FLA. SOLAR ENERGY CENTER LIBRA

energy

from the desert

energy from the desert

Practical Proposals for Very Large Scale Photovoltaic Systems

Edited by Kosuke Kurokawa
Keiichi Komoto
Peter van der Vleuten
David Faiman

London • Sterling, VA

LIBRARY
Florida Solar Energy Center
1679 Clearlake Road
Cocoa FL 32922-5703

013858 ✓

First published by Earthscan in the UK and USA in 2007

Copyright © Photovoltaic Power Systems Executive Committee of the International Energy Agency, 2007

All rights reserved

ISBN 978-1-84407-363-4

Typeset by MapSet Ltd, Gateshead, UK
Printed and bound by Gutenberg Press, Malta
Cover design by Susanne Harris

For a full list of publications please contact:

Earthscan
8–12 Camden High Street
London, NW1 0JH, UK
Tel: +44 (0)20 7387 8558
Fax: +44 (0)20 7387 8998
Email: earthinfo@earthscan.co.uk
Web: **www.earthscan.co.uk**

22883 Quicksilver Drive, Sterling, VA 20166-2012, USA

Earthscan is an imprint of James and James (Science Publishers) Ltd and publishes in association with the International Institute for Environment and Development

A catalogue record for this book is available from the British Library

Library of Congress Cataloging-in-Publication Data has been applied for

Neither the authors nor the publisher make any warranty or representation, expressed or implied, with respect to the information contained in this publication, nor assume any liability with respect to the use of, or damages resulting from, this information.

Printed on elemental chlorine-free paper

Please note: in this publication a comma has been used as a decimal point, according to the ISO standard adopted by the International Energy Agency.

Contents

Foreword	*x*
Preface	*xi*
Task 8 Participants	*xii*
List of Contributors	*xiii*
Acknowledgements	*xiv*
List of Figures and Tables	*xv*
List of Acronyms and Abbreviations	*xxv*
EXECUTIVE SUMMARY	**xxix**
1. Introduction	**xxix**
1.1 Objectives	xxix
1.2 Concept of very large scale photovoltaic power generation (VLS-PV) systems	xxix
1.3 Practical approaches to realize VLS-PV	xxxi
2. The Mediterranean Region: Case study of very large scale photovoltaics	**xxxiii**
3. The Middle East region: A top-down approach for introducing VLS-PV plants	**xxxv**
4. The Asian region: project proposals of VLS-PV on the Gobi Desert	**xxxviii**
4.1 Demonstrative research project for VLS-PV in the Gobi Desert of Mongolia	xxxviii
4.2 Feasibility study on 8 MW large scale PV system in Dunhuang, China	xxxix
5. The Oceania region: Realizing a VLS-PV power generation system at Perenjori	**xlii**
6. Desert region community development	**xliv**
7. Conclusions and recommendations	**xlvii**
7.1 General conclusions	xlvii
7.2 Requirements for a project development	xlviii
7.3 Recommendations	xlviii
1. INTRODUCTION	1
1.1 Objectives	1
1.2 Concept of very large scale photovoltaic power generation (VLS-PV) systems	1
1.2.1 Availability of desert area for PV technology	1
1.2.2 VLS-PV concept and definition	2
1.2.3 Potential of VLS-PV advantages	3
1.2.4 The viability of VLS-PV's development	4
1.3 Practical approaches to realize VLS-PV	4
1.3.1 Required approaches for PV and renewables	4
1.3.2 Preliminary recommendations on a policy level for VLS-PV	5
1.3.3 Approaches for practical project proposal	6

2. THE MEDITERRANEAN REGION: CASE STUDY OF VERY LARGE SCALE PHOTOVOLTAICS 8
2.1 Introduction 8
2.2 General conditions 8
2.2.1 Country profiles 8
2.2.2 Overview of most important country data 20
2.2.3 Evaluation with respect to VLS-PV 20
2.3 Economic feasibility of VLS-PV systems 21
2.3.1 PV electricity-generation cost 21
2.3.2 Main results and discussion 22
2.4 A closer look at Portugal and Spain 23
2.4.1 VLS-PV in Portugal 23
2.4.2 VLS-PV in Spain 23
2.5 Summary and conclusions 23

3. THE MIDDLE EAST REGION: A TOP-DOWN APPROACH FOR INTRODUCING VLS-PV PLANTS 26
3.1 Introduction 26
3.2 Potential of solar electricity based on land area 26
3.3 Electricity requirement 27
3.4 Suggested technology 28
3.4.1 Concentrator photovoltaic (CPV) technology 28
3.4.2 The individual CPV units and their expected performance 28
3.4.3 Storage 29
3.4.4 Transmission 29
3.5 The rate of VLS-PV introduction 29
3.6 The cost of VLS-PV plants 30
3.6.1 The cost of a 1 GW plant 30
3.6.2 The cost of storage batteries 30
3.6.3 The cost of a manufacturing facility 30
3.7 Israel as a case study, and a suggested investment algorithm 30
3.7.1 Projected revenue 30
3.7.2 A suggested investment algorithm 31
3.7.3 Results for the Israel example 32
3.7.4 Discussion of the Israel results 32
3.8 Other major electricity consumers in the Middle East 34
3.8.1 Saudi Arabia 36
3.8.2 Iran 38
3.8.3 Turkey 40
3.8.4 Egypt 42
3.8.5 United Arab Emirates 44
3.8.6 Iraq 46
3.8.7 Kuwait 48
3.8.8 Syria 50
3.9 Summary and conclusions 52

4. THE ASIAN REGION: PROJECT PROPOSALS OF VLS-PV ON THE GOBI DESERT 54
4.1 Demonstrative research project for VLS-PV in the Gobi Desert of Mongolia 54
4.1.1 Introduction 54
4.1.2 General information on Mongolia and specified areas 55
4.1.3 Overview of the project proposed 57
4.1.4 Preliminary demonstrative research project for VLS-PV in Mongolia 62
4.1.5 Recommendations 66
4.2 Feasibility study on an 8 MW large scale PV system in Dunhuang, China 67
4.2.1 Introduction 67
4.2.2 General information on the site 67
4.2.3 Technical profile 68
4.2.4 Financial feasibility 70
4.2.5 Recommendations 70

5. THE OCEANIA REGION: REALIZING A VLS-PV POWER GENERATION SYSTEM AT PERENJORI 71
5.1 Introduction 71
5.2 Potential of VLS-PV project at Perenjori 71
5.2.1 Power requirement and creating extra load in the region 71
5.2.2 Basic meteorological data for designing and sizing the project 71
5.2.3 Possible scenario for realizing VLS-PV project 72
5.2.4 Grid connections versus stand-alone PV power-generation systems 72
5.2.5 Funding of VLS-PV power-generation projects 73
5.2.6 Pathway to setting up a local PV manufacturing facility 73
5.2.7 Rainwater collection from PV modules 73
5.2.8 Greenhouse gas emission abatement 74
5.3 Pilot 10 MW PV power-generation project at Perenjori 74
5.3.1 Estimated project cost 74
5.3.2 Levelled cost of power generation 74
5.3.3 Estimated annual power generation 75
5.3.4 Economics of rainwater collection from PV modules 76
5.4 Analysis of government subsidy programmes 77
5.4.1 Renewable Remote Power-Generation Programme (RRRGP) 77
5.4.2 RRRGP sub-programmes 77
5.4.3 Private investment, superannuation funds and ethical trust funds 78
5.4.4 Ownership and management issues 79
5.4.5 Power purchase agreements with potential end users 79
5.5 Expected effects induced by VLS-PV at Perenjori 79
5.5.1 Greenhouse gas emissions abatement 79
5.5.2 Economical and social development of Perenjori region 79
5.5.3 Local industry development 79
5.5.4 Future investment potential in the region 80
5.6 Future option: Hydrogen production from VLS-PV 80
5.7 Recommendations 80

6. DESERT REGION COMMUNITY DEVELOPMENT 82
6.1 Introduction 82
6.2 Approaches to establishing a desert community 82
6.2.1 Sustainable PV stations 82
6.2.2 Regional society 83
6.2.3 Suggestion for development modelling of solar photovoltaic generation and greening model society with stability 85
6.3 Sustainable agriculture development using PV systems 87
6.3.1 Basic characteristics of desert soil and countermeasures against over-cropping 87
6.3.2 Electricity needs for sustainable irrigated agriculture 92
6.3.3 Effects of VLS-PV on the local environment 94
6.4 Recommendations 99

7. CONCLUSIONS AND RECOMMENDATIONS 102
7.1 General conclusions 102
7.1.1 Background 102
7.1.2 Technical aspects 102
7.1.3 Economic aspects 102
7.1.4 Social aspects 103
7.1.5 Environmental aspects 103
7.2 Requirements for a project development 103
7.3 Recommendations 103

APPENDIX A: CASE STUDIES OF VLS-PV SYSTEMS IN DESERTS 105
A.1 A life-cycle analysis for various very large scale photovoltaic power generation (VLS-PV) systems in world deserts 105
A.1.1 Introduction 105
A.1.2 General assumptions 105
A.1.3 VLS-PV system design 108
A.1.4 Evaluation results 113
A.1.5 Conclusions 117
A.2 Resource analysis of solar energy by using satellite images 131
A.2.1 Introduction 131
A.2.2 Subject area 131
A.2.3 Method and results 132
A.2.4 A resource evaluation of PV system 135
A.2.5 Ground truth 135
A.2.6 Conclusions 136
A.3 The highest efficiency PV module by practical concentrator technologies, performance, reliability and applications to deserts 138
A.3.1 Introduction 138
A.3.2 Technology overview 138
A.3.3 Reliability in the desert 142
A.3.4 Outdoor characteristics 143
A.3.5 Comparison with the crystalline silicon flat-plate module 144
A.3.6 Land utilization for PV generation 145
A.3.7 Possibility of 'plant breeding' PV 150
A.3.8 Conclusions 150
A.4 Recent developments in low-concentration photovoltaics (LCPV) 151
A.4.1 Introduction 151
A.4.2 The Day4™ Electrode 152
A.4.3 Concentrator optics 152
A.4.4 Cost estimates 153
A.4.5 Leveraging current production 153
A.5 Application of bifacial PV systems to VLS-PV 154
A.5.1 Introduction 154
A.5.2 Characteristics of bifacial PV cells and modules 154
A.5.3 Preliminary examination of application of bifacial PV cells with vertical installation to VLS-PV 157
A.5.4 Conclusions 162
A.6 Large scale PV plant experience in the Arizona Desert and evaluation of the energy payback time 163
A.6.1 Introduction 163
A.6.2 Photovoltaic power plant experience at Tucson Electric Power 163
A.6.3 Energy payback and life-cycle CO_2 emissions of the balance of system (BOS) in an optimized 3,5 MW PV installation 168
A.6.4 Conclusions 173
A.7 Photovoltaic and solar thermal systems: Similarities and differences 176
A.7.1 Introduction 176
A.7.2 Solar thermal (ST) systems 176
A.7.3 Parabolic trough ST systems 176
A.7.4 Comparisons of potential large scale ST and PV plants 177
A.7.5 Conclusions 178
A.8 Impact assessment of VLS-PV on the global climate 179
A.8.1 Introduction 179
A.8.2 Early investigations on climatological changes by land surface property 179
A.8.3 Preliminary modelling of VLS-PV installation 180
A.8.4 Choice of model 181
A.8.5 Methodology 181
A.8.6 Model results and discussion 182

APPENDIX B: FUTURE POSSIBILITIES TOWARDS WORLD MAJOR ENERGY 185
B.1 Long-term perspective of energy and renewables 185
B.1.1 Introduction 185
B.1.2 Long-term future energy perspective 185
B.1.3 Conclusions 190
B.2 Integrating VLS-PV systems within an electrical grid 190
B.2.1 Introduction 190
B.2.2 Renewable energy-dominated grids 191
B.2.3 Blended generation 191
B.2.4 Shaping the power in renewable energy-dominated grids 191
B.2.5 Large scale energy storage 192
B.2.6 'Smart' grids 192
B.2.7 Conclusions 193
B.3 Solar hydrogen scenario 194
B.3.1 Introduction 194
B.3.2 Theoretical possibilities of hydrogen production 195
B.3.3 International Energy Agency (IEA) hydrogen programme 196
B.3.4 A new approach: Concentrator photovoltaics (CPV) combined with high temperature electrolyser 197
B.3.5 Conclusions 200

II

Foreword

This book deals with the concept of very large scale photovoltaic power generation (VLS-PV) systems and represents the second report that has been prepared on this subject within the International Energy Agency (IEA) Photovoltaic Power Systems Programme (PVPS), Task 8: Very Large Scale Photovoltaic Power Systems. Under the leadership of Professor K. Kurokawa, and based on the previous conceptual work, the international project team has elaborated upon the subject with concrete case studies.

The current rapid growth of the photovoltaic market is dominated by grid-connected applications and the average size of these projects has recently increased. Starting from small scale system sizes in the kilowatt (kW) power range, systems of tens to hundreds of kW are now regularly realized. These photovoltaic systems are often mounted on or integrated within buildings and have reached system sizes up to a few megawatts (MW). In parallel, ground-based arrays of the order of 10 MW and more have been recently built. Through these projects, the photovoltaic sector has gained important experience in the design, construction and operation of photovoltaic systems in the MW range.

Surface and land availability will limit the greater size of photovoltaic systems in densely populated areas of the planet. On the other hand, enormous surfaces of high solar irradiation exist in arid or desert areas. These offer a large potential for photovoltaic systems beyond the size currently known, for which the term VLS-PV systems has been introduced. Such systems may seem unrealistic due to the current state of technology. However, this is precisely the origin of this report, aimed at exploring the detailed conditions for VLS-PV systems in selected regions.

The report provides a detailed analysis of technical, socio-economic and environmental aspects of VLS-PV systems for case studies in the Mediterranean, the Middle East, and the Asian and Oceania regions. Moreover, it includes a brief comparison with concentrated solar (thermal) power (CSP) systems and a first assessment on global climate impact.

I am confident that through its careful, detailed and thorough analysis, this report forms a comprehensive assessment of the concept of very large scale photovoltaic power systems. It should help to rationalize the discussion of this concept and to provide the basis for concrete feasibility studies and project proposals.

I would like to thank Professors K. Kurokawa and K. Komoto, as well as the whole Task 8 expert team for their dedicated effort in making this publication possible. I hope that this unique analysis will contribute to the advance of photovoltaics in new applications.

Stefan Nowak
Chairman, IEA PVPS

Preface

It is already known that the world's very large deserts present a substantial amount of energy-supplying potential. Given the demands on world energy in the 21st century, and when considering global environmental issues, the potential for harnessing this energy is of huge import and has formed the backbone and motive for this report.

The work on very large scale photovoltaic power generation (VLS-PV) systems, upon which this report is based, first began under the umbrella of the International Energy Agency (IEA) Photovoltaic Power System Programme (PVPS) Task 6 in 1998. The new Task 8 – Very Large Scale Photovoltaic Power Generation Utilizing Desert Areas – was established in 1999.

The scope of Task 8 is to examine and evaluate the potential of VLS-PV systems, which have a capacity ranging from several megawatts to gigawatts, and to develop practical project proposals for demonstrative research towards realizing VLS-PV systems in the future.

For this purpose, in the first phase (1999–2002), key factors that enable the feasibility of VLS-PV systems were identified, and the benefits of VLS-PV applications for neighbouring regions were clarified.

In order to disseminate these possibilities, we published the first book entitled *Energy from the Desert: Feasibility of Very Large Scale Photovoltaic Power Generation (VLS-PV) Systems* in 2003. At the same time, we started the second phase of activity.

In the second phase (2003–2005), in-depth case studies on VLS-PV systems were carried out, and practical proposals for demonstrative research projects on pilot photovoltaic systems suitable for selected regions were outlined, which should enable sustainable growth in VLS-PV systems in the future. Practical projects for large scale PV systems were also discussed.

'It might be a dream, but ...' has been a motive for continuing our beloved study of VLS-PV. Now we have become confident that this is no longer a dream.

This report deals with one of the promising recommendations for sustainable development in terms of solving world energy and environmental problems in the 21st century.

Professor Kosuke Kurokawa
Editor
Operating Agent, Alternate, Task 8

Keiichi Komoto
Editor, Task 8

II

Task 8 Participants

Kazuhiko Kato, OA
Research Centre for Photovoltaics, National Institute of Advanced Industrial Science and Technology (AIST), Japan

Kosuke Kurokawa, OA–Alternate
Tokyo University of Agriculture and Technology (TUAT), Japan

Masanori Ishimura, Secretary
Photovoltaic Power Generation Technology Research Association (PVTEC), Japan

Claus Beneking
ErSol Solar Energy AG, Germany

Matthias Ermer
SolarWorld Industries Deutschland GmbH, Germany

David Faiman
Department of Solar Energy and Environmental Physics, Jacob Blaustein Institute for Desert Research, Ben-Gurion University of the Negev, Israel

Thomas N. Hansen
Tucson Electric Power Company, US

Herb Hayden
Arizona Public Service, US

Keiichi Komoto
Mizuho Information and Research Institute, Inc (MHIR), Japan

John S. MacDonald
Day4Energy, Inc, Canada

Fabrizio Paletta
Centro Electtrotecnico Sperimentale Italiano (CESI), Italy

Angelo Sarno
Ente per le Nuove Technologie, l'Energia el'Ambiente (ENEA), Italy

Jinsoo Song
Korea Institute of Energy Research (KIER), Korea

Leendert A. Verhoef
New-Energy-Works, The Netherlands

Peter van der Vleuten
Free Energy International, The Netherlands

Namjil Enebish, Observer
National Renewable Energy Center, Mongolia

List of Contributors

Ichiro Araki	Appendix A.5
Kenji Araki	Appendix A.3
Miguel Arrarás	Chapter 2
Andreas Beneking	Chapter 2
Claus Beneking	Chapter 2
Brahim Bessaïs	Chapter 2
Tomoki Ehara	Appendix B.1
C. Elam	Appendix B.3
Namjil Enebish	Chapter 4.1
Matthias Ermer	Chapter 1
David Faiman	Chapter 3, Appendix A.7
Vasilis M. Fthenakis	Appendix A.6
José Pedro Ganiguer	Chapter 2
Thomas N. Hansen	Appendix A.6
Masakazu Ito	Chapter 6, Appendices A.1, A.2
Carlos Itoiz	Chapter 2
Guido Kalenbach	Chapter 2
Kazuhiko Kato	Chapters 1, 4.1, Appendix A.1
Makoto Kato	Appendix A.8
H. C. Kim	Appendix A.6
Keiichi Komoto	Chapters 1, 4.1, 4.2, 7, Appendices A.1, B.1
Sanjay Kumar	Appendix A.8
Kosuke Kurokawa	Chapters 1, 4.2, 6, 7, Appendices A.2, A.8, B.3
J. B. Lasich	Appendix B.3
Shenghong Ma	Chapter 4.2
John S. MacDonald	Appendices A.4, B.2
J. E. Mason	Appendix A.6
R. D. McConnell	Appendix B.3
Larry Moore	Appendix A.6
Terry Mysak	Appendix A.6
Taku Nishimura	Chapter 6, Appendix A.8
Thomas Nowicki	Chapter 2
Mauricio Olite Ariz	Chapter 2
Kenji Otani	Chapter 4.1
Doreen Otto	Chapter 2
Hal Post	Appendix A.6
Dov Raviv	Chapter 3
Roy Rosenstreich	Chapter 3
Christoph Schneider	Chapter 2
Dilawar Singh	Chapter 5
Sangeeta Sinha	Appendix A.8
Jinsoo Song	Chapters 4.1, 4.2
Peter van der Vleuten	Chapter 2
Leendert A. Verhoef	Chapter 2
Sicheng Wang	Chapter 4.2
Karsten Weltzien	Chapter 2

II

Acknowledgements

This report was accomplished with the kind support of various organizations and people around the world.

Our activity, Task 8, started in 1999 and was concluded in 2002 with its first phase of activity. In 2003, we held an international symposium, Energy from the Desert, in Osaka, Japan, as a side event of the Third World Conference on Photovoltaic Energy Conversion (WCPEC-3); during this event, we obtained various kinds of expert advice and new memberships. We began our current second phase of activity in 2003.

We thank Ichiro Hashimoto and Yukao Tanaka (NEDO, Japan), who provided us with a great deal of support as the operating agent country member of the International Energy Agency (IEA) PVPS Executive Committee.

We acknowledge Byanma Jigjid (Mongolia) and Noboru Yumoto (Japan), who gave impressive and attractive presentations in the international symposium in Osaka in 2003.

We thank Pietro Menna and David Collier, who were members of PVPS Task 8 in the early stage of this second phase of activity.

The project proposal for the Mediterranean region comprised contributions from Deutsche Energie Agentur (DENA), assistance from Birgit Rekow, and financial support from the German Federal Ministry of Environment (BMU), granted through PTJ.

The project proposals for the Asian region and some case studies of very large scale photovoltaic power generation (VLS-PV) systems in deserts were developed thanks to the members of the Japanese domestic committee for VLS-PV: Kenji Otani (AIST); Masakazu Ito (TITEC); Taku Nishimura (The University of Tokyo); Ichiro Araki (Hitachi); Kenji Araki (Daido Steel); Tetsuo Kichimi (Resources Total System); Hiroyuki Sugihara (Kandenko); Tetsu Nishioka (GETC); Tetsuya Fukunaga (UFJI); and Tomoki Ehara (MHIR). We also thank Yukihoko Kimura, Tatsuo Yasui, Hidemi Mitsuyasu (NEDO) and Tsunehisa Harada (PVTEC) for supporting and managing the committee.

The research employed in preparing Chapter 3 and Appendix A, section A.7 'Photovoltaic and solar thermal systems: Similarities and differences', was partially funded by the Israel Ministry of National Infrastructure.

Appendix B, section B.2 'Integrating VLS-PV systems within an electrical grid', was greatly assisted by Keith McPherson of the Cascadia Institute in Seattle, USA.

We greatly thank Isaburo Urabe, who was a former secretary of Task 8 and strongly supported all Task 8 activities.

Finally, the Task 8 members thank the IEA PVPS Executive Committee and the participating countries of Task 8 for giving them valuable opportunities for performing this task.

List of Figures and Tables

FIGURES

E1.1	Solar pyramid	xxx
E1.2	Image of a VLS-PV system in a desert area	xxxi
E2.1	The Mediterranean regions within the case study	xxxiii
E3.1	Land requirement for providing 80 % of each country's electrical requirements	xxxvii
E3.2	Total solar generating and storage capacity at the end of 36 years	xxxvii
E3.3	Maximum credit line required at 3 % real interest	xxxvii
E3.4	Total number of jobs created	xxxvii
E4.1	Location of VLS-PV systems	xxxix
E4.2	Location of Dunhuang, Gansu	xl
E4.3	1 MW PV sub-station	xl
E5.1	Location map of Perenjori	xlii
E5.2	Possible scenario to realize VLS-PV at Perenjori	xliii
E6.1	Framework of desert community development and research topic	xlv
E6.2	Desalinated drip irrigation system	xlv
1.1	World deserts	1
1.2	Solar pyramid	2
1.3	Image of a VLS-PV system in a desert area	3
1.4	Global network image	4
1.5	Very large scale PV system deployment scenario	4
1.6	Check-flow of recommendations	5
2.1	Physical map of Morocco	8
2.2	Average annual horizontal solar radiation in Morocco in kWh/m^2	9
2.3	Primary energy supply in Morocco in 2002 (without energy imports)	9
2.4	Physical map of Tunisia	11
2.5	Average annual horizontal solar radiation in Tunisia in kWh/m^2	12
2.6	Primary energy supply in Tunisia in 2002	12
2.7	Physical map of Portugal	14
2.8	Average annual horizontal solar radiation in Portugal and Spain in kWh/m^2	15
2.9	Primary energy supply in Portugal in 2002	15
2.10	Physical map of Spain	18
2.11	Primary energy supply in Spain 2002	18
2.12	Relation of GDP per capita to energy consumption per capita	21
3.1	A typical large dense array CPV collector, in which an array of cells is illuminated by a single light-gathering element (in this case, a parabolic reflecting dish)	28
3.2	A typical large individual cell CPV collector, in which each individual cell in the array is illuminated by its own light-gathering element (in this case, a Fresnel lens)	28

3.3 (a) Projected annual growth of electrical energy generation in Israel for the first 36 years of a top-down VLS-PV programme, starting from an all-fossil 44 TWh per year in 2004 and reaching 107 TWh, of which 87 % is solar 33
3.3 (b) Projected annual growth of percentage solar electricity generation in Israel for the first 36 years of a top-down VLS-PV programme, starting from 0 % in 2004–2008 (years 1 to 5) and reaching 87 % in year 36 33
3.3 (c) Annual investments required for the top-down VLS-PV programme proposed for Israel and subsequent income from solar plants 33
3.3 (d) Annual cash flow associated with investment, at 3 % interest, in a top-down VLS-PV programme in Israel 33
3.4 Real estate requirements for a top-down VLS-PV programme that would provide typically 80 % of each country's electrical requirements within 36 years for the nine largest electricity-producing countries in the Middle East 34
3.5 Percentage of each country's total land area 34
3.6 Size of proposed VLS-PV plants and storage units that would be erected annually in a top-down programme for the nine largest electricity-producing countries in the Middle East 34
3.7 Total solar-generating and storage capacity at the end of 36 years of a top-down VLS-PV programme for the nine largest electricity-producing countries in the Middle East 34
3.8 Maximum credit line required (at 3 % real interest) for carrying out a top-down VLS-PV programme for the nine largest electricity-producing countries in the Middle East 34
3.9 The numbers of years required for the credit line to reach its maximum and subsequently be completely liquidated by revenues from the VLS-PV plants for the nine largest electricity-producing countries in the Middle East 34
3.10 Total number of jobs created by a top-down VLS-PV programme for the nine largest electricity-producing countries in the Middle East 34
3.11(a) Projected annual growth of electrical energy generation in Saudi Arabia for the first 36 years of a top-down VLS-PV programme, starting from 142,8 TWh per year in 2004 and reaching 444,8 TWh, of which 70 % is solar 36
3.11(b) Projected annual growth of the solar electricity-generation percentage in Saudi Arabia for the first 36 years of a top-down VLS-PV programme, starting from 0 % during 2004–2008 (years 1 to 5) and reaching 70 % in year 36 37
3.11(c) Annual investments required for the top-down VLS-PV programme proposed for Saudi Arabia and subsequent income from solar plants 37
3.11(d) Annual cash flow associated with investment, at 3 % interest, in a top-down VLS-PV programme in Saudi Arabia 37
3.12(a) Projected annual growth of electrical energy generation in Iran for the first 36 years of a top-down VLS-PV programme, starting from 130,4 TWh per year in 2004 and reaching 191,3 TWh, of which 81 % is solar 38
3.12(b) Projected annual growth of solar electricity-generation percentage in Iran for the first 36 years of a top-down VLS-PV programme, starting from 0 % during 2004–2008 (years 1 to 5) and reaching 81 % in year 36 39
3.12(c) Annual investments required for the top-down VLS-PV programme proposed for Iran and subsequent income from solar plants 39
3.12(d) Annual cash flow associated with investment, at 3 % interest, in a top-down VLS-PV programme in Iran 39
3.13(a) Projected annual growth of electrical energy generation in Turkey for the first 36 years of a top-down VLS-PV programme, starting from 124,8 TWh per year in 2004 and reaching 189,4 TWh, of which 82 % is solar 40
3.13(b) Projected annual growth of solar electricity-generation percentage in Turkey for the first 36 years of a top-down VLS-PV programme, starting from 0 % during 2004–2008 (years 1 to 5) and reaching 82 % in year 36 41
3.13(c) Annual investments required for the top-down VLS-PV programme proposed for Turkey and subsequent income from solar plants 41
3.13(d) Annual cash flow associated with investment, at 3 % interest, in a top-down VLS-PV programme in Turkey 41
3.14(a) Projected annual growth of electrical energy generation in Egypt for the first 36 years of a top-down VLS-PV programme, starting from 82,8 TWh per year in 2004 and reaching 160,1 TWh, of which 78 % is solar 42

3.14(b) Projected annual growth of solar electricity-generation percentage in Egypt for the first 36 years of a top-down VLS-PV programme, starting from 0 % during 2004–2008 (years 1 to 5), and reaching 78 % in year 36 43
3.14(c) Annual investments required for the top-down VLS-PV programme proposed for Egypt and subsequent income from solar plants 43
3.14(d) Annual cash flow associated with investment, at 3 % interest, in a top-down VLS-PV programme in Egypt 43
3.15(a) Projected annual growth of electrical energy generation in the United Arab Emirates for the first 36 years of a top-down VLS-PV programme, starting from 39,9 TWh per year in 2004 and reaching 69,6 TWh, of which 80 % is solar 44
3.15(b) Projected annual growth of solar electricity-generation percentage in the UAE for the first 36 years of a top-down VLS-PV programme, starting from 0 % during 2004–2008 (years 1 to 5) and reaching 80 % in year 36 45
3.15(c) Annual investments required for the top-down VLS-PV programme proposed for the UAE and subsequent income from solar plants 45
3.15(d) Annual cash flow associated with investment, at 3 % interest, in a top-down VLS-PV programme in the UAE 45
3.16(a) Projected annual growth of electrical energy generation in Iraq for the first 36 years of a top-down VLS-PV programme, starting from 35 TWh per year in 2004 and reaching 91,9 TWh, of which 88 % is solar 46
3.16(b) Projected annual growth of solar electricity-generation percentage in Iraq for the first 36 years of a top-down VLS-PV programme, starting from 0 % during 2004–2008 (years 1 to 5) and reaching 88 % in year 36 47
3.16(c) Annual investments required for the top-down VLS-PV programme proposed for Iraq and subsequent income from solar plants 47
3.16(d) Annual cash flow associated with investment, at 3 % interest, in a top-down VLS-PV programme in Iraq 47
3.17(a) Projected annual growth of electrical energy generation in Kuwait for the first 36 years of a top-down VLS-PV programme, starting from 33,5 TWh per year in 2004 and reaching 104,3 TWh, of which 77 % is solar 48
3.17(b) Projected annual growth of solar electricity-generation percentage in Kuwait for the first 36 years of a top-down VLS-PV programme, starting from 0 % during 2004–2008 (years 1 to 5) and reaching 77 % in year 36 49
3.17(c) Annual investments required for the top-down VLS-PV programme proposed for Kuwait and subsequent income from solar plants 49
3.17(d) Annual cash flow associated with investment, at 3 % interest, in a top-down VLS-PV programme in Kuwait 49
3.18(a) Projected annual growth of electrical energy generation in Syria for the first 36 years of a top-down VLS-PV programme, starting from an all-fossil 26,8 TWh per year in 2004 and reaching 63,5 TWh, of which 73 % is solar 50
3.18(b) Projected annual growth of solar electricity-generation percentage in Syria for the first 36 years of a top-down VLS-PV programme, starting from 0 % during 2004–2008 (years 1 to 5) and reaching 73 % in year 36 51
3.18(c) Annual investments required for the top-down VLS-PV programme proposed for Syria and subsequent income from solar plants 51
3.18(d) Annual cash flow associated with investment, at 3 % interest, in a top-down VLS-PV programme in Syria 51

4.1 Trends in population by area 55
4.2 Composition of gross domestic product (GDP) by industry 55
4.3 Location of the existing power plants 55
4.4 Electrification map of Mongolia 56
4.5 Trends in electricity consumption by sector 57
4.6 Transportation base map of Mongolia 58
4.7 Trends in amounts of freight transportation (t.km) 58
4.8 Trends in amounts of freight transportation (t) 58
4.9 Location of VLS-PV systems 59
4.10 Image of PV system layout of the first phase 60

4.11	2 MW sodium-sulphur battery and 170 kW vanadium redox flow battery with 1 MW PV facilities at the National Institute of Advanced Industrial Science and Technology (AIST), Japan	61
4.12	Installed instrument	64
4.13	Performance ratio of PV module 1 (poly-Si)	66
4.14	Performance ratio of PV module 2 (mono-Si)	66
4.15	Location of Dunhuang, Gansu	67
4.16	Gobi area in Dunhuang	67
4.17	The share of electricity consumption in Dunhuang, 2003	67
4.18	Irradiation data in Dunhuang	68
4.19	200 kW PV channel	68
4.20	1 MW PV sub-station	69
4.21	Configuration of PV sub-array	69
4.22	Field design for 1 MW PV sub-station	70
4.23	Field design for 8 MW LS-PV system	70
5.1	Location map of Perenjori	71
5.2	Solar radiation in Australia	72
5.3	Possible scenario to realize VLS-PV at Perenjori	72
5.4	Possible scenario for developing a local PV manufacturing industry	73
5.5	Rainwater collection from PV modules	74
5.6	Cost composition of 10 MW PV power-generation system project	75
5.7	Levelled life-cycle generation cost of 10 MW PV power plant	76
5.8	Effective generation cost considering revenues from rainwater collection	77
5.9	CO_2 and coal saving by 10 MW PV plants	79
5.10	Schematic flowchart of the hydrogen production from PV system	80
6.1	Image of VLS-PV system	83
6.2	Estimated results for a suitable area within the Gobi Desert in Mongolia	83
6.3	A flow diagram of an electro-dialysis desalination system using PV	84
6.4	Example of a desert area village's material, energy and currency flow	84
6.5	Framework of desert community development and research topic	86
6.6	Expansion of Sahara Desert	87
6.7	Irrigation canal during the dry season, Gansu Province, China	89
6.8	Pivot irrigation	89
6.9	Sodium adsorption ratio (SAR), concentration, moisture content and soil permeability	91
6.10	Electrolyte concentration and hydraulic conductivity	91
6.11	Downward leaching of salt at the Loess Plateau in Gansu Province, China	92
6.12	Irrigation–drainage combined system	93
6.13	Subsurface irrigation	93
6.14	Desalinated drip irrigation system	94
6.15	Interrelationship among atmosphere, vegetation and soil in the water cycle	95
6.16	Regional water circulation	96
A.1	Six deserts are selected for case studies	106
A.2	Image of basic array and 500 kW array unit	106
A.3	Conceptual image of a 1 GW VLS-PV system	107
A.4	Thematic circuit diagram of a 100 MW VLS-PV	107
A.5	Annual average estimated day's cable and transmission loss curve with 30° tilt angle in the Sahara Desert	110
A.6	Cable and transmission losses in world deserts (not including transformer and SVC loss)	111
A.7	Transmission losses of TACSR 240 sq, 410 sq and 680 sq	111
A.8	Design drawings of 30° tilted basic array when using m-Si	112
A.9	Foundation for tracking system	112
A.10	Tracking equipment design	112
A.11	Example of tracking system design	113
A.12	Transportation method for China	113
A.13	Initial cost of a 100 MW VLS-PV system in the Sahara Desert (Nema)	114

A.14 Initial cost of a 100 MW VLS-PV system in the Sahara Desert (Ouarzazate) 114
A.15 Initial cost of a 100 MW VLS-PV system in the Negev Desert 115
A.16 Initial cost of a 100 MW VLS-PV system in the Thar Desert (Jodhpur) 115
A.17 Initial cost of a 100 MW VLS-PV system in the Sonoran Desert (Chihuahua) 115
A.18 Initial cost of a 100 MW VLS-PV system in the Great Sandy Desert (Port Headland) 115
A.19 Initial cost of a 100 MW VLS-PV system in the Gobi Desert (Huhhot) 115
A.20 Initial cost of a 100 MW VLS-PV system in the Gobi Desert (Sainshand) 115
A.21 Generation cost of 100 MW VLS-PV systems in eight desert regions at optimal tilt angle as a function of annual global horizontal irradiation 116
A.22 Initial cost of a 100 MW VLS-PV system using m-Si PV modules 116
A.23 Initial cost of a 100 MW VLS-PV system using a-Si PV modules 116
A.24 Initial cost of a 100 MW VLS-PV system using CdTe PV modules 116
A.25 Initial cost of a 100 MW tracking VLS-PV system using m-Si PV modules 116
A.26 Generation cost with m-Si PV modules 117
A.27 Generation cost with a-Si PV modules 117
A.28 Generation cost with CdTe PV modules 117
A.29 Generation cost with sun-tracking systems 117
A.30 Total energy requirement and energy payback time of a 100 MW VLS-PV system using m-Si PV modules 118
A.31 Total energy requirement and energy payback time of a 100 MW VLS-PV system using a-Si PV modules 118
A.32 Total energy requirement and energy payback time of a 100 MW VLS-PV system using CdTe PV modules 118
A.33 Life-cycle CO_2 emissions and life-cycle CO_2 emission rate of a 100 MW PV system using m-Si PV modules 118
A.34 Life-cycle CO_2 emissions and life-cycle CO_2 emission rate of a 100 MW PV system using a-Si PV modules 119
A.35 Life-cycle CO_2 emissions and life-cycle CO_2 emission rate of a 100 MW PV system using CdTe PV modules 119
A.36 Comparison of VLS-PV systems of m-Si, a-Si and CdTe modules 119
A.37 Examples of the surface of the Gobi Desert 130
A.38 Six major world deserts as subject area 131
A.39 Examples of reflecting features and radiation characteristics 131
A.40 Example of vegetation classification by vegetation index 132
A.41 Suitable vegetation levels for VLS-PV systems 132
A.42 Seasonal changes in vegetation levels 133
A.43 Parameters of the maximum likelihood classifier 134
A.44 Result of extracted edge line by satellite image 135
A.45 Estimated results of part of the Sahara Desert 135
A.46 Estimated area percentage as suitable land for a VLS-PV system using Landsat-7 136
A.47 Examining the accuracy of estimated results 136
A.48 PV potential world map 136
A.49 550X and 400X on an open-loop tracker 138
A.50 The receiver, a key component of concentrator technology 138
A.51 Efficiency of the cell versus number of suns 139
A.52 I-V curve of the three-junction cell at 498 sun irradiation 139
A.53 Structure of the three-junction concentrator cell 139
A.54 A bare III-V three-junction concentrator chip 139
A.55 Heat spreading concept for a concentrator module 140
A.56 A kaleidoscope homogenizer: an indispensable optical device for multi-junction concentrator solar cells 140
A.57 Ray-tracing for concentrator optics 140
A.58 Generations of injection-moulded Fresnel lenses 141
A.59 Inside of the 400X concentrator module with 36 receivers connected in series 141
A.60 Module temperature versus ambient temperature 142
A.61 Comparison between the accelerated test and the outside exposure test 142
A.62 Accelerated test (equivalent to 20 years of installation in the desert area) by compound stress of simultaneous concentrated UV irradiation and water condensation 143

A.63 Off-axis beam test (thanks to the optical characteristics of the non-imaging Fresnel lens, etc., fingers will not be burned by an off-axis beam) 143
A.64 Typical outdoor I-V curve of the 550X module 143
A.65 Histogram of the corrected module efficiency for a clear summer day 144
A.66 Comparison to the mc-Si flat-plate module (rated efficiency is 14,06 %) 145
A.67 Spectrum calculated by the sunshine model used at solstice in June and December 145
A.68 Direct beam spectrum: Measured versus model 146
A.69 Relative efficiency by the change of spectrum of direct normal irradiance: Ge-based spectrum-matched three-junction cell 146
A.70 Short-circuit current of concentrator module versus direct normal irradiance (DNI) 146
A.71 Efficiency of concentrator module versus DNI 147
A.72 Power generation of spectrum-sensitive concentrator array with a space factor of 0,9 in an east–west direction and space factor of 0,5 in a north–south direction 147
A.73 Power generation of isolated spectrum-sensitive concentrator module without shading effect 148
A.74 Contour plot of area efficiency as a function of space factor in east–west and north–south directions 148
A.75 Contour plot of the ratio of integrated irradiance based on the observed DNI in Matsumoto, Japan 149
A.76 Concentrator photovoltaics (CPV) for 'breeding plants', which collects a direct beam and provides diffused sunlight to plants 149
A.77 Soft shadow by rejection of direct beam but transmittance of diffused sunlight 149
A.78 Measurement results, Day4™ Cell and unmodified cell 152
A.79 Non-imaging Fresnel lens with receiver plane 152
A.80 Simulation of Fresnel lens acceptance angles for incident radiation from the angles θ and φ 153
A.81 Pilot concentrator 153
A.82 Concentrator-based power system: Conceptual view 153
A.83 Cross-section of a bifacial cell 154
A.84 Photograph of a bifacial cell 155
A.85 Typical I–V and P–V curves of a bifacial PV module HB3M-48 155
A.86 Photograph of PV module HB3M-48 155
A.87 Photograph of a bifacial vertical fence-integrated PV array 155
A.88 Simulation results on daily generated power distribution 156
A.89 Simulation results on annually generated power with various azimuth angles and bifacialities 156
A.90 Overview of the PV array on the campus of the Chitose Institute of Science and Technology 156
A.91 Layout of bifacial fence-type PV array 156
A.92 Electrical diagram of the PV system 157
A.93 Measured generated power on a fine day without snow on the ground 157
A.94 Measured generated power on a fine day with snow on the ground 157
A.95 Electrical diagram of a VLS-PV system with bifacial PV modules 158
A.96 A schematic image of a bifacial PV fence-type installation 158
A.97 A schematic assembly process of a fence-type bifacial PV 158
A.98 A schematic layout of a 500 kW system (layout no 1) 159
A.99 A schematic layout of a 500 kW system (layout no 7) 159
A.100 A schematic layout of a 500 kW system (layout no 6) 159
A.101 Monthly global irradiation in case A 161
A.102 Monthly global irradiation in case B 161
A.103 Annual power output of bifacial PV system (simulation) 161
A.104 Springerville PV generating plant 164
A.105 Typical 135 kW DC system 164
A.106 Xantrex PV-150 system inverter 164
A.107 Average monthly final yield for all systems over operating history 165
A.108 Monthly reference yield (sun hours) for 2004 165
A.109 Average monthly performance ratio for all systems in 2004 166
A.110 Unscheduled maintenance events by component 166
A.111 Unscheduled maintenance costs by component 166
A.112 Average monthly capacity factor for Springerville systems 168
A.113 Photograph of TEP's Springerville PV plant 168
A.114 Installed balance-of-system (BOS) costs for the mc-Si PV installations 169
A.115 Life-cycle energy consumption of BOS: Reference case 171
A.116 Life-cycle GHG emissions of BOS: Reference case 171

A.117 Comparison of life-cycle energy use between assessments 1 and 2 172
A.118 Comparison of life-cycle energy use between assessments 1 and 2 172
A.119 Life-cycle energy consumption of BOS: Impact of inverters' life expectancy 173
A.120 Life-cycle GHG emissions of BOS: Impact of inverters' life expectancy 173
A.121 Life-cycle energy consumption of BOS: Impact of recycled aluminium 173
A.122 Life-cycle GHG emissions of BOS: Impact of recycled aluminium 173
A.123 Life-cycle GHG emissions of BOS: Impact of Life Cycle Inventory (LCI) data 174
A.124 Schematic parabolic trough collector 177
A.125 Heat balance at land surface 180
A.126 Water balance at land surface 180
A.127 An example of land surface disturbances induced by a decrease in albedo 181

B.1 Trends in total energy consumption by fuel 185
B.2 World primary energy supply projection by fuel (*World Energy Outlook*), 1971–2030 186
B.3 World primary energy supply projection by region (*World Energy Outlook*), 1971–2030 186
B.4 Regional energy and economy per capita 186
B.5 Comparison of energy supply and CO_2 emissions from 1971–2020 186
B.6 World primary energy supply by fuel (*International Energy Outlook* reference case), 1990–2025 187
B.7 Oil price projections, 1970–2025 187
B.8 World primary energy supply by fuel (*World Energy, Technology and Climate Policy Outlook* reference scenario), 1990–2025 187
B.9 World primary energy supply by fuel (sustainable development scenario), 1990–2050 188
B.10 World primary energy supply and CO_2 emissions (sustainable development scenario), 1990–2025 188
B.11 World energy supply by fuel (Shell, dynamic as usual), 1975–2050 188
B.12 World energy supply by fuel (Shell, spirit of the coming age), 1975–2050 188
B.13 World primary energy supply by fuel (German Advisory Council on Global Change), 2000–2100 189
B.14 Annual CO_2 emissions and carbon sequestration (German Advisory Council on Global Change), 2000–2100 189
B.15 World renewable energy supply by fuel (European Renewable Energy Council), 2001–2040 189
B.16 Growth in renewable energy technologies in each period 189
B.17 Comparison of world energy scenarios, 1980–2050 189
B.18 Generation sources in a renewable energy-dominated grid 191
B.19 Blended generation: Wind and solar 191
B.20 Comparison of different hydrogen systems driven by fossil fuel and renewable energy, respectively 195
B.21 Solar farm of dish concentrators on aboriginal lands near Alice Springs, Australia; each dish is nominally 20 kW (note the visitors for scale) 198
B.22 Characteristics of a spectral splitter mirror 198
B.23 Schematic of system shows sunlight reflected and focused on the receiver, with reflected infrared directed to a fibre-optics waveguide for transport to a high temperature solid oxide electrolysis cell; solar electricity is sent to the same electrolysis cell that uses both heat and electricity to split water 198

TABLES

E2.1 Solar irradiation, energy yield and PV electricity-generation cost data compared with the conventional electricity price level and local feed-in tariff rates for stationary systems at two representative sites in four Mediterranean countries xxxiv
E3.1 Land area, electricity requirements and solar electricity potential in the Middle East xxxv
E3.2 Expected economic benefits to Israel of VLS-PV plant introduction during the first 36 years xxxvi
E4.1 Proposed projects for VLS-PV development in Mongolia xxxviii
E4.2 Capital investment for 8 MW LS-PV system xl
E5.1 Estimated project cost for 10 MW PV power generating system xliii

1.1 Summary of a scenario for VLS-PV development 6

2.1 Electricity tariffs in Morocco in 2002 9
2.2 Installed capacity of renewable energy in Morocco in 2002 9
2.3 Planned capacity of renewable energy in Morocco in 2002 10
2.4 Moroccan institutions on energy, renewable energy and PV 10

2.5	Overview of Moroccan PV companies	11
2.6	Electricity tariffs in Tunisia in 2002 (tax not included)	13
2.7	Tunisian institutions on energy, renewable energy and PV	13
2.8	Expected share of different energy sources for electricity generation in 2010 and 2020	14
2.9	Installed capacity of renewable energy in Portugal (not including the Azores and Madeira)	15
2.10	Electricity tariff in Portugal in 2002	16
2.11	Expected installed capacity of renewable energy in Portugal in 2010	16
2.12	Portuguese institutions on energy, renewable energy and PV	16
2.13	PV companies in Portugal	17
2.14	Installed capacity of renewable energy in Spain	18
2.15	Planned electricity capacity from renewable energy in Spain in 2011	19
2.16	Spanish institutions on energy, renewable energy and PV	19
2.17	Spanish feed-in prices: Fixed prices in 2004 and bonus system in comparison	19
2.18	PV companies in Spain	20
2.19	Key data of the Mediterranean countries	21
2.20	Cost structure for stationary VLS-PV systems in the MW range	22
2.21	Annual cost per kW for a total system cost of 4 000 EUR/kW	22
2.22	Solar irradiation, energy yield and PV electricity-generation cost data compared with the conventional electricity price level and local feed-in tariff rates for stationary systems at two representative sites in four Mediterranean countries	22
3.1	Middle East statistics for 2002	26
3.2	Potential electrical yield of VLS-PV technologies in the Negev	27
3.3	Estimated annual solar electricity potential of countries in the Middle East	27
3.4	Expected economic benefits to Israel of VLS-PV plant introduction during the first 36 years	32
3.5	Expected economic benefits to Saudi Arabia of VLS-PV plant introduction during the first 36 years	36
3.6	Expected economic benefits to Iran of VLS-PV plant introduction during the first 36 years	38
3.7	Expected economic benefits to Turkey of VLS-PV plant introduction during the first 36 years	40
3.8	Expected economic benefits to Egypt of VLS-PV plant introduction during the first 36 years	42
3.9	Expected economic benefits to the United Arab Emirates of VLS-PV plant introduction during the first 36 years	44
3.10	Expected economic benefits to Iraq of VLS-PV plant introduction during the first 36 years	46
3.11	Expected economic benefits to Kuwait of VLS-PV plant introduction during the first 36 years	48
3.12	Expected economic benefits to Syria of VLS-PV plant introduction during the first 36 years	50
4.1	Existing power plants in Mongolia (as of 2000)	56
4.2	Trends in electricity balance in Mongolia (GWh)	57
4.3	Electricity load demand 2005–2020	57
4.4	Population and industry of cities along the railway	57
4.5	Proposed projects of VLS-PV development in Mongolia	59
4.6	Installed instrument for research in Sainshand	64
4.7	Measurement parameters	65
4.8	Measured meteorological data in Sainshand	65
4.9	Electricity consumption in Dunhuang in 2003 (GWh/year)	68
4.10	Electricity price by sector (yuan/kWh)	68
4.11	Irradiation data in Dunhuang	68
4.12	Specifications of proposed PV module	69
4.13	Specifications of proposed DC/AC inverter	69
4.14	Requirements of PV system components	69
4.15	Capital investment for 8 MW LS-PV system	70
5.1	Mining projects anticipated in the region	72
5.2	Estimated project cost for 10 MW PV power-generation system (AUD)	75
5.3	Assumptions for 10 MW pilot PV power plant	75
5.4	Estimated power generation from 10 MW pilot PV power plant	75
5.5	Annual cost for 10 MW pilot PV power plant (AUD)	76
5.6	Generation cost of 10 MW pilot PV power plant	76
5.7	Estimated additional revenues by rainwater collection from PV modules	76

5.8 Effective generation cost considering revenues from rainwater collection 76

6.1 Summary of community proposed 85
6.2 Area of major arid regions 88
6.3 Salt tolerance of plants 89
6.4 Salt content of soils and reduction of yield 92

A.1 Variety of deserts 120
A.2 Global irradiation in the world's deserts 120
A.3 Balance of system (BOS) equipment for a 100 MW system 120
A.4 Case studies in this research 120
A.5 PV module specifications 120
A.6 Local unit prices of steel and concrete (USD/tonne) 120
A.7 Unit prices of cable and components 120
A.8 Annual operation and maintenance costs 121
A.9 Energy data for Japan and China 121
A.10 Energy and CO_2 contents of products used in this study 121
A.11 Data for transportation of system components 121
A.12 Data for system construction in China 121
A.13 Economic data used in this study 121
A.14 Dividing beam and diffuse irradiation model 121
A.15 Estimated irradiation in desert areas (kWh/day) 122
A.16 Assumptions of efficiency 122
A.17 Loss ratios in six deserts with 30° tilt angle structure 122
A.18 Requirements of 100 MW PV system components 122
A.19 Annual cost of a 100 MW VLS-PV system in the Sahara Desert (Nema) 123
A.20 Annual cost of a 100 MW VLS-PV system in the Sahara Desert (Ouarzazate) 123
A.21 Annual cost of a 100 MW VLS-PV system in the Negev Desert 124
A.22 Annual cost of a 100 MW VLS-PV system in the Thar Desert (Jodhpur) 124
A.23 Annual cost of a 100 MW VLS-PV system in the Sonoran Desert (Chihuahua) 125
A.24 Annual cost of a 100 MW VLS-PV system in the Great Sandy Desert (Port Headland) 125
A.25 Annual cost of a 100 MW VLS-PV system in the Gobi Desert (Huhhot) 126
A.26 Annual cost of a 100 MW VLS-PV system in the Gobi Desert (Sainshand) 126
A.27 Generation cost of a 100 MW VLS-PV system (US cents/kWh) 127
A.28 System components (for 30°) 127
A.29 Annual cost for a 100 MW VLS-PV system with m-Si modules 128
A.30 Annual cost for a 100 MW VLS-PV system with a-Si modules 128
A.31 Annual cost for a 100 MW VLS-PV system with CdTe modules 129
A.32 Annual cost for a 100 MW sun-tracking VLS-PV system 129
A.33 Generation cost of 100 MW VLS-PV systems at optimal tilt angle (30°) 129
A.34 Total energy requirement and energy payback time of a 100 MW sun-tracking VLS-PV system using m-Si PV modules 129
A.35 Life-cycle CO_2 emissions and life-cycle CO_2 emission rate of a sun-tracking 100 MW PV system with m-Si PV modules 129
A.36 Main characteristics of satellite images 131
A.37 The resource potential of six world deserts 137
A.38 Uncorrected peak efficiency measurement 144
A.39 Low-concentrator photovoltaics (LCPV) sun concentrator system cost estimates 153
A.40 Range of solar irradiation, state of Arizona 154
A.41 Specifications of a bifacial PV module HB3M-48 155
A.42 Electrical performance of the sub-arrays 157
A.43 Comparison of irradiances and generated energies 157
A.44 Evaluation results of wind pressure resistance of fence-type PV arrays 160
A.45 Layout summary of fence-type bifacial PV 160
A.46 Global irradiation on the inclined plane for vertical installation with an inclination angle of 90° (kWh/m2/day) 160
A.47 Calculation conditions used for simulations of power generated 162

A.48 List of Springerville crystalline silicon systems 164
A.49 Maintenance costs as a percentage of capital investment 166
A.50 Cost breakdown for Springerville systems 167
A.51 System costs for the future (USD/Wdc) 167
A.52 Material inventory, in kilograms, of the BOS components for a 1 MW field PV plant 170
A.53 Energy use and greenhouse gas (GHG) emissions for BOS production for reference case (33 % secondary aluminium and a 30-year lifetime of inverters, with 10 % part replacement every ten years) 171
A.54 Potential electrical yield of types of VLS-PV technology in the Negev 177
A.55 Ordinary ranges of ground albedo 182
A.56 Assumed ground albedo for case studies 182

B.1 Proved reserves of fossil fuels and uranium 185
B.2 Comparison of results between reference and carbon abatement (CA) cases 187
B.3 Characteristics of renewable energy sources 190
B.4 Electrical energy storage technologies 192
B.5 A chronology of the main hydrogen and fuel cell events since 1766 194
B.6 Various theoretical means of hydrogen production 195
B.7 Possible means of solar hydrogen production 196
B.8 Electrolysis cell measurements at beginning and end of 17 minutes of system operation 198
B.9 Component and system costs for 10 MW hybrid concentrator photovoltaic (CPV) project for electrolytic production of hydrogen 199
B.10 Hydrogen production data for mature 10 MW plant 199
B.11 Cost comparison for the hybrid CPV production of electrolytic hydrogen 199

List of Acronyms and Abbreviations

°	degree
μm	micrometre
Ω	ohm
AC	alternating current
ADENE	National Energy Agency of Portugal
AEI	Australian Ethical Investments Ltd
AGO	Australian Greenhouse Office
Ah	ampere hour
AIST	National Institute of Advanced Industrial Science and Technology (Japan)
Al	aluminium
AMISOL	Association Marocaine des Industries Solaires
ANER	Tunisian National Agency for Renewable Energies
AOD	aerosol density
APISOLAR	Associação Portuguesa da Indústria Solar (Portugal)
APPA	Asociación de Productores de Energías Renovables (Spain)
APS	Arizona Public Service
a-Si	amorphous silicon
ASIF	Asociación para la Indústria Fotovoltáica (Spain)
Au	gold
AVHRR-NDVI	Advanced Very High Resolution Radiometer Normalized Difference Vegetation Index
BATS	Biosphere–Atmosphere Transfer Scheme
bbl	barrel of oil (equivalent to 42 US gallons or approximately 159 L)
BMU	German Federal Ministry of Environment
BOO	build, operate and own
BOS	balance of system
BOT	build, operate and transfer
BP	British Petroleum
BSF	back surface field
C	Celsius
Ca	calcium
CA	carbon abatement
$CaCl_2$	calcium chloride
$CaSO_4$	calcium sulphate
CAPE	convective available potential energy
CCC	critical coagulation concentration
CCM3	Community Climate Model version 3.0 (of NCAR)
CDER	Centre for the Development of Renewable Energy Sources
CDM	clean development mechanism
CdTe	cadmium telluride
CEC	cation exchange capacity
CEReS	Centre for Environmental Remote Sensing
CES	Central Energy System (Mongolia)
CESI	Centro Electtrotecnico Sperimentale Italiano
CFC	chlorofluorocarbon
CIGS	copper indium gallium diselenide, $Cu(In,Ga)Se_2$
CIPIE	Commission Interdépartementale de la Production Indépendante d'Electricité (Tunisia)
CITET	Centre International des Technologies de l'Environnement
CLM	Community Land Model
cm	centimetre
cm^2	square centimetre
CNE	Comision Nationál de Energía – Spanish Regulation Authority
CNI	Confederación Nacional de Instaladores (Spain)
CO_2	carbon dioxide
CPC	compound parabolic concentrator
CPV	concentrator photovoltaic(s)
c-Si	crystalline silicon
CSP	concentrated solar (thermal) power
CSPIE	Commission Supérieure de la Production Indépendante d'Electricité (Tunisia)

Cu	copper
CV cable	cross-linked polyethylene insulated vinyl sheath cable
DAS	data acquisition systems
DC	direct current
DENA	Deutsche Energie Agentur
DER	Portuguese Ministry of Economy
DGCL	Direction Générale des Collectivités Locales (Morocco)
DGE	General Directorate for Energy
DOE	US Department of Energy
DNI	direct normal irradiance
EC	European Commission
EC	electrical conductivity
ED	electro-dialysis
EDF	Electricité de France
EDP	Eletricidade de Portugal
EES	East Energy System (Mongolia)
EIA	Energy Information Administration (US)
EJ	exajoule (10^{18} J)
ENDESA	Empresa Energetica Espanola
ENEA	Ente per le Nuove Tecnologie, l'Energia el'Ambiente (Italy)
ENI	Ente Nazionale Idrocarburi
EPS	Environmental Portfolio Standard programme
EPT	energy payback time
EREC	European Renewable Energy Council
ERSE	Entidade Reguladora dos Serviços Energéticos (Portugal)
ESP	exchangeable sodium percentage
EU	European Union
F	Fahrenheit
FACTS	flexible AC transmission system
FAO	United Nations Food and Agriculture Organization
FC	fuel cell
Fe	iron
FF	fill factor
FRP	fibre-reinforced plastic
g	gram
GaAs	gallium arsenide
g-C	gram carbon
GCM	general circulation model
GDP	gross domestic product
GEF	Global Environment Facility
GEWEX/GAME	Global Energy and Water Cycle Experiment – Asian Monsoon Experiment
GHG	greenhouse gas
GIS	geographic information systems
GJ	gigajoules
GPS	global positioning system
GW	gigawatt
GWh	gigawatt hour
H_2	hydrogen
ha	hectare
HCPV	high concentration photovoltaic(s)
HDPE	high-density polyethylene
HTSC	high temperature superconducting
Hz	hertz
IAHE	International Association of Hydrogen Energy
IDAE	Instituto para la Diversificación y Ahorro de la Energía (Spain)
IEA	International Energy Agency
IEC	International Electrotechnical Commission
IEEE	Institute of Electrical and Electronics Engineers
IFC	International Finance Corporation
IFCO	Iranian Fuel Conservation Organization
ILO	International Labour Organization
IMET	Italian Territorial and Environment Ministry
IPCC	Intergovernmental Panel on Climate Change
Isc	short-circuit current
ISO	International Organization for Standardization
J	joule
JIS	Japan Industrial Standards
JSES	Japan Solar Energy Society
JWA	Japan Weather Association
K	kelvin
k	kilo
K	potassium
kg	kilogram
KIER	Korea Institute of Energy Research
KJ	kilojoule
Kt	kilotonne
km	kilometre
kV	kilovolt
kVA	kilovolt ampere
kW	kilowatt
kWh	kilowatt hours
kWh/m^2	kilowatt hours per square metre
kWp	kilowatt peak
L	litre
LAI	leaf area index
LCA	life-cycle analysis/assessment
LCI	Life Cycle Inventory
LCM	Land Surface Model
LCPV	low-concentrator photovoltaics
LEC	levelled energy cost
LEO	low energy office
LH_2	liquid hydrogen
LNG	liquid natural gas
LSM	Land Surface Model
LS-PV	large scale photovoltaic power generation
m	metre

m^2	square metre
m^3	cubic metre
MAPE	Medida de Apoio ao Aproveitamento do Potencial Energético e Racionalização de Consumos
MEDREC	Mediterranean Renewable Energy Centre
met	metabolic rate
METPV	Meteorological Test Data for Photovoltaic System
Mg	magnesium
MHIR	Mizuho Information and Research Institute, Inc (Japan)
MIBEL	Mercado Ibérico de Electricidade
MJ	megajoule
ml	millilitre
MLC	maximum likelihood classifier
mm	millimetre
MOVPE	metal organic vapour-phase epitaxy
mph	miles per hour
MPP	maximum power point
MPPT	maximum power point tracking
MRET	Mandatory Renewable Energy Target (Australia)
m/s	metres per second
MSAVI	Modified Soil Adjusted Vegetation Index
m-Si	monocrystalline silicon
Mtoe	million tonne oil equivalent
MW	megawatt
MWh	megawatt hour
MWp	megawatt peak
N	newton
Na	sodium
NASA	National Aeronautics and Space Administration
NCAR	National Centre for Atmospheric Research
NEDO	New Energy and Industrial Technology Development Organization
NHA	National Hydrogen Association
NIR	near infrared ray
NOx	nitrogen oxide
NOAH	National Centres for Environmental Prediction/Oregon State University/Air Force/Hydrologic Research Laboratory
NREL	National Renewable Energy Laboratory
NUM	National University of Mongolia
O_2	oxygen
OA	operating agent
OECD	Organisation for Economic Co-operation and Development
ODA	official development assistance
O&M	operation and maintenance
OMEL	Companía Operadora del Mercado Espanol de Electricidad
ONE	Office National de l'Electricité (Morocco)
OPEC	Organization of Petroleum Exporting Countries
PAEE	*Plan de Ahorro y Eficiencia Energético*
PEM	proton exchange membrane
PER	*Plan de Fomento para Energías Renovables*
PERG	Programme for Decentralized Electrification
PERL	passivated emitter, rear locally diffused
PFC	perfluorocarbon
PFER	*Plan de Fomento de las Energías Renovables*
PMMA	polymethylmethacrylate
PNR	Programme National de Recherche
PNR	Tunisian National Research Program
POE	Programa Operacional de Economia (Portugal)
PPA	power purchase agreement
ppm	parts per million
PPP	power purchasing party
p-Si	polycrystalline silicon
PTJ	Projektträger Jülich (funding agency of BMU)
PV	photovoltaic(s)
PVMaT	Photovoltaic Manufacturing Technology programme
PVMTI	Photovoltaic Market Transformation Initiative
PVTEC	Photovoltaic Power Generation Technology Research Association (Japan)
RAMS	Regional Atmospheric Modeling System
RAPS	Remote Area Power Supply programme
R&D	research and development
RB	random barrier
REN	Rede Eléctrica National
RES	renewable energy sources
RH	relative humidity
RISE	Research Institute for Sustainable Energy (Australia)
RMS	root mean square
RO	reverse osmosis
RP	reference price
R/P	reserves/production ratio
RRRGP	Renewable Remote Power-Generation Programme
SAI	stem area index
SAR	sodium adsorption ratio
SCE-t	standard coal equivalent tonne

SEDO	Sustainable Energy Development Office (Australia)
SEI	Sistema Eléctrico Independente (Portugal)
SEIA	Solar Energy Industries Association
SEN	Sistema Eléctrico Nacional (Portugal)
SEP	Sistema Eléctrico Publico (Portugal)
SHS	solar home system
Si	silicon
SINES	Société Internationale de l'Energie et des Sciences (Tunisia)
SO_2	sulphur dioxide
SOEC	solid oxide electrolysis cell
SOLTEN	Sol Tenerife (Project name of Tenerife PV project)
SPES	Sociedade Portuguesa de Energia Solar (Portugal)
SPM	SunLight Power Maroc
ST	solar thermal
STC	standard test conditions
STEG	Société Tunisienne d'Electricité et du Gaz (Morocco)
SVC	static var compensator
t	tonne
TACSR	thermal-resistant aluminium conductor
t-C	tonne carbon
TEP	Tucson Electric Power
TJ	terajoule
t.km	tonne per kilometre
toe	tonne oil equivalent
TUAT	Tokyo University of Agriculture and Technology
TWh	terawatt hour
UAE	United Arab Emirates
UK	United Kingdom
UL	Underwriters Laboratories
UNEP	United Nations Environment Programme
US	United States
UV	ultraviolet
V	volt
VIC	Variable Infiltration Capacity
VLS-PV	very large scale photovoltaic power generation
W	watt
WBGU	German Advisory Council on Global Change
WCPEC-3	Third World Conference on Photovoltaic Energy Conversion
WE-NET	World Energy Network
WES	Western Energy System (Mongolia)
W/m^2	watts per square metre
WWF	World Wide Fund for Nature
YSZ	yttria-stabilized zirconia

1. Introduction

1.1 OBJECTIVES

The scope of this study is to examine and evaluate the potential of very large scale photovoltaic power generation (VLS-PV) systems, which have capacities ranging from several megawatts to gigawatts, and to develop practical project proposals for demonstrative research towards realizing VLS-PV systems in the future.

Our previous report, *Energy from the Desert: Feasibility of Very Large Scale Photovoltaic Power Generation (VLS-PV) Systems*, identified the key factors enabling the feasibility of VLS-PV systems and clarified the benefits of these systems' applications for neighbouring regions, the potential contribution of system application to global environmental protection, and renewable energy utilization in the long term. It is apparent from the perspective of the global energy situation, global warming and other environmental issues that VLS-PV systems can:

- contribute substantially to global energy needs;
- become economically and technologically feasible;
- contribute considerably to the environment; and
- contribute considerably to socio-economic development.

This report will reveal virtual proposals of practical projects that are suitable for selected regions, and which enable sustainable growth of VLS-PV in the near future, and will propose practical projects for realizing VLS-PV systems in the future.

1.2 CONCEPT OF VERY LARGE SCALE PHOTOVOLTAIC POWER GENERATION (VLS-PV) SYSTEMS

Solar energy is low-density energy by nature. To utilize it on a large scale, a massive land area is necessary. One third of the land surface of the Earth is covered by very dry desert and high-level insolation (in-coming solar radiation), where there is a lot of available space. It is estimated that if a very small part of these areas, approximately 4 %, was used for installing photovoltaic (PV) systems, the resulting annual energy production would equal world energy consumption.

A rough estimation was made to examine desert potential under the assumption of a 50 % space factor for installing PV modules on the desert surface as the first evaluation. The total electricity production would be 2 081 x 10^3 TWh (= 7,491 x 10^{21} J), which means a level of almost 17 times as much as the world primary energy supply (0,433 x 10^{21} J in 2002). These are hypothetical values, ignoring the presence of loads near the deserts. Nevertheless, these values at least indicate the high potential of developing districts located in solar energy-rich regions as primary resources for energy production.

Figure E1.1 shows that the Gobi Desert area in western China and Mongolia can generate as much electricity as the current world's primary energy supply. Figure E1.2 depicts an image of a VLS-PV system in a desert area

Three approaches are under consideration to encourage the spread and use of PV systems:

1 Establish small scale PV systems independent of each other. There are two scales for such systems: installing stand-alone, several hundred watt-class PV systems for private dwellings, and installing 2 to 10 kW-class systems on the roofs of dwellings, as well as 10 to 100 kW-class systems on office buildings and schools. Both methods are already being used. The former is employed to furnish electric power in developing countries and is known as the solar home system (SHS), and the latter is used in industrialized countries.

2 Establish 100 to 1 000 kW-class mid-scale PV systems on unused land on the outskirts of urban areas. The IEA Photovoltaic Power Systems Programme (PVPS)/Task 6 studied PV plants for this scale of power generation. Systems of this scale are in practical use in about a dozen sites in the world at

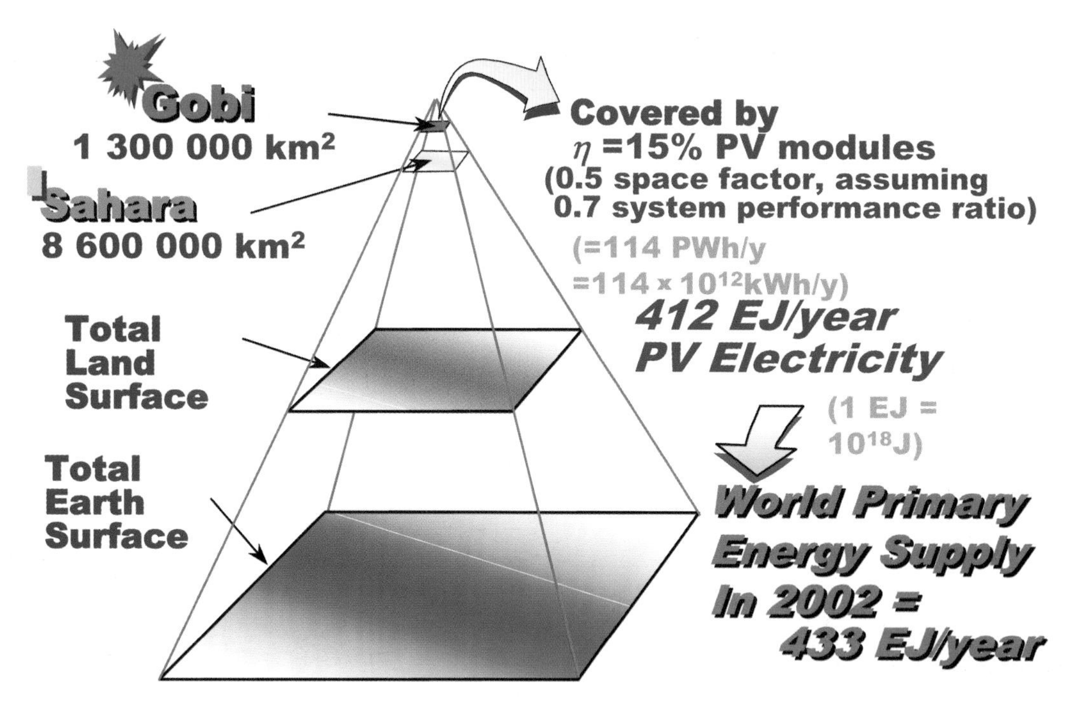

Figure E1.1 Solar pyramid

the moment; but their number is expected to increase rapidly in the early 21st century. This category can be extended up to multi-megawatt size.

3 Establish PV systems larger than 10 MW on vast barren, unused lands that enjoy extensive exposure to sunlight. In such areas, a total of even more than 1 GW of PV system aggregation can be easily realized. This approach enables installing a large number of PV systems quickly. When the cost of generated electric power is lowered sufficiently in the future, many more PV systems will be installed. This may lead to a drastically lower cost of electricity, creating a positive cycle between cost and consumption. In addition, this may become a feasible solution for future energy need and environmental problems across the globe, and ample discussion of this possibility is worthwhile.

The third category corresponds to VLS-PV. The definition of VLS-PV may be summarized as follows:

- The size of a VLS-PV system may range from 10 MW to 1 or several GW, consisting of one plant, or an aggregation of plural units distributed in the same district operating in harmony with each other.
- The amount of electricity generated by VLS-PV can be considered significant for people in the district, nation or region.
- VLS-PV systems can be classified according to the following concepts, based on their locations:
  - land based (arid to semi-arid deserts);
  - water based (lakes, coastal and international waters);
  - locality options: developing countries (lower-, middle- or higher-income countries; large or small countries) and Organisation for Economic Co-operation and Development (OECD) (countries).

Although the concept of VLS-PV includes water-based options, much discussion is required. It is not neglected here, but is viewed as a future possibility outside the major emphasis of this report.

The advantages of VLS-PV are summarized as follows:

- It is very easy to find land in or around deserts appropriate for large energy production with PV systems.
- Deserts and semi-arid lands are normally high insolation areas.
- The estimated potential of such areas can easily supply the estimated world energy needs by the middle of the 21st century.
- When large-capacity PV installations are constructed, step-by-step development is possible through utilizing the modularity of PV systems. According to regional energy needs, plant capacity can be increased gradually. It is an easier approach for developing areas.
- Even very large installations are quickly attainable in order to meet existing energy needs.
- Remarkable contributions to the global environment can be expected.
- When VLS-PV is introduced to some regions, other types of positive socio-economic impacts may be induced, such as technology transfer to regional PV

Figure E1.2 Image of a VLS-PV system in a desert area

industries, new employment and growth of the economy.

- The VLS-PV approach is expected to have a drastic influence on the chicken-and-egg cycle in the PV market. If this does not happen, the goal of achieving VLS-PV may move a little further away.

VLS-PV systems are already feasible. PV systems with a capacity of more than 10 MW were installed and connected to the public grid in 2006. Currently, a land-based VLS-PV concept is in its initial stage.

These advantages make VLS-PV a very attractive option and worthy of being discussed in the context of the 21st century's global energy needs.

1.3 PRACTICAL APPROACHES TO REALIZE VLS-PV

Solar energy resources, PV technologies and renewable energy will help to realize important economic, environmental and social objectives in the 21st century, and will be a critical element in achieving sustainable development.

In order to advance the transition to a global energy system for sustainable development, it is very important to orient large and increasing investments towards renewable energy. If investments continue with business as usual, mostly in conventional energy, societies will be further locked into an energy system incompatible with sustainable development and one that further increases the risks of climate change.

In order to promote renewable energy and its diversity of challenges and resource opportunities (as well as the financial and market conditions among and within regions and countries) different approaches are required. Establishing policies for developing markets, expanding financing options and developing the capacity required are crucial in incorporating the goals of sustainable development within new policies.

As a result, increasing renewable energy use is a policy, as well as a technological, issue that applies directly to achieving VLS-PV.

VLS-PV is a major project that has not yet been experienced. Therefore, in order to realize VLS-PV, it is necessary to identify issues that should be solved and to discuss a practical development scenario.

When thinking about a practical project proposal for achieving VLS-PV, a common objective is to find the best sustainable solution. System capacity for suitable development is dependent upon each specific site with its own application needs and available infrastructure, especially access to long-distance power transmission, and human and financial resources.

From the technological viewpoint, it is important to start with the research and development (R&D) or pilot stage when considering overall desert development. In terms of commercial operation, the pilot or demonstration stage is more crucial.

The proposals developed in this report may motivate stakeholders to realize a VLS-PV project in the near future. Moreover, the practical project proposals, with their different viewpoints and directions, will provide essential knowledge for developing detailed instructions for implementing VLS-PV systems.

II

2. The Mediterranean region: Case study of very large scale photovoltaics

The economic conditions for VLS-PV systems in the Mediterranean region were examined. Originally focusing on the Sahara Desert-bordering countries of Morocco and Tunisia, Portugal and Spain were also included in order to compare the impact of recently approved PV feed-in tariffs with the less-supportive framework environments in Northern Africa. Two sites were selected for each country, one more affected by marine climate influences with lower irradiation, and one representing a higher irradiated desert-like location. The study was performed from a professional project developer's perspective by determining PV electricity generation cost and potential revenues from electricity sales from VLS-PV systems to customers, either to consumers on a standard electricity price level or to grid-operating entities on a feed-in tariff basis.

Experience with already realized MW systems shows that a stationary (non-tracking, flat-plate) large scale PV installation can now be realized at around 4 010 EUR/kW. The value serves as a fair approximation for the following calculations, including a limited overhead cost of 8 %. Note that this overhead does not yet include a further 6 to 8 % capital acquisition cost, which is typically required if the project is sold to private or fund investors, a frequently encountered way of project financing at present. Three-quarters of the system cost is for the PV modules, the module prices thus being the main parameter for future cost reduction. For annual cost, 20 years' linear depreciation and 100 % loan financing at a 5 % interest rate serve as model parameters, which, of course, need to be adapted for concrete project proposals. No investment for land was considered here, and the estimated land rental cost is instead included in the 2 % annual operation and maintenance cost. The total annual cost per kW was assumed to be equal to 380 EUR/kW at all locations.

PV electricity generation costs in the analysed Mediterranean countries are between 23,9 and 37,7 EUR cents/kWh, as shown in Table E2.1. As expected, the generation cost for PV calculated is higher than the price level of conventional electricity drawn from the grid in all places. In this context, it is important to note

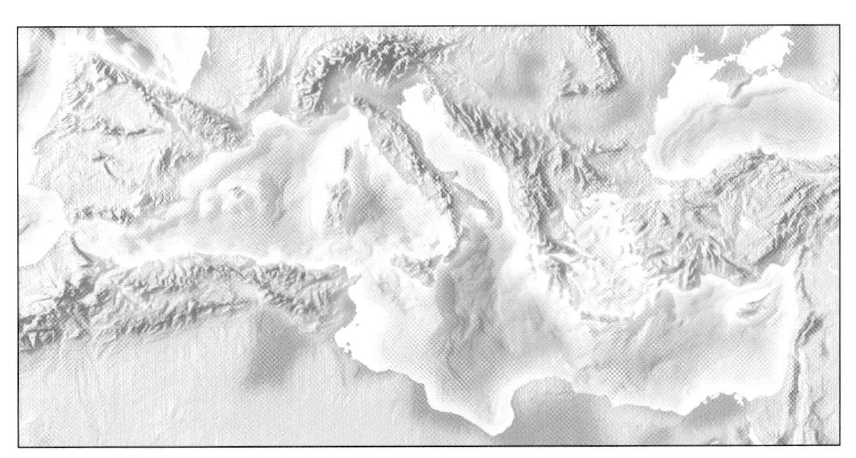

Figure E2.1 The Mediterranean regions within the case study

Table E2.1 Solar irradiation, energy yield and PV electricity-generation cost data compared with the conventional electricity price level and local feed-in tariff rates for stationary systems at two representative sites in four Mediterranean countries

Country	Site	Annual global irradiation ($kWh/m^2/year$)	Annual energy yield (kWh/kW/year)	Generation cost for PV (EUR cents/kWh)	Conventional grid electricity price level (EUR cents/kWh)	Feed-in tariff rate (EUR cents/ kWh)
Morocco	Casablanca	1 772	1 337	28,4	~8–12	None
	Quarzazate	2 144	1 589	23,9		
Tunisia	Tunis	1 646	1 219	31,2	~2–5	None
	Gafsa	1 793	1 339	28,4		
Portugal	Porto	1 644	1 312	29,0	~12	~55 <5 kW
	Faro	1 807	1 360	27,9		~31–37 >5 kW
Spain	Oviedo	1 214	1 008	37,7	~9	41,44 <100 kW
	Almería	1 787	1 372	27,7		21,62 >100 kW

that the assumed 100 % loan financing makes up a substantial proportion of the generation cost. Generation costs well below 20 EUR cents/kWh result for almost all sites, without including the financing cost and the 7 % safety margin in the annual energy yield. This confirms that PV generation costs are not too far above the conventional price line and could reach or even fall below this line after a price decrease of PV modules, which is already anticipated by foreseeable advances in technology and economy of scale with increasing mass production.

Although the lowest generation cost of 23,9 EUR cents/kWh is reached in Quarzazate, this is not low enough to become attractive for a buy-back scheme in Morocco, even considering that the general electricity price level is comparatively high there. Tunisia has a centralized electricity industry with a low price level, making the situation for PV even more difficult. Morocco and Tunisia have no specific legal framework to support PV electricity generation and no existing feed-in tariff. Therefore, the economic feasibility for VLS-PV is low in these Northern African countries if based on achieving income from electricity sales to consumers or on the grid alone – that is, not considering any investment subsidies.

Portugal and Spain also have much lower prices for conventional electricity than the calculated PV electricity-generation cost. In these countries, however, smaller systems appear to be economically feasible with the available feed-in tariffs at higher irradiation sites. The exciting question for VLS-PV is whether large systems can also be economically operated under special circumstances. Answering this question requires a closer look at the conditions in these Southern European countries.

In summary, we expect the best conditions for VLS-PV to develop in Spain on an intermediate time scale of two to five years, even though there are now several larger projects proposed in Portugal. Concrete realization of VLS-PV projects depends upon successful negotiation between project developers, PV and electricity industries, and politicians with regard to acceptance, sustainability and incentives in every single project.

Generally, VLS-PV as a centralized electricity source needs to compete with conventional electricity sources and other centralized renewable energy sources (RES), such as solar thermal and wind energy, which are also proposed and implemented strongly in the studied region, in addition to decentralized small scale PV. Lower investment costs, additional support and/or higher feed-in tariffs for large systems are required, in addition to intelligent financing schemes, in order to make VLS-PV economically feasible in the considered Mediterranean region on a larger scale.

II

3. The Middle East region: A top-down approach for introducing VLS-PV plants

A top-down approach to providing solar electricity to any given region must address the following five questions:

1. How much land area is available for the harvest of sunshine, and how much electricity could this resource provide?
2. How much electricity is required?
3. What kind of technology should be used, and how much of it would be needed for the task?
4. At what rate should the technology be introduced?
5. What monetary resources would be required and how could these resources be provided?

This study provides a set of answers for the principal electricity-consuming countries in the Middle East.

First, we studied the current electricity requirements and land availability of all countries in the region, with the specific aim of being able to provide some 80 % of their total electricity needs with solar energy within 36 years. For all of the major electricity-producing countries, it was concluded that land area considerations should present no obstacles to such aims.

Second, we studied *existing* concentrator photovoltaic (CPV) technology at the system component level, considering the expected costs involved in their mass production. These costs included the VLS-PV plants and the necessary mass production facilities for

Table E3.1 Land area, electricity requirements and solar electricity potential in the Middle East

Country	Land area (km^2)	Electricity production in 2002 (TWh)	Solar electricity potential by technology: Static, 30° tilt ($TWh\ y^{-1}$)	One-axis tracking ($TWh\ y^{-1}$)	Two-axis tracking ($TWh\ y^{-1}$)	Concentrator photovoltaic (CPV) ($TWh\ y^{-1}$)
Bahrain	665	6,9	43,5	46,3	22,1	32,4
Cyprus	9 240	3,6	604,3	643,1	306,8	450,0
Egypt	995 450	81,3	65 102,4	69 283,3	33 048,9	48 478,4
Iran	1 636 000	129,0	106 994,4	113 865,6	54 315,2	79 673,2
Iraq	432 162	34,0	28 263,4	30 078,5	14 347,8	21 046,3
Israel	20 330	42,7	1 329,6	1 415,0	675,0	990,1
Jordan	91 971	7,3	6 014,9	6 401,2	3 053,4	4 479,0
Kuwait	17 820	32,4	1 165,4	1 240,3	591,6	867,8
Lebanon	10 230	8,1	669,0	712,0	339,6	498,2
Oman	212 460	9,8	13 894,9	14 787,2	7 053,7	10 346,8
Qatar	11 437	9,7	748,0	796,0	379,7	557,0
Saudi Arabia	1 960 582	138,2	128 222,1	136 456,5	65 091,3	95 480,3
Syria	184 050	26,1	12 036,9	12 809,9	6 110,5	8 963,2
Turkey	770 760	123,3	50 407,7	53 644,9	25 589,2	37 536,0
UAE	82 880	39,3	5 420,4	5 768,4	2 751,6	4 036,3
Yemen	527 970	3,0	34 529,2	36 746,7	17 528,6	25 712,1

Table E3.2 Expected economic benefits to Israel of VLS-PV plant introduction during the first 36 years

Interest rate	3 % /year
Yearly added solar power	1,5 GW
Yearly added six-hour storage power	0,5 GW
Credit line capacity required for the entire project	9 781 MUSD
Interest paid	3 397 MUSD
Loan repaid after	21 years
Total solar power installed	46,5 GW
Total storage power installed	15,5 GW
Electricity price after five years, when solar electricity sales start	9 US cents/kWh
Electricity price after 22 years, when all debts are paid off	5,5 US cents/kWh
Land area required for installation	558 km^2
Fraction of total national land area	2,7 %
Yearly manpower requirements for solar production	4 500 jobs
Yearly manpower requirements for solar operation	11 625 jobs
Yearly manpower requirements for storage production	1 500 jobs
Yearly manpower requirements for storage operation	3 875 jobs
Headquarters and engineering	1 395 jobs
Total number of jobs after 36 years	22 895 jobs

their manufacture. It was concluded that, in Israel, VLS-PV plants would cost no more than 850 USD/kW, and that production facilities, capable of an annual throughput of 1,5 GW collectors and 0,5 GW storage, would cost approximately 1 170 MUSD.

Third, we studied the kind of investment that would be necessary to create a production facility in four years, the first VLS-PV during the fifth year, and one successive new VLS-PV plant every year thereafter. Assuming an open credit line being made available by the government (or investors) at a 3 % real rate of interest, it was concluded that, in the Israeli case:

- the credit line would reach its maximum value in the 13th year;
- the maximum required credit would be equal to the cost of approximately ten fossil-fuelled plants;
- the credit line *plus interest* would be fully paid off by electricity revenues after 21 years;
- by that time, revenues would be sufficiently high to enable both the continued annual production of VLS-PV plants with no further investment, *and* the decommissioning and replacement of old plants after 30 years of service.

It is important to point out that after the initial investment has been paid off, the price of electricity no longer depends upon any factors related to its generation. It becomes a purely arbitrary figure that can be fixed at any desired level. For our examples, we arbitrarily fixed it at 5,5 US cents/kWh. However, if it is deemed desirable to continue installing VLS-PV plants at the rate of one per year, then the price of electricity can be lowered to a figure enabling the annual net revenue from sales to precisely cover the cost of one new VLS-PV plant. Similarly, if it becomes necessary to replace old plants after 30 years of service, it is sufficient to fix the electricity price during the 29th year at a level covering the cost of constructing two new VLS-PV plants the following year, etc. Simple arithmetic shows that in both of these examples, the required electricity price will be less than the 9 US cents/kWh that we have adopted for our calculations.

In the fourth part of this study, we repeated the Israeli calculations for the other major electricity producers in the region, making certain simplifying assumptions that were specified in each case. Given uncertainties surrounding local electricity prices, labour costs, and production/consumption growth rates, our results for these countries should be regarded as indicative rather than definitive.

A number of far-reaching conclusions can be drawn from the results of this study. First, VLS-PV plants can yield electricity at costs fully competitive with fossil fuel. Second, one may think in terms of typically 80 % of a country's entire electricity requirements coming from solar energy within a period of 30 to 40 years. Third, VLS-PV plants turn out to be *triply renewable*. In addition to the normal sense in which solar is deemed to be a renewable energy, the revenues from this top-down approach would be sufficient to completely finance the continued annual construction of VLS-PV plants *and* the replacement of 30-year-old VLS-PV plants with new ones *without* the need for any further investment.

In conclusion, the present top-down study strongly indicates that VLS-PV could directly compete with fossil fuels as the principal source of electricity for any country in the Middle East, and an investment scheme has been suggested for implementation.

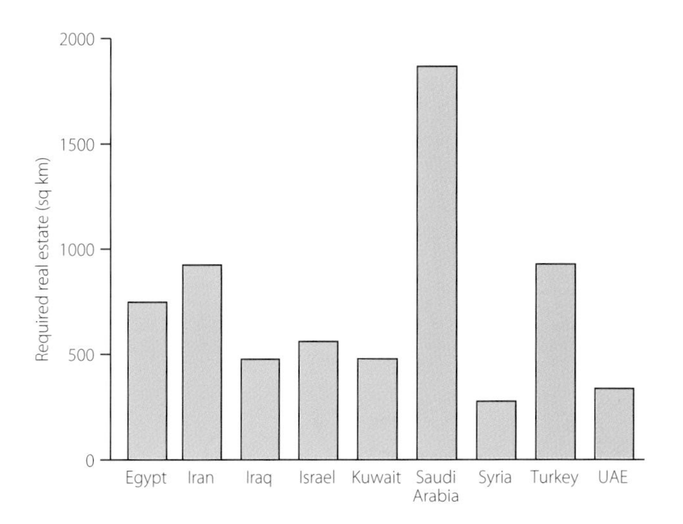

Figure E3.1 Land requirement for providing 80 % of each country's electrical requirements

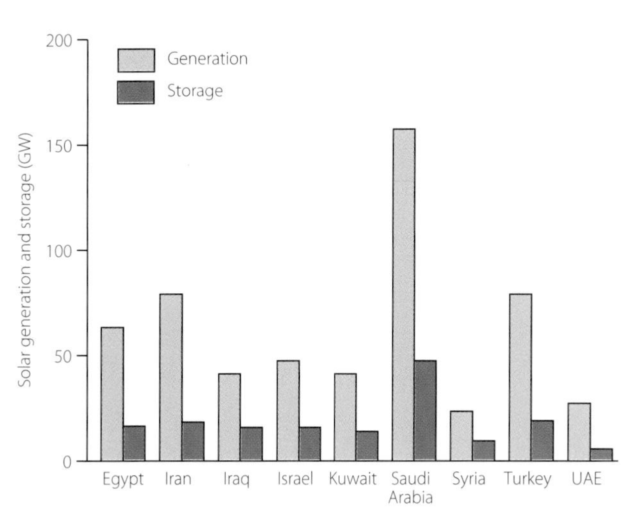

Figure E3.2 Total solar generating and storage capacity at the end of 36 years

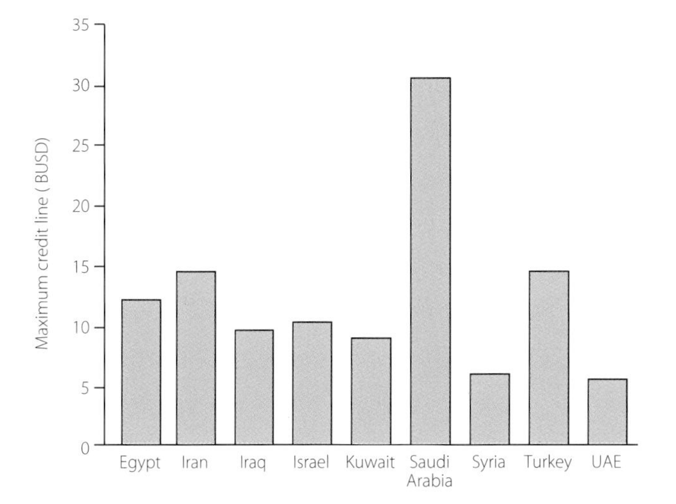

Figure E3.3 Maximum credit line required at 3 % real interest

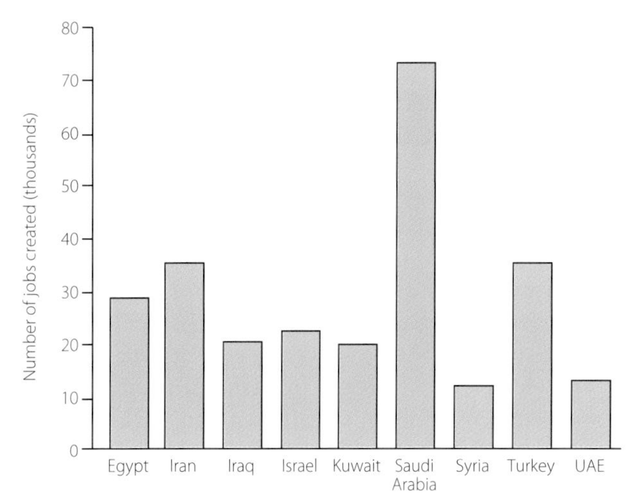

Figure E3.4 Total number of jobs created

II

4. The Asian region: Project proposals of VLS-PV on the Gobi Desert

4.1 DEMONSTRATIVE RESEARCH PROJECT FOR VLS-PV IN THE GOBI DESERT OF MONGOLIA

Mongolia has the vast Gobi Desert area in the southern and south-east parts. There are two types of electricity users in Mongolia, nomadic families and users of the electricity network. While electrification using PV for nomadic families has occurred, an existing electricity network supports Mongolian economic activity.

The electricity networks (transmission lines) have been constructed only in specific regions, such as those centring on Ulaanbaatar, the capital of Mongolia. The transmission lines have basically been constructed along a railway connecting Atlanbulug with Zumiin Uud through Ulaanbaatar – the borders in the north and south-east. The railway is playing a very important role in Mongolian economic activity. Therefore, these areas along the railway and transmission lines are expected to further develop in the future. However, electricity for the areas is generated by coal at Ulaanbaatar, worsening the atmospheric environment around Ulaanbaatar. As a result, installing large scale carbon-free renewable electricity such as the VLS-PV system may contribute both to protecting against air pollution and supporting regional development.

The VLS-PV scheme is a project that has not been carried out before. In order to achieve VLS-PV, a sustainable development scheme will be required. There are many technical and non-technical aspects that should be considered. Therefore, we will propose a demonstrative research project in the areas along the railway and discuss a future possibility for VLS-PV in the Gobi Desert, in Mongolia. The proposed project will include three phases as follows (see Table E4.1 and Figure E4.1). The potential sites in the Gobi Desert area along the railway were identified using long-term meteorological observation data conducted over the last

Table E4.1 Proposed projects for VLS-PV development in Mongolia

	Location	Capacity	Demands
First phase: R&D/pilot stage	Sainshand	1 MW	Households and public welfare (significant level compared to the peak demand and electricity usage in Sainshand city)
Second phase: demonstration stage	Four sites along the railway: 1 Sainshand 2 Zumiin Uud 3 Choir 4 Bor-Undur	10 MW/site (total: 40 MW)	Industry (surpasses the peak demand and almost equivalent to electricity usage around these locations)
Third phase deployment stage	Five sites along the railway: 1 Sainshand 2 Zumiin Uud 3 Choir 4 Bor-Undur 5 Mandalgobi One site between Oyu Tolgoi and Tsagaansuvrage	100 MW/site (sub-total: 500 MW) and 500 MW (total: 1 GW)	Power supply (almost double the peak demand and significant level compared to electricity usage in Mongolia)

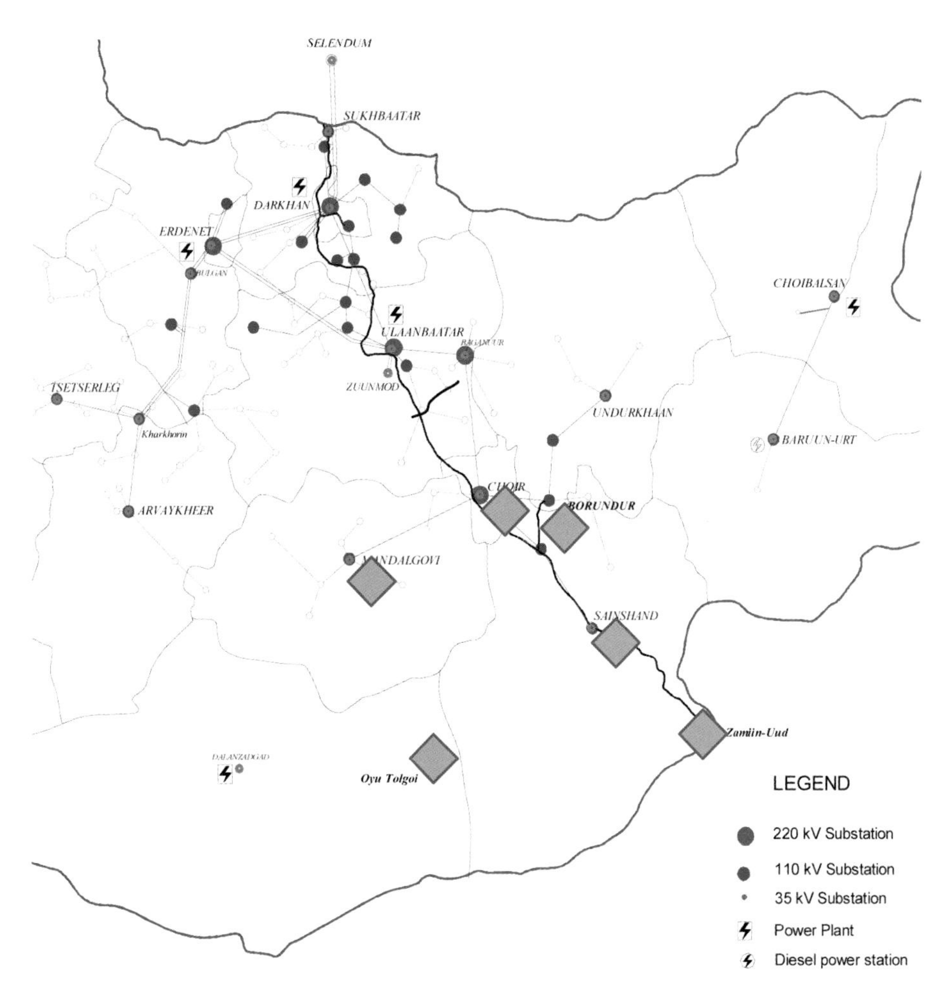

Figure E4.1 Location of VLS-PV systems

30 years. Grid access, as well as favourable market, economic, climatic and weather conditions, prevail in southern Mongolia – hence the choice of the candidate sites for the development of the VLS-PV system in the Gobi.

It is expected that the first phase will take four to five years. The project site will be Sainshand and the capacity of the PV system will be 1 MW. The assumed demands are households and public welfare needs in the region. The project has benefits beyond electricity. Apart from the creation of jobs and employment, the tourism industry will also benefit. In the second phase, 10 MW PV systems will be installed in Sainshand, Zumiin Uud, Choir and Bor-Undur, which are located along the railway lines. These sites are important cities and the scale is classified as medium-large scale in Mongolia. The total capacity of PV systems installed will reach 40 MW, and the demands assumed are to supply industry sectors, such as mining, located in the sites' neighbourhoods. The project will then be shifted to the third phase, which is the deployment phase. In this stage, 10 MW PV systems will be enhanced to 100 MW VLS-PV systems, and one new 100 MW system will be constructed in Mandalgobi. Besides these 100 MW VLS-PV systems, another 500 MW VLS-PV system will be constructed in between Oyu Tolgoi and Tsagaansuvraga, which are located in Umnugobi and Dornogobi provinces.

Renewable energy development is a promising way for social development and is one of the most important policies in Mongolia. Two documents, the Law for the Promotion of Renewable Energy and a proposal for a Utilization and National Renewable Energy Programme, have recently been drafted and submitted to the government for the approval of parliament. Final approval of these two documents will positively affect taxes and other funding that will assist in the development of VLS-PV systems.

4.2 FEASIBILITY STUDY ON 8 MW LARGE SCALE PV SYSTEM IN DUNHUANG, CHINA

Energy shortages and environmental pollution have become the bottleneck of social and economic development in China. Improving the current structure of energy supply and promoting utilization of renewable energy are effective solutions for these problems. The photovoltaic power generation system involves clean energy without greenhouse gas emissions. In China, there are huge lands in the Gobi Desert and elsewhere

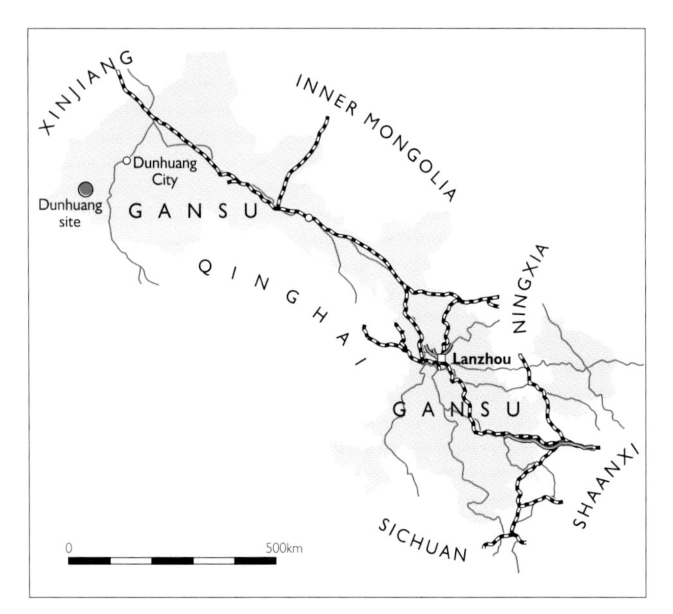

Figure E4.2 Location of Dunhuang, Gansu

Table E4.2 Capital investment for 8 MW LS-PV system

	Investment (million yuan)	Share (percentage)
Equipment	277,02	85,91
PV module	236,8	73,43
Inverter	34,2	10,61
Transformer	3,52	1,09
Test and monitoring	2,5	0,78
Civil construction	15,56	4,83
Transportation and installation	7,35	2,34
Feasibility study and preliminary investment	7,0	2,17
Miscellaneous	15,36	4,65
Total	322,47	100

Note: 1 yuan = approximately 0,12 USD

that provide the possibility of large scale PV systems on very large scale applications. Only when PV is used for large scale applications can costs be reduced to the level of those associated with traditional electric power.

The Gobi area in Gansu is about 18 000 km^2. This area can be used to build 500 GW VLS-PV, which is more than the whole power capacity in China today. The targeted place for 8 MW large scale photovoltaic power generation (LS-PV) in the Gobi Desert is at Qiliying, 13 km from Dunhuang city. The latitude is N40° 39', with a longitude of E94° 31' and an elevation of 1 200 m. It is only 5 km from Qiliying to the 6 000 kVA/35 kV transformer station, so it will not cost that much to build a high voltage transmission line.

The 8 MW PV system will be divided into eight sub-stations of 1 MW each. Each 1 MW sub-station will feed the generated electricity to a high voltage grid (35 000 V) through a 1 000 kVA transformer. Each 1 MW sub-station will be divided into five channels with 200 kW each, as shown in Figure E4.3. Each 200 kW PV channel will be equipped with a grid-connected inverter to convert the DC power from the PV into three-phase AC power for the primary of the 1 000 kVA transformer.

Each 1 MW sub-station and each 200 kW channel will be independent. Such design offers the advantages of being easier for troubleshooting and maintenance, being flexible for potential investors, and allowing various types of PV systems to be installed and compared.

The system efficiency is assumed to be 0,77. Using the efficiency and the annual in-plain irradiation facing south with a 40° tilted angle, the annual output is calculated to be 13 761 MWh/year.

Total capital investment is 322,47 million yuan (approximately 38,7 MUSD), and 86 % of total investment is for PV system equipment, such as PV modules, inverters and transformers, as shown in Table E4.2. However, it is expected that 96,74 million yuan (approximately 11,6 MUSD; 30 % of the total capital) will be a grant provided by the central government of China, and the real required capital will be 225,73 million yuan (approximately 27,1 MUSD).

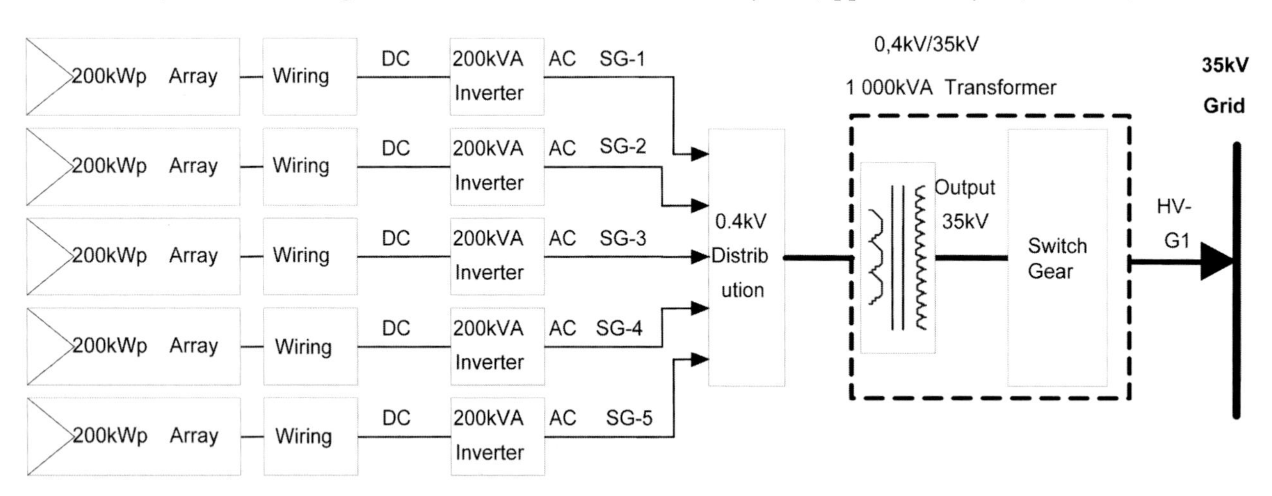

Figure E4.3 1 MW PV sub-station

The Gansu grid company will guarantee 1,683 yuan/kWh (0,202 USD/kWh) as the feed-in tariff and the annual income for the PV system will be 23,16 million yuan (approximately 2,78 MUSD): 13,761 MWh/year x 1,683 yuan/kWh. The tariff purchased by the grid company will be added on to all the electricity consumed in Gansu Province, which was 340 x 10^8 kWh in 2002. Therefore, the additional tariff will be 0,000 68 yuan/kWh (0,000 082 USD/kWh): 23,16 million yuan/340 x 10^8 kWh. For a family consuming 2 kWh/day, the annual consumption will be 730 kWh, and they will only need to pay 0,5 yuan/year (0,06 USD/year) in addition.

The proposed 8 MW LS-PV plant in Dunhuang city is considered the first pilot project in China of the Great Desert Solar PV Programme proposed by the World Wide Fund for Nature (WWF) and an expert group. The development of further large scale PV systems in other regions is also being discussed. It has been proposed that 30 GW of solar PV power generation capacity could be developed by 2020 if government incentive policies are developed and are in place. This could enable China to become a leading country in solar power development in the world.

II

5. The Oceania region: Realizing a VLS-PV power generation system at Perenjori

Perenjori is a small township approximately 350 km north-east of Perth in the wheat belt of Western Australia and situated at S-29° latitude and E116,2° longitude. The required land to set up a VLS-PV power generation project can be obtained at a reasonably low price or leased for 30 to 50 years from local farmers. The land is flat and suitable for mounting the structure or installing solar PV power generation projects.

Several issues arise in terms of achieving a very large scale solar photovoltaic power generation project at Perenjori. Although the Great Sandy Desert receives more solar radiation than Perenjori, the overall economy of setting up the project and generation costs will be less at Perenjori due to its proximity to enough loads and the availability of the local grid.

At present, the load is almost negligible; but there is a strong interest in promoting the mining industry in the region, provided that sufficient and good quality power is available for mining activities. A number of mining companies have also shown interest in setting up their mining operations in the Perenjori regions, and there will be a load of the order of 1 GW over the next 10 to 15 years.

The size of a VLS-PV system may range from 10 MW (pilot) to 1 or several GW (commercial), consisting of one plant or an aggregation of a number of units, distributed in the same region and operating in harmony with one another. Figure E5.2 gives a rough idea of how a VLS-PV project can be realized in a radius of 100 km of Perenjori over the next 15 years, aggregating to a capacity of over 1 GW.

What will be feasible is to install several stand-alone solar PV systems as per the load requirement of each individual mining operation in the region. Then, when three to four projects have been set up in the region, they can be interconnected by creating a small local grid.

A project for installing a 10 MW pilot power generation system will be proposed as the first step for a VLS-PV system at Perenjori. The estimated project cost of a 10 MW PV power generation project at Perenjori will be of the order of 60 MAUD (approximately 45,6 MUSD), with the following cost breakdowns as shown in Table E5.1. Almost 70 % of the project cost comprises PV modules only.

The generation cost of the pilot project of 10 MW, after availing of a 50 % subsidy from the government, will be approximately 14 AUD cents/kWh (0,11 USD/kWh) under the Mandatory Renewable Energy

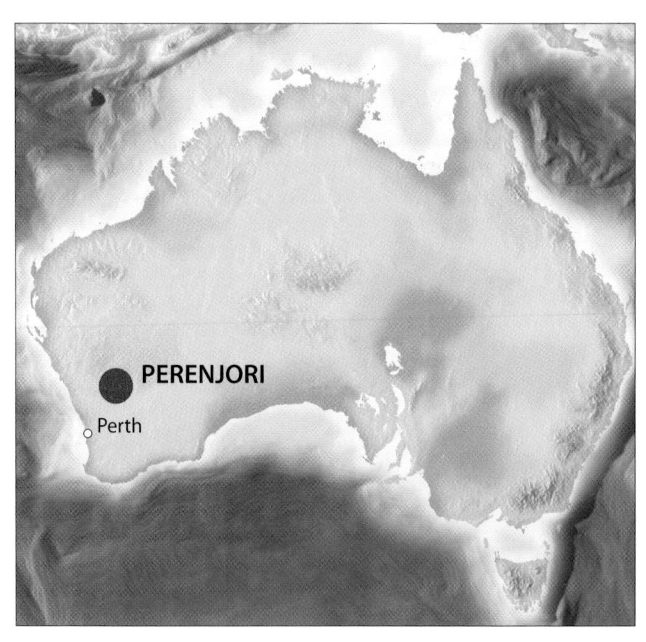

Figure E5.1 Location map of Perenjori

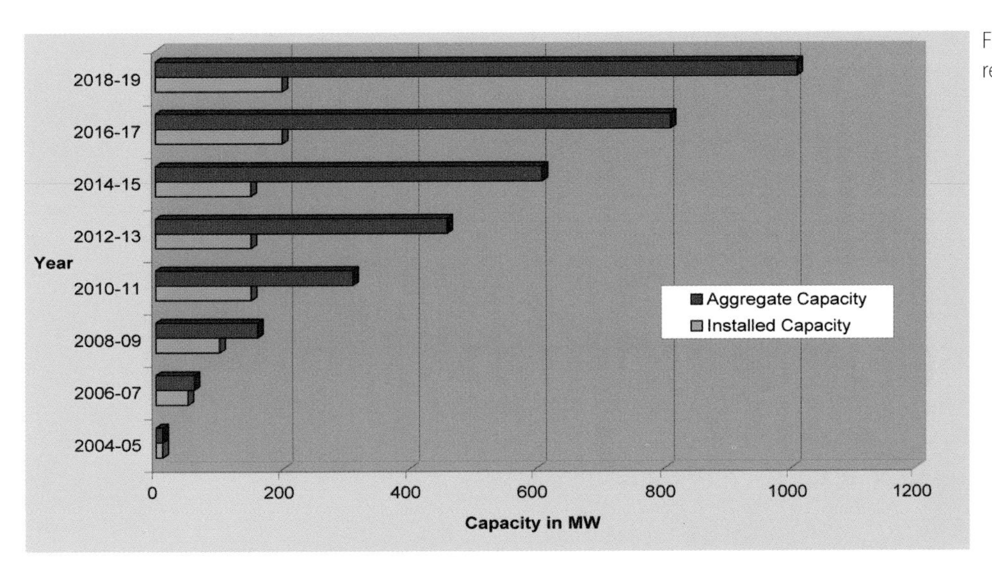

Figure E5.2 Possible scenario to realize VLS-PV at Perenjori

Table E5.1 Estimated project cost for 10 MW PV power generating system

Components	Unit cost	Total cost (AUD)
PV modules	4,4 AUD/Watt	44 000 000
Mounting structure with single-axis tracking	10 % of modules	4 400 000
Inverter(s)	125 000 AUD	5 000 000
Transformers and cabling	4 % of modules	1 760 000
Installation and commissioning	7 % of modules	3 080 000
Land	Lump sum	500 000
Miscellaneous, including transportation to site	Lump sum	1 260 000
Total		60 000 000

Note: 1 AUD = approximately 0,76 USD

Target (MRET) of the federal government of Australia, which is very much comparable with the cost of power generation from a diesel power project. At a price of 36 AUD cents/litre (0,27 USD/litre), diesel fuel is available to the mining companies in Perenjori; the cost of power generation from diesel power projects comes to about 12 AUD cents/kWh (0,09 USD/kWh). Therefore, the mining companies would be interested in purchasing power from the proposed pilot project of 10 MW, and the installation of a diesel-based power project as a back-up has been suggested.

Prior to setting up the pilot project, arrangements must be made by the project developers to sell the power to the mining operators. The power purchase agreements (PPAs) should be signed for the whole lifetime of the project. The Shire of Perenjori and Mid West Development Commission would play an important role in negotiating the terms and conditions of the PPAs, and their assistance would be necessary to attract project developers for this project, as well as for other projects to be installed later on.

II

6. Desert region community development

In this study, the following issues are researched and investigated:

- the possibility of utilizing VLS-PV in desert areas;
- a sustainable scenario for installing VLS-PV systems;
- modelling of a sustainable society with solar photovoltaic generation and greening in the desert;
- specific problems of VLS-PV systems in desert areas;
- the use of solar PV generation with regard to elemental technology/systems for sustainable agriculture; and
- sustainable agriculture utilizing other regional renewable energy options.

Figure E6.1 depicts a desert community development that aims to achieve an ideal community. Agriculture and tree planting can be facilitated with plentiful renewable energy, and electricity is used by neighbouring cities. The community must have the potential to appeal to other people through the following themes:

- *Sustainable PV stations: sustainable energy production.* Here, the VLS-PV system is the main feature, using sunshine, along with wind power and other renewable energy. At night, electricity comes from battery storage instead of from the grid.
- *Sustainable farming.* Utilizing soil and water conservation technology, we will conserve and rehabilitate the landscape. This may not be difficult to achieve with PV support because the main reason for desertification is human activity relating to energy needs. Conservation and rehabilitation will take plant and animal ecology into account.
- *Sustainable community.* Statistical and scenario analyses are used to develop an ideal community. In order to sustain regional society, in addition to the facilities and technology needed, education and training are provided.
- *Remote sensing.* Remote sensing technology has the potential to find suitable places in which to implement VLS-PV and wind power systems. It can generate data on soil and water required for sustainable agricultural production.
- *Desalination.* Renewable energy power can operate a desalination system, which will supply drinking and irrigation water.
- *Effect of PV station on forest, grassland and farmland.* Proper operation of a water pump and desalination system can provide good quality water and remove salt from the groundwater. This can enhance crop yields and reduce the use of fuelwood. Renewable energy-driven greenhouse agriculture can produce high quality and high product yields.
- *Effect of PV station on the local community.* PV stations can go beyond supplying electricity to local communities to producing more employment in the region. This would also increase local incomes through selling electricity. PV structures can also protect houses from strong winds.
- *Effect of forest, grassland and farmland benefits on local community.* Agriculture can supply food to communities and leads to wind protection effects and employment. People in local communities can obtain more income by selling products.
- *Technology.* Implementing a subsurface drainage system may save groundwater quality, and desalination equipment protects soil from damage due to salinization.

In sum, this virtual community is ideal, but is based on the existence of a good groundwater supply and a power grid near the community.

Nevertheless, agriculture in the desert is difficult even though arid and semi-arid regions are better suited for solar irradiation. While there is occasionally more than enough sunshine, it diminishes easily. Rain-fed

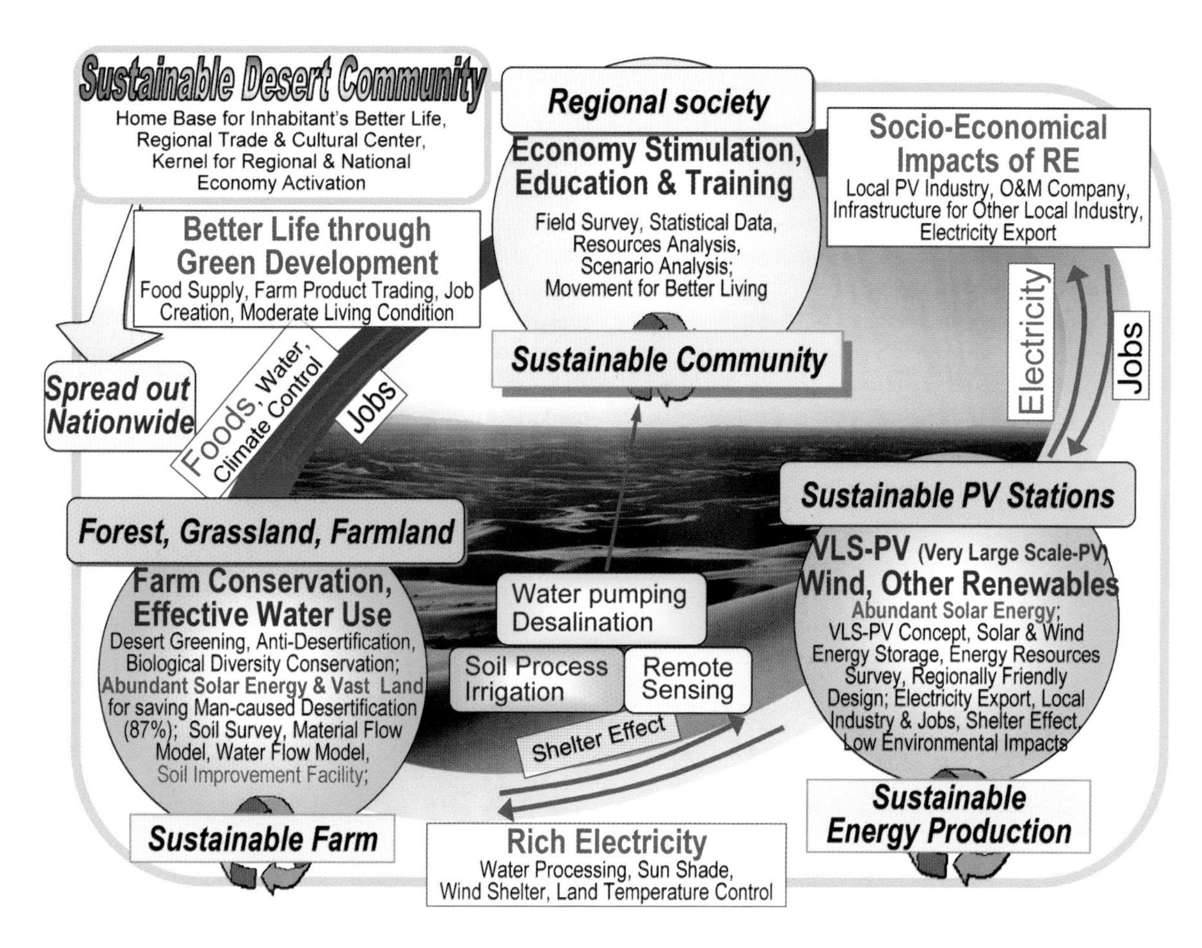

Figure E6.1 Framework of desert community development and research topic

agriculture is limited by low precipitation. Although irrigated agriculture has been maintained in some areas, it often causes soil degradation, largely through improper water distribution, and inadequate irrigation practice and quality.

Figuring out the appropriate amount and quality of irrigation water, including groundwater behaviour, is important. Controlling water quality for irrigation, as well as controlling the amount of irrigation using pumps, is also necessary. Although this is easy to achieve in developed countries, arid and semi-arid regions still have problems due to the cost of improving water distribution through pumping. In this study we suggest realistic proposals for using PV systems to provide the energy necessary to control the quality and quantity of irrigation water.

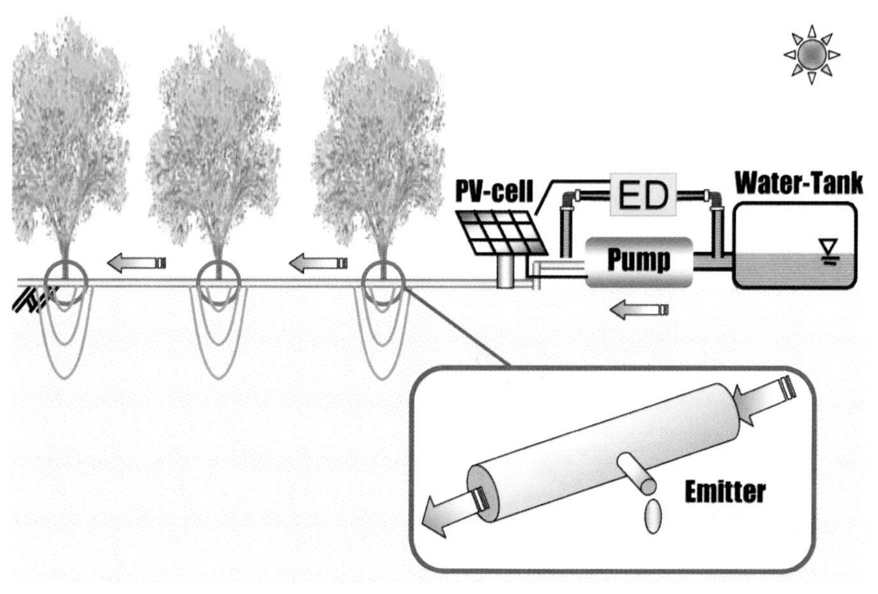

Figure E6.2 Desalinated drip irrigation system

Implementing a PV system in an arid or semi-arid region also has the potential of introducing electricity as an alternative energy source to wood used as fuel. This is significant for forest conservation purposes. PV systems also protect water courses by reusing desalinized groundwater. In the past, water was desalinated through distillation, using solar heat; but this required a huge area to desalinate salty water because the distillation rate of solar heat is quite low. Considerable amounts of low saline water are available when we use a PV system to generate electricity to drive a desalination system, such as reverse osmosis (RO) or electro-dialysis (ED). Figure E6.2 depicts a drip irrigation system with a desalination system that uses PV as energy source.

In order to achieve the sustainable development of desert region communities, it is important to consider technology innovation and to protect the community and its environment. Maintaining community food production is a further important issue.

Agriculture is necessary for food supply, and lack of appropriate knowledge of agriculture often leads to soil degradation and wasting water, thus worsening desertification. In this study, we propose different types of irrigation systems using PV to save soil and water in arid and semi-arid areas. In addition, it is advisable to provide an impact assessment of the climatic results of introducing VLS-PV into a desert region since a PV system will affect quality of life and food production in arid and semi-arid regions.

II

7. Conclusions and recommendations

7.1 GENERAL CONCLUSIONS

The scope of this report is to examine and evaluate the potential of VLS-PV systems, which have a capacity ranging from several megawatts to gigawatts, and to develop practical project proposals for demonstrative research towards realizing VLS-PV systems in the future.

It is apparent that VLS-PV systems can contribute substantially to global energy needs, can become economically and technologically feasible, and can contribute considerably to environmental and socio-economic development. With the objective of ensuring these contributions and outlining the actions necessary for realizing VLS-PV in the future, this report has developed concrete project proposals.

When thinking about a practical project proposal for achieving VLS-PV, a common objective is to find the best sustainable solution. System capacity for suitable development is dependent upon each specific site with its own application needs and available infrastructure, especially access to long-distance power transmission, and human and financial resources

Although the expected impacts of VLS-PV will differ in each region, depending upon the local situation, and upon any options and concerns, there are several conclusions in common.

Background

- World energy demands will increase as world energy supplies diminish due to trends in world population and economic growth in the 21st century. In order to save conventional energy supplies while supporting growth in economic activity, especially within developing countries, new energy resources and related technologies must be developed.
- The potential amount of world solar energy that can be harnessed is sufficient for global population needs during the 21st century. It has been forecast that photovoltaic technology shows promise as a major energy resource for the future.
- Much potential exists in the world's desert areas. If appropriate approaches are found, they will provide solutions to the energy problem of those countries that are surrounded by deserts.

Technical aspects

- It has already been proven that various types of PV systems can be applied for VLS-PV. PV technologies are now considered to have reached the necessary performance and reliability levels for examining the feasibility of VLS-PV.
- To clarify technical reliability and sustainability, a step-by-step approach will be required to achieve GW scale. This approach is also necessary for the sustainable funding scheme.
- Technological innovation will make VLS-PV in desert areas economically feasible in the near future.
- Global energy systems, such as hydrogen production and high temperature superconducting technology, as well as the higher conversion efficiency of PV cells and modules, will make VLS-PV projects more attractive.

Economic aspects

- The economic boundary conditions for VLS-PV are still unsatisfactory at current PV system prices without supporting schemes, so that lowering the investment barrier, additional support, and/or higher feed-in tariffs for large systems are still required.

- VLS-PV using flat-plate PV modules or CPV modules produced using mass production technologies are expected to be an economically feasible option for installing large central solar electric generation plants in or adjacent to desert areas in the future as a replacement for central fossil-fuelled generating plants.
- Intelligent financing schemes such as higher feed-in tariffs in developed countries, and international collaboration schemes for promoting large scale PV or renewable energy in developing countries will make VLS-PV economically feasible.
- For the sustainable development of VLS-PV, initial financial support programmes should be sufficient to completely finance the continued annual construction of VLS-PV plants and the replacement of old VLS-PV plants by new ones, eliminating the need for any further investment after the initial financing is repaid.

Social aspects

- Renewable energy development is a promising way of ensuring social development and is one of the most important policies in developing countries.
- VLS-PV projects can provide a pathway to setting up PV industry in the region, while the existence of a traditional and strong PV industry will provide an additional factor contributing to an economic and political environment that favours the application of PV.
- International collaboration and institutional and organizational support schemes will lead to the success of VLS-PV projects in developing countries.
- Developing VLS-PV will become an internationally attractive project, making the region and/or country a leading country/region in the world.
- When we think about developing VLS-PV, we should be careful to consider the social sustainability of the region – for instance, in terms of complementing agricultural practices and desert community development.

Environmental aspects

- In light of global environmental problems, there is a practical necessity for renewable energy, and it is expected to be an important option for the mid 21st century.
- A proper green area coupled with VLS-PV would positively impact upon convective rainfall, rather than produce a negative impact due to a rise in sensible and surface heat flux. It would also decrease dust storms, as well as reduce and mitigate CO_2 emissions.

7.2 REQUIREMENTS FOR A PROJECT DEVELOPMENT

VLS-PV is anticipated to be a very globally friendly energy technology, contributing to the social and economic development of the region in an environmentally sound manner.

However, concrete realization of VLS-PV projects depends upon successful negotiation between project developers and PV and electricity industries, and must have the support of government regarding its acceptance, sustainability and economic incentives.

In order to make negotiations successful and to establish a VLS-PV project, approaches required for project development are as follows:

- Clarify critical success factors on both technical and non-technical aspects.
- Demonstrate technical capability and extendibility.
- Demonstrate economic and financial aspects.
- Show local, regional, and global environmental and socio-economic effects.
- Assess, analyse and allocate project risks such as political and commercial risks.
- Find available institutional and organizational schemes.
- Provide training programmes for installation, operation and maintenance.
- Develop instructions or a guideline for these approaches.

7.3 RECOMMENDATIONS

There are strong indications that VLS-PV could directly compete with fossil fuel and with existing technology as the principal source of electricity for any country that has desert areas. This could be accomplished by finding an investment scheme and by getting institutional and organizational support for its implementation.

In addition, the technology innovations regarding PV and global energy systems in the future will make VLS-PV economically and technologically attractive and feasible.

The following recommendations are outlined to support the sustainable growth of VLS-PV in the near future:

- Discuss and evaluate future technical options for VLS-PV, including electricity network, storage and grid management issues, as well as global renewable energy systems.
- Analyse local, regional and global environmental and socio-economic effects induced by VLS-PV systems from the viewpoint of the whole life cycle.
- Clarify critical success factors for VLS-PV projects, on both technical and non-technical aspects, based on experts' experiences in the field of PV and large

scale renewable technology, including industry, project developers, investors and policy-makers.

- Develop available financial, institutional and organizational scenarios, and general instruction for practical project proposals to realize VLS-PV systems.
- The International Energy Agency (IEA) PVPS community will continue Task 8 activities. Experts from the fields of grid planning and operation, desert environments, agriculture, finance and investment should be involved.
- The IEA PVPS community welcomes non-member countries to discuss the possibility of international collaboration in IEA PVPS activities.

CHAPTER ONE

Introduction

1.1 OBJECTIVES

The scope of this study is to examine and evaluate the potential of very large scale photovoltaic power generation (VLS-PV) systems, which have capacities ranging from several megawatts to gigawatts, and to develop practical project proposals for demonstrative research towards realizing VLS-PV systems in the future.

Our previous report, *Energy from the Desert: Feasibility of Very Large Scale Photovoltaic Power Generation (VLS-PV) Systems*, identified the key factors enabling the feasibility of VLS-PV systems and clarified the benefits of these systems' applications for neighbouring regions, the potential contribution of system application to global environmental protection, and renewable energy utilization in the long term.[1] It is apparent from the perspective of the global energy situation, global warming and other environmental issues that VLS-PV systems can:

- contribute substantially to global energy needs;
- become economically and technologically feasible;
- contribute considerably to the environment; and
- contribute considerably to socio-economic development.

In order to secure these contributions, a long-term perspective and a consistent policy are necessary regarding technological, organizational, and financial issues. Action is now required to unveil the giant potential of VLS-PV systems in deserts.

This report will reveal virtual proposals of practical projects that are suitable for selected regions, and which enable sustainable growth of VLS-PV in the near future, and will propose practical projects for realizing VLS-PV systems in the future.

1.2 CONCEPT OF VERY LARGE SCALE PHOTOVOLTAIC POWER GENERATION (VLS-PV) SYSTEMS

1.2.1 Availability of desert area for PV technology

Solar energy is low-density energy by nature. To utilize it on a large scale, a massive land area is necessary. One third of the land surface of the Earth is covered by very dry desert and high-level insolation, where there is a lot of available space. It is estimated that if a very small part of these areas, approximately 4 %, was used for installing photovoltaic (PV) systems, the resulting annual energy production would equal world energy consumption.

Rough estimations were made to examine desert potential under the assumption of a 50 % space factor for installing PV modules on the desert surface as the first evaluation. The total electricity production would be 2 081 x 10^3 TWh (= 7,491 x 10^{21} J), which means a level of almost 17 times as much as the world primary energy supply (0,433 x 10^{21} J in 2002).[2] These are hypothetical values, ignoring the presence of loads near the deserts. Nevertheless, these values at least indicate the high potential of developing districts located in solar energy-rich regions as primary resources for energy production.

Figure 1.2 shows that the Gobi Desert area in western China and Mongolia can generate as much

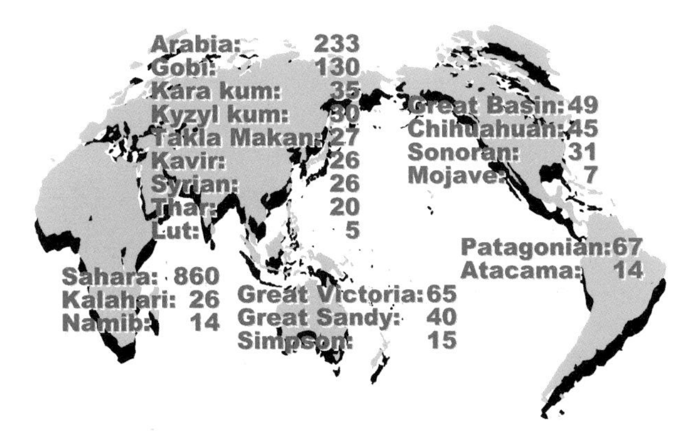

Figure 1.1 World deserts (unit: 10^4 km^2)

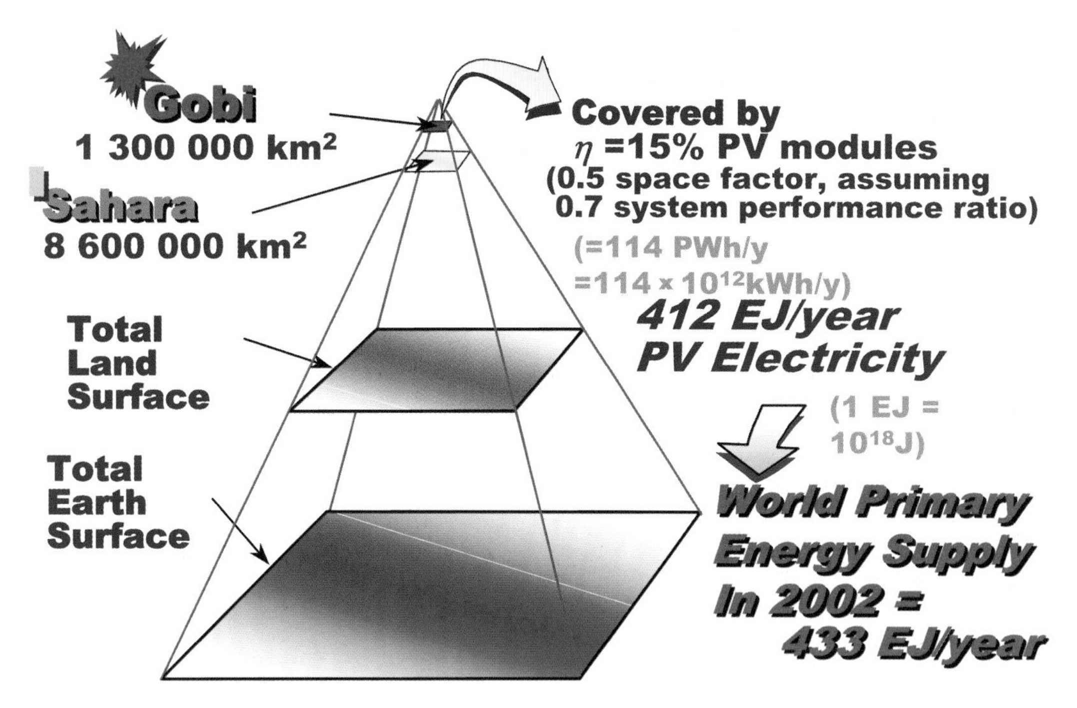

Figure 1.2 Solar pyramid

electricity as the current world's primary energy supply. Figure 1.3 depicts an image of a VLS-PV system in a desert area

1.2.2 VLS-PV concept and definition

Three approaches are under consideration to encourage the spread and use of PV systems:

1 Establish small scale PV systems independent of each other. There are two scales for such systems: installing stand-alone, several hundred watt-class PV systems for private dwellings, and installing 2 to 10 kW-class systems on the roofs of dwellings, as well as 10 to 100 kW-class systems on office buildings and schools. Both methods are already being used. The former is employed to furnish electric power in developing countries and is known as the solar home system (SHS), and the latter is used in industrialized countries. This seems to be used extensively in areas of short- and medium-term importance.
2 Establish 100 to 1 000 kW-class mid-scale PV systems on unused land on the outskirts of urban areas. The PVPS/Task 6 studied PV plants for this scale of power generation. Systems of this scale are in practical use in about a dozen sites in the world at the moment; but their number is expected to increase rapidly in the early 21st century. This category can be extended up to multi-megawatt size.
3 Establish PV systems larger than 10 MW on vast barren, unused lands that enjoy extensive exposure to sunlight. In such areas, a total of even more than 1 GW of PV system aggregation can be easily realized. This approach enables installing a large number of PV systems quickly. When the cost of generated electric power is lowered sufficiently in the future, many more PV systems will be installed. This may lead to a drastically lower cost of electricity, creating a positive cycle between cost and consumption. In addition, this may become a feasible solution for future energy need and environmental problems across the globe, and ample discussion of this possibility is worthwhile.

The third category corresponds to VLS-PV. The definition of VLS-PV may be summarized as follows:

- The size of a VLS-PV system may range from 10 MW to 1 or several GW, consisting of one plant, or an aggregation of plural units distributed in the same district operating in harmony with each other.
- The amount of electricity generated by VLS-PV can be considered significant for people in the district, nation or region.
- VLS-PV systems can be classified according to the following concepts, based on their locations:
  - land based (arid to semi-arid deserts);
  - water based (lakes, coastal and international waters);
  - locality options: developing countries (lower-, middle- or higher-income countries; large or small countries) and Organisation for Economic Co-operation and Development (OECD) (countries).

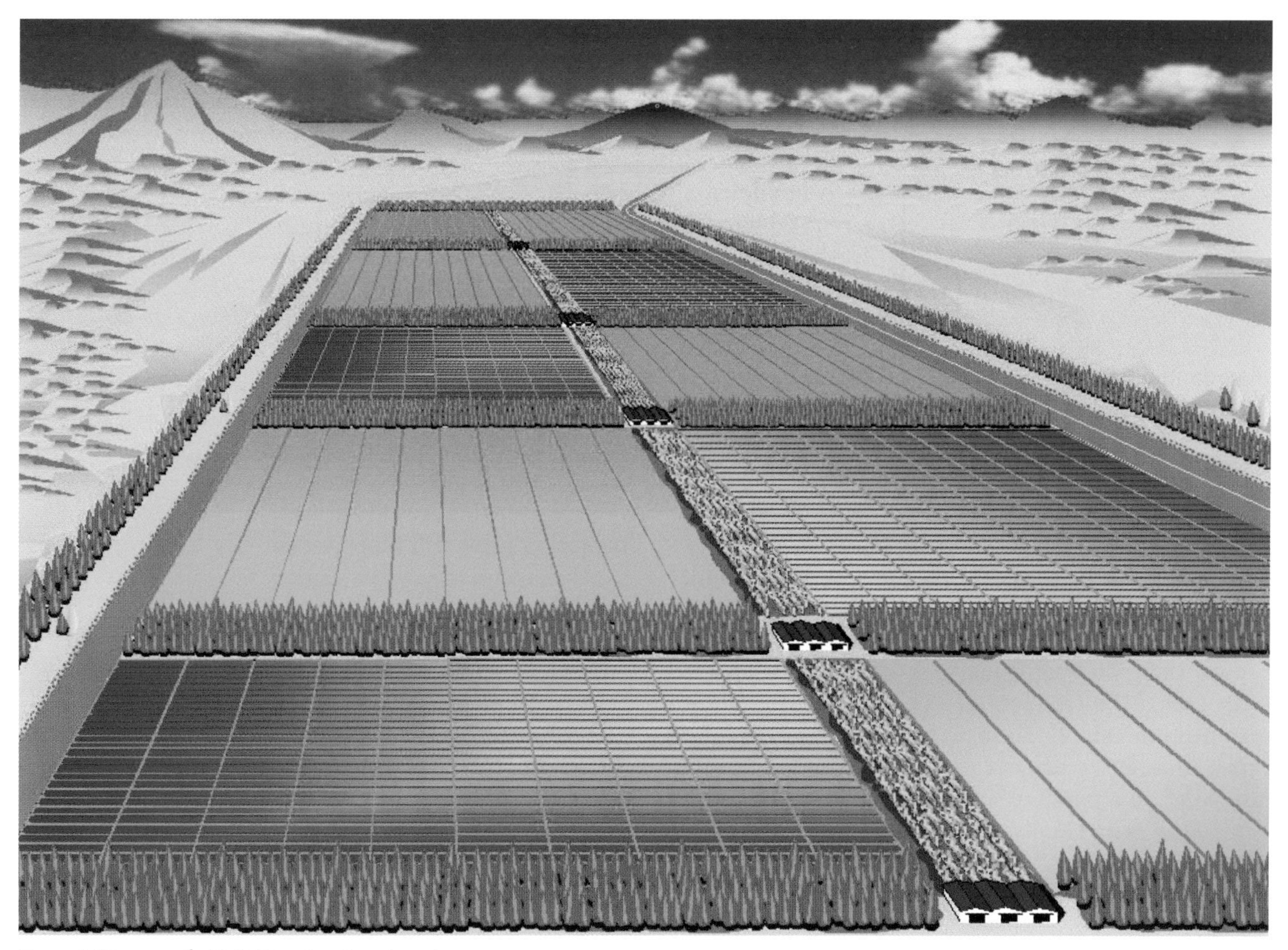

Figure 1.3 Image of a VLS-PV system in a desert area

VLS-PV systems are already feasible. PV systems with a capacity of more than 10 MW were installed and connected to the grid in 2006. Currently, a land-based VLS-PV concept is in its initial stage.

Although the concept of VLS-PV includes water-based options, much discussion is required. It is not neglected here, but is viewed as a future possibility outside the major emphasis of this report.

1.2.3 Potential of VLS-PV advantages

The advantages of VLS-PV are summarized as follows:

- It is very easy to find land in or around deserts appropriate for large energy production with PV systems.
- Deserts and semi-arid lands are normally high insolation areas.
- The estimated potential of such areas can easily supply the estimated world energy needs by the middle of the 21st century.
- When large-capacity PV installations are constructed, step-by-step development is possible through utilizing the modularity of PV systems. According to regional energy needs, plant capacity can be increased gradually. It is an easier approach for developing areas.
- Even very large installations are quickly attainable in order to meet existing energy needs.
- Remarkable contributions to the global environment can be expected.
- When VLS-PV is introduced to some regions, other types of positive socio-economic impacts may be induced, such as technology transfer to regional PV industries, new employment and growth of the economy.
- The VLS-PV approach is expected to have a drastic influence on the chicken-and-egg cycle in the PV market. If this does not happen, the goal of achieving VLS-PV may move a little further away.

These advantages make VLS-PV a very attractive option and worthy of being discussed in the context of the 21st century's global energy needs. This concept is illustrated in Figure 1.4.

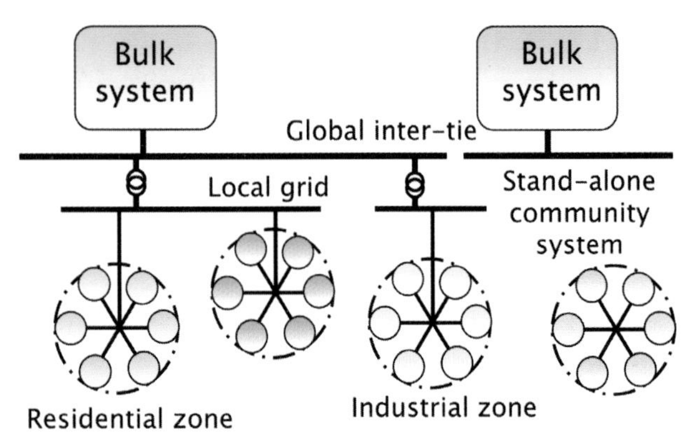

Figure 1.4 Global network image

1.2.4 The viability of VLS-PV's development

Basic case studies were reported concerning regional energy supply using VLS-PV systems in desert areas where solar energy is abundant. According to this report, the following scenario is suggested to achieve large scale PV introduction. At first, bulk systems are installed individually in some locations. They are then interconnected with each other through a power network and are incorporated with regional electricity demand growth. Finally, such a district would become a large power source. This scenario is summarized in Figure 1.5 as follows:

- *Stage 1:* a stand-alone bulk system is introduced to supply electricity for surrounding villages or anti-desertification facilities in the vicinity of deserts.
- *Stage 2:* remote, isolated networks germinate. Plural systems are connected by a regional grid. This contributes to load levelling and reduced power fluctuation.
- *Stage 3:* the regional network is connected to a primary transmission line. Generated energy can be supplied to a load centre and industrial zone. Total use combined with other power sources and storage becomes important for matching the demand pattern and improving the capacity factor of the transmission line. Furthermore, around the time that stage 3 is reached, in the case of a south-to-north tie, seasonal differences between demand and supply can be adjusted. An east-to-west tie can shift peak hours.
- *Stage 4:* finally, a global network is developed. Most of the energy consumed by human beings can be supplied through solar energy.

For the last stage, a breakthrough in advanced energy transportation will be expected on a long-term basis, such as a superconducting cable, flexible AC transmission system (FACTS) or chemical media.

1.3 PRACTICAL APPROACHES TO REALIZE VLS-PV

1.3.1 Required approaches for PV and renewables

Solar energy resources, PV technologies and renewable energy will help to realize important economic, environmental and social objectives in the 21st century, and will be a critical element in achieving sustainable development.

Recent rapid growth of renewable energy in some countries and regions is strongly supported by policies for deploying renewables. Developing countries work to expand and modernize their energy systems, and industrialized countries work to replace ageing systems and to meet rising demand. Their positive policies are accelerating the acceptance of renewables and are

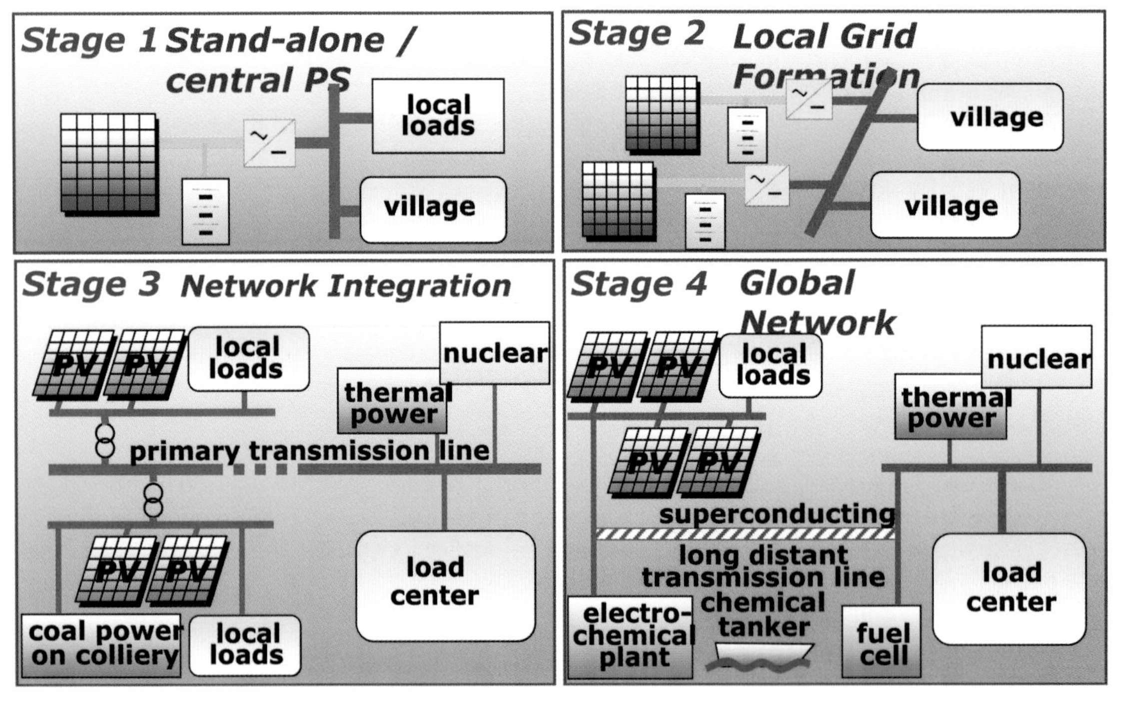

Figure 1.5 Very large scale PV system deployment scenario

leading their investments.

In order to advance the transition to a global energy system for sustainable development, it is very important to orient large and increasing investments towards renewable energy. If investments continue with business as usual, mostly in conventional energy, societies will be further locked into an energy system incompatible with sustainable development and one that further increases the risks of climate change.

As a result, increasing renewable energy use is a policy, as well as a technological, issue that applies directly to achieving VLS-PV.

In order to promote renewable energy and its diversity of challenges and resource opportunities (as well as the financial and market conditions among and within regions and countries), different approaches are required. With the aim of adopting policy changes and incorporating the goals of sustainable development within their policies, *Policy Recommendations for Renewable Energies*[3] was prepared by the International Conference for Renewable Energies, held in Bonn, Germany, in 2004, and the following policy priorities were discussed and declared.[3,4]

Establish policies for developing markets

The need for coherent regulatory and policy frameworks that support the development of thriving markets for renewable energy technologies and that recognize the important role of the private sector should be emphasized. This includes removing barriers and allowing for fair competition in energy markets, while considering the concept of internalizing external costs for all energy sources. Such frameworks are essential if the potential for renewable energy technologies is to be realized effectively and efficiently, if favourable conditions for public and private investments in renewable energies are to be created, and if modern energy services are to be extended to populations currently without access.

Expand financing options

Enhanced international cooperation for capacity-building and technology transfer, effective institutional arrangements at all levels, corporate responsibility, micro-financing, public–private partnerships, and advanced policies by export credit agencies are crucial in expanding finance for renewable energies. Financial incentives and higher shares of official development assistance (ODA) as catalytic funding should also be considered. International financial institutions should significantly expand their investments in renewables and energy efficiency, establishing clear objectives for renewable energies in their portfolios.

Develop the capacity required

Strengthening human and institutional capacities for renewable energies should be supported. This includes:

- building capacity for policy analysis and technology assessment, as well as strengthening educational efforts, gender mainstreaming and the role of women;
- raising the awareness of government decision-makers and financiers of the benefits of renewable energies;
- promoting consumer demand for renewable energy technologies;
- supporting development of marketing, maintenance and other service capacities; and
- strengthening regional and international collaboration, as well as stakeholder participation, to facilitate access to, and sharing of, relevant information and good practice.

1.3.2 Preliminary recommendations on a policy level for VLS-PV

In our previous report[1], we concluded that VLS-PV systems would have a positive impact. We defined a flowchart for recommendations, as shown in Figure 1.6 and gave the following recommendations on a policy level:

- National governments and multinational institutions should adopt VLS-PV in desert areas as a viable energy generation option in global, regional and local energy scenarios.
- The International Energy Agency (IEA)–PVPS community should continue Task 8 to expand the study and refine the research and development (R&D) and pilot phase, and should involve

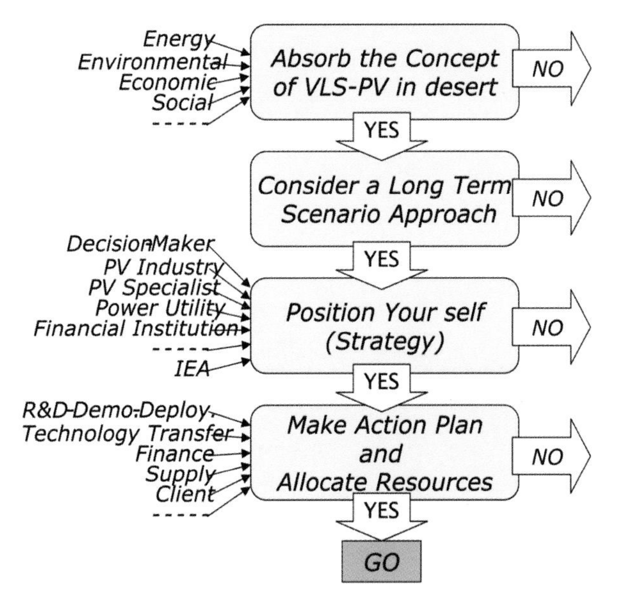

Figure 1.6 Check-flow of recommendations

participation by desert experts and financial experts, as well as collect further feedback information from existing PV plants.

- International organizations and national governments of industrialized countries should provide financing to generate feasibility studies in plural desert areas around the world and to implement the pilot and demonstration phase.
- Countries with deserts should (re)evaluate their deserts not as potential problem areas, but as vast and profitable (future) resources for sustainable energy production. The positive influence on local economic growth, regional anti-desertification and global warming should be recognized.

Although the degree of impacts will differ among and within regions, depending upon the local situation, options and concerns, VLS-PV is expected to enhance the security of energy supply; reduce the threat of climate change; stimulate economic growth; create jobs; raise incomes; reduce poverty; improve social equity; and protect the environment at all levels.

In order to realize VLS-PV, however, stakeholders involved must be willing to consider the third and fourth steps depicted in Figure 1.6, and decision-makers and financial institutions should be included in the discussion. An effective approach considered for this purpose is to develop practical project proposals for realizing VLS-PV and to develop more concrete recommendations.

1.3.3 Approaches for practical project proposal

Possible scenario for VLS-PV development

VLS-PV will be a major project that has not been implemented before. Therefore, in order to achieve VLS-PV, it is necessary to identify issues that should be solved and to discuss a practical development scenario.

Focusing on technical development, a basic concept for VLS-PV development, consisting of four stages (see Table 1.1), was proposed.

S-0: R&D stage (may be unnecessary in advanced countries)

In order to verify the basic characteristics of the PV system in a desert area, two main technical issues should be examined: the reliability of PV in a desert area and the feasibility of grid connection. Non-technical issues to be examined include the conditions for site selection and planning cooperation frameworks, including training engineers and funding schemes. Recently, PV systems with a capacity of around 10 MW have been installed and operated in technologically advanced countries. It is very important to consider social conditions for the whole VLS-PV project in regions that have no experience of large scale PV systems.

S-1: Pilot stage

A large scale PV system that has a capacity of tens of megawatts will be constructed and operated to evaluate and verify the preliminary characteristics of a large scale PV system. Technical issues here are of a higher level and shift to concentrated and grid-connected PV systems for building VLS-PV technical standards. PV module facilities will be constructed and operated, although the PV modules will be imported. PV module production will be introduced in the next stage. Preventing desertification through such activities as ensuring adequate vegetation cover and establishing tree plantations will also be started at this stage.

Table 1.1 Summary of a scenario for VLS-PV development

	Sub-stage	
	Technical issues	Non-technical issues
S-0: R&D stage *(may be unnecessary in advanced countries)*	• Examination of the reliability of PV systems in a desert area • Examination of the required ability of PV systems for grid connection	• Site selection for VLS-PV based on conditions • Planning project formation, including training engineers and funding
S-1: Pilot stage *Scale of system: 25 MW*	• Development of the methods of operation and maintenance (O&M) for VLS-PV • Examination of the control of power supply from PV systems to the grid line	• Development of the area around VLS-PV in order to prevent desertification • Training engineers for PV module production on site
S-2: Demonstration stage *Scale of system: 100 MW*	• Development of the technical standards for O&M of VLS-PV, including grid connection	• Training engineers for mass production of PV modules and for balance of systems (BOS) production on site • Preparation for industrialization by private investment
S-3: Deployment stage *Scale of system: 1 GW*	• Developing the concept of 'solar breeder' from the viewpoint of technical and non-technical issues	

S-2: Demonstration stage

A 100 MW PV system will be constructed and operated to research methods of grid-connected operation and maintenance for when VLS-PV actually takes on a part of the local power supply. Information on grid-connecting VLS-PV to the existing grid line will involve technical standards for deploying VLS-PV. However, non-technical issues will also be covered. Mass production of PV modules will be carried out and BOS production on site will occur at the deployment stage.

S-3: Deployment stage

The deployment stage verifies the capability of VLS-PV as a power source. In this stage, technologies for generating and supplying electricity will be nearly complete. However, for deployment of VLS-PV in the future, some options, such as demand control, electricity storage and recycling of components, will be required. These will contribute to building the concept of a 'solar breeder' and business plans for VLS-PV will be proposed.

Direction and requirement for practical project proposals

Implementation of the VLS-PV project will activate the world's PV industries in a wide range of technologies, from solar cells production to system construction. However, stages S-0 to S-2 (see Table 1.1) are necessary to achieve the final stage, and developments in each stage should be steadily carried out.

When thinking about a practical project proposal for VLS-PV development, a common objective is to find the best sustainable solution. System capacity for suitable development depends upon each specific site and with its own application needs, infrastructure, and human and financial resources.

From the technological viewpoint, it is important to start with the R&D or pilot stage when considering overall desert development. In terms of commercial operation, the pilot or demonstration stage is more crucial.

Project proposals depend upon the type or stage of the project; however, at a minimum, the following items should be considered:

- background and objective;
- targeted capacity of VLS-PV systems;
- location;
- project period;
- technical aspects;
- economic aspects;
- social aspects;
- environmental aspects; and
- future perspective.

Location should be specified based on various conditions, such as land area, meteorological conditions, energy and electricity issues, integration with existing infrastructure, and social development in the region. Technical aspects should be discussed from the viewpoints of PV technology for electricity supply, and suitable technology for the desert should also be covered. Furthermore, it is important to consider the technology transfer to the region. Economic aspects should include a possible financial plan. Social and environmental aspects should focus on both positive and negative impacts.

The proposals developed in this report may motivate stakeholders to realize a VLS-PV project in the near future. Moreover, the practical project proposals, with their different viewpoints and directions, will provide essential knowledge for developing detailed instructions for implementing VLS-PV systems

Based on these viewpoints, this report will provide examples of practical project proposals.

REFERENCES

1. Kurokawa, K. (ed) (2003) *Energy from the Desert: Feasibility of Very Large Scale Photovoltaic Power Generation (VLS-PV) Systems*, James and James, London
2. IEA (2004) *World Energy Outlook*, IEA, Paris, France
3. 'Policy recommendations for renewable energies', International Conference for Renewable Energies, Bonn, Germany, 4 June 2004
4. 'Political declaration', International Conference for Renewable Energies, Bonn, Germany, 4 June 2004

CHAPTER TWO

The Mediterranean region: Case study of very large scale photovoltaics

2.1 INTRODUCTION

Very large scale photovoltaic power generation (VLS-PV) systems can contribute considerably to global energy needs, the environment and socio-economic development, while simultaneously becoming more economically and technologically feasible.[1] This chapter examines the economic conditions for VLS-PV systems in the Mediterranean region. Originally focusing on the Sahara Desert-bordering countries of Morocco and Tunisia, Portugal and Spain were also included to compare the impact of recently approved photovoltaic (PV) feed-in tariffs with less supportive frameworks in Northern Africa.

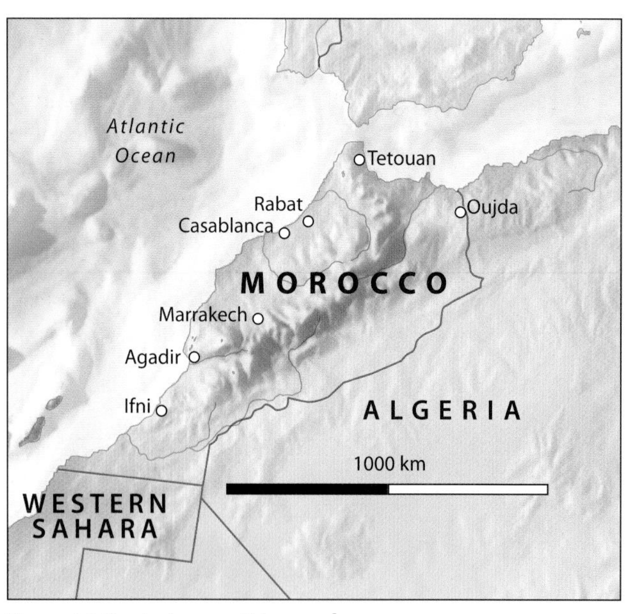

Figure 2.1 Physical map of Morocco[3]

Two sites were selected for each country, one more affected by marine climate influences with lower irradiation, and one representing a higher irradiated desert-like location. The study was performed from a professional project developer's point of view by determining PV electricity generation cost and potential revenues from the electricity sale of VLS-PV systems to customers, either to consumers at a standard electricity price level or to grid-operating entities on a feed-in tariff basis.

2.2 GENERAL CONDITIONS

2.2.1 Country profiles

Morocco

Morocco, located in North-Western Africa, as shown in Figure 2.1, has a population of 29,7 million people living mainly in the fertile north-western coastal regions. Casablanca, with 3,1 million inhabitants, is the leading industrial, commercial and financial centre of Morocco and includes an important port. The area of Morocco is 446 550 km^2. The south-eastern boundary, in the Sahara, is not precisely defined. The population density is 67 persons/km^2. The income per capita was 1 532 EUR in 2002.[2]

The annual horizontal global radiation input of Morocco increases from north to south and from west to east. The lowest intensities of about 1 700 kWh/m^2/year are found in the north. In the lee of the Grand Atlas Mountains and in the Sahara Desert, the average annual radiation input is above 2 100 kWh/m^2/year – for example, in Quarzazate, as shown in Figure 2.2.[4]

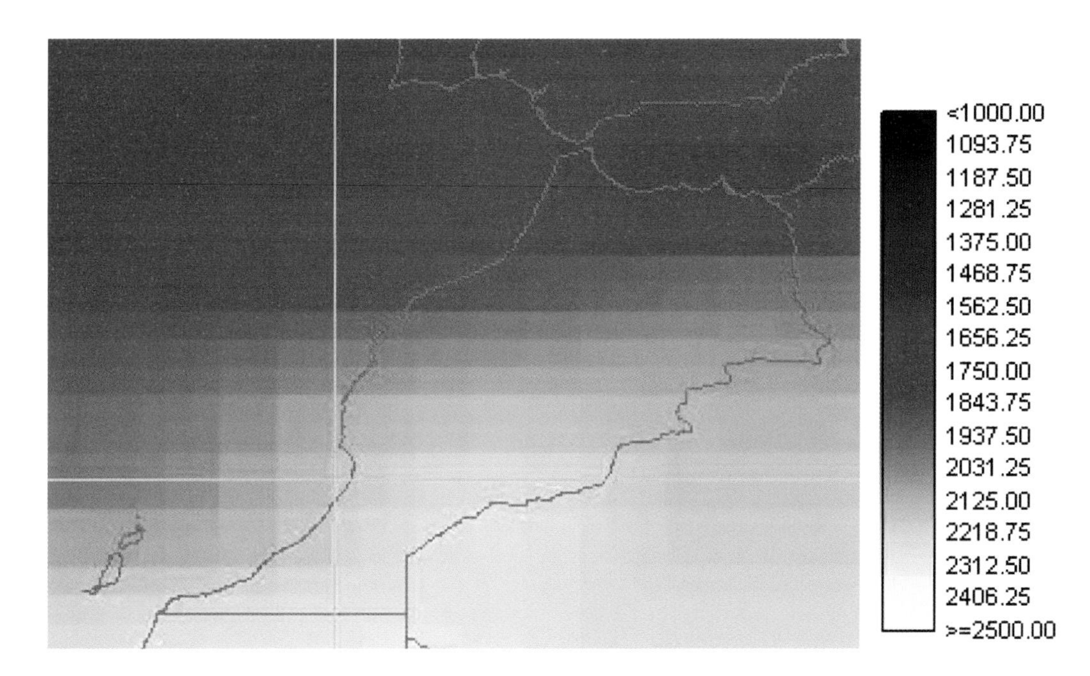

Figure 2.2 Average annual horizontal solar radiation in Morocco in kWh/m^2 [4]

The primary energy supply in Morocco as shown in Figure 2.3, is based on the fossil energy sources of oil (62,0 %) and coal (32,6 %). Approximately 97 % of these two sources had to be imported in 2002.

The total electricity production in 2002 was about 17 213 GWh, with 70,0 % from coal, 23,9 % from oil, 4,9 % from hydro and 1,1 % from other sources.[5] In 2002, 14 147 GWh of electricity were injected into the Moroccan power grid. Most electricity was generated by plants fired with imported coal primarily from South Africa or oil from Saudi Arabia, Iran, Iraq and Nigeria, while some electricity was imported from Spain (1 355 GWh) and Algeria (37 GWh). Hydropower and wind energy contributed 5 % and 1 % to the country's power supply.[6] Morocco is expecting a rise in total power consumption up to 25 000 GWh by 2010. According to existing plans, the expansion of power imports over the next few years[7] and an energy policy based on diversification, market competition and demand management will handle this.[5,8]

In Morocco, electricity tariffs depend upon the voltage: a low voltage of 220 V or a high voltage of 380 V. The billing is based on realistic electricity consumption, as shown in Table 2.1.[9] The price of electricity for end users, excluding independent power providers, is fixed by the government of Morocco.

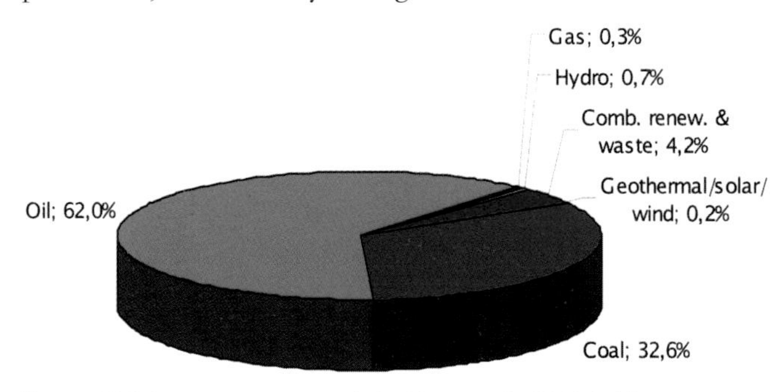

Figure 2.3 Primary energy supply in Morocco in 2002 (without energy imports)[5]

Table 2.1 Electricity tariffs in Morocco in 2002[9]

Consumption per month	Price in EUR cents/kWh
0–100 kWh	7,6
101–200 kWh	8,2
201–500 kWh	8,8
500 kWh	12,1

Between 1997 and 2000, the tariffs for medium voltage and high voltage electricity were lowered by 28 % overall.[7]

Of all the Maghreb countries, Morocco has done the most to develop its renewable energy sources for generating electricity. Morocco has a high potential for hydropower. The technically exploitable hydroelectric power potential is estimated at 2 500 MW, of which about 37 % has been developed to date. Morocco has very good wind conditions with mean wind speeds exceeding 11 m/s in some places, creating substantial exploitable wind potential. The government of Morocco is planning to increase the share of electricity generated by wind power to 4 % by 2010. To date, only 0,24 % of the total energy comes from renewable energy sources. Table 2.2 provides an overview of the installed capacity of the renewable energies in 2002.

In 2001, the Moroccan Ministry of Energy and Mining defined the energy policy guidelines for promoting renewable energy sources in Morocco in the

Table 2.2 Installed capacity of renewable energy in Morocco in 2002[7, 10, 11]

	Installed capacity
Photovoltaic	6 MW
Solar thermal	45 000 m^2
Wind power	54 MW
Biomass	3 000 m^3
Hydropower	1 170 MW

Table 2.3 Planned capacity of renewable energy in Morocco in 2002[7, 11]

	Capacity in MW	Political aim by
Hydropower (big installations)	n.a.	n.a.
Wind power	1 200	2010
Hydropower (small installations)	2 000	2010
Biomass + biogas	n.a.	n.a.
Photovoltaic (solar home systems)	10	2006
Solar thermal	200	2008

Note: n.a. = data not available.

Table 2.4 Moroccan institutions on energy, renewable energy and PV[7]

	Function	Task
ONE	Central Energy Corporation	Development of programmes for renewable energies (e.g. PV)
CDER	Centre for the Development of Renewable Energy Sources	Studies: expert knowledge and information; quality education for experts in the field of renewable energies
AMISOL	Association for Solar Energy	Support and development of solar technologies
PVMTI	Photovoltaic Market Transformation Initiative	Support for and commercialization of PV technologies

National Strategy Plan for the Development of Renewable Energy Sources. According to the plan, the share of power generated from renewable energy sources should increase to 10 % by 2010 and 20 % by 2020. By promoting the use of renewable energy sources, the government intends to reduce Morocco's dependence upon imported energy, while completing the electrification of rural areas, creating jobs and lowering the country's carbon dioxide (CO_2) emissions. Morocco adopted the Kyoto Protocol in January 2002, through which new sources of project financing geared to promoting renewable energy sources opened up.[7] The planned capacity of varying renewable energy sources until 2008 is shown in Table 2.3.

Morocco began to liberalize its energy market during the 1990s. The Office National de l'Électricité (ONE), a public-law company answering to the Ministry of Energy and Mining, has been responsible for the generation and transmission of electricity in Morocco since 1963. ONE operates as a single buyer. Since 1994, however, power plants with ratings above 10 MW can also be built and operated by private enterprises if the project is subject to open tendering and all power produced is sold to ONE. Depending upon the region, the actual supply of electricity to the ultimate consumers is attended to either by ONE, local government or private enterprise. The government of Morocco is planning to further liberalize the electricity market in the years to come. In connection with this, ONE is to be unbundled into independent enterprises responsible for the generation, transmission and distribution of electricity. The power grid will also be opened up for independent power producers who will then be able to market their electricity without necessarily having individual contracts with ONE.[7] Today, around 70 % of the electricity-supplying companies are private. The electricity market will be completely liberalized by 2007.[10]

In 1982, the Centre for the Development of Renewable Energy Sources (CDER) was established and subordinated to the Ministry of Energy and Mining. CDER's work includes conducting studies, disseminating knowledge, performing quality-control checks on equipment (PV systems, in particular), and training specialists in renewable energy sources. ONE and CDER together have planned and developed a PV programme for rural electrification. The most important association for solar energy is Association Marocaine des Industries Solaires (AMISOL). Its work aims at supporting alternative energies such as photovoltaic power, pump systems and wind farms. In addition, the Photovoltaic Market Transformation Initiative (PVMTI), an important solar energy programme devoted to developing national markets for PV systems in India, Kenya and Morocco, was launched by the Global Environment Facility (GEF) and the International Finance Corporation (IFC) in 1998.[7] The most important Moroccan institutions responsible for renewable energy are shown in Table 2.4.

Morocco has no law prescribing a legal obligation to purchase and pay for electricity generated from renewable energy sources. The prices for electricity from privately operated power plants are negotiated between ONE and each operator, and fixed in individual contracts. In these contracts, ONE pledges to purchase the electricity produced by the power plants.[7]

There are few programmes that support the development of renewable energy sources, in general, and PV, in particular. One is the Programme for Decentralized Electrification (PERG), which promotes solar home systems (SHSs) for the electrification of rural areas. From 1995 until 2002, PERG raised the degree of rural electrification from 18 % to 55 %. It is expected to reach 100 % by 2008, which means that even villages situated far from the power grid benefit from solar power. The rural electrification programme is financially supported by the local administration, the Direction Générale des Collectivités Locales (DGCL), the communities and by Germany, France, Japan, Spain and the European Union (EU). In addition, PERG also

Table 2.5 Overview of Moroccan PV companies[12]

Company	Activity
Afrisol	Sales; rural electrification; telecommunication; street lighting
Casabloc	Sales; PV systems; lighting
Getradis Energies Renouvables	Import; sales; distribution; installation of PV solar heating; pumps
Noor Web	Design; installation of PV systems
Phototherm Electronique	Sales; distribution; installation of solar thermal and PV systems and production of electronic equipment
Sehi	Sales of PV and solar pumping systems
Setel-Siemens VE-E	Sales; distribution of PV products
Sicotel	Sales of PV systems, specializing in lighting and pumping
Spolyten	Sales of PV systems
UmaSolar sa	Sales of PV systems, solar pool and water heating systems

includes the off-grid electrification of communities and villages situated too far from the existing distribution networks for coverage. Financially, ONE is continuing to subsidize each installed SHS with contributions ranging from 397 EUR to 1 634 EUR, the one precondition being that each company enters into an appropriate contract with ONE.[7]

Currently, while rural electrification with small PV systems (SHS) has made good progress, with 7 MW of electricity capacity installed in 2002, large grid-connected PV systems do not exist. Only a 1 kW photovoltaic plant has been realized as a pilot and model installation. CDER forecasts the market volume for PV systems will be about 10 MW for the years 2001 to 2006. In 2010, the share of total electricity generation by PV installations is expected to be about 0,8 %, which shows that Morocco does not expect very fast progress.[7]

In support of harnessing renewable energy sources, the import tax on components for the generation of renewable energy has been reduced to 2,5 %. No other tax incentives or remissions are planned for the time being.[7]

The Photovoltaic Market Transformation Initiative (PVMTI) envisages an investment volume of 5 MUSD for Morocco. Part of these resources has already been allocated to two companies: the Moroccan financing company Salafin SA, which in June 2002 received 1 MUSD for a lending programme designed to promote solar energy systems to be supplied, installed and maintained by Afrisol SA; and Association Al Amana, to which the IFC is providing 720 000 USD in the form of guarantees and loans for a micro-financing project offering small scale loans for the purchase of solar energy systems in the Taroudant region. The Moroccan PV enterprise Noor Web is supplying the solar energy equipment for PVMTI.[7]

The leading solar enterprise in Morocco is Afrisol SA. Established in 1987, Afrisol sells American-made Solarex solar energy systems. Since 1998, Afrisol has been operating as Solarex's main distributor in the Maghreb/West Africa region and other African countries. At the time of writing, Afrisol was in the process of implementing a solar project with a financial volume of 1 MUSD for PVMTI. Afrisol is also planning to build a factory for solar modules in the vicinity of Tangier within the next few years. Other major solar enterprises in Morocco, as shown in Table 2.5, are SunLight Power Maroc (SPM), Noor Web (also involved in PVMTI) and Total Energie Maroc. In Morocco, no module producer exists. However, there are about 20 companies assembling simple PV systems from imported components, or importing PV systems and selling them on the Moroccan market.[7]

Tunisia

Tunisia, located in Northern Africa as shown in Figure 2.4, has a population of 9,9 million people, mostly concentrated in a narrow strip along the Mediterranean

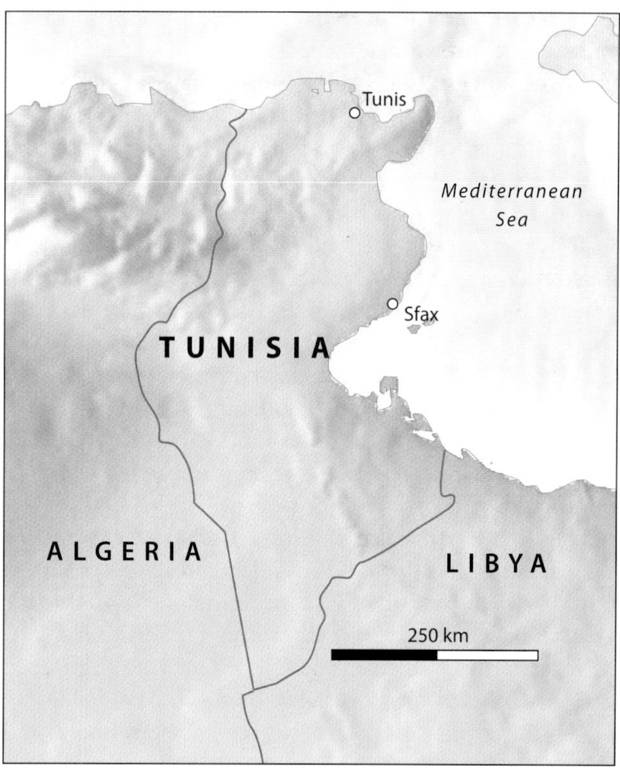

Figure 2.4 Physical map of Tunisia[3]

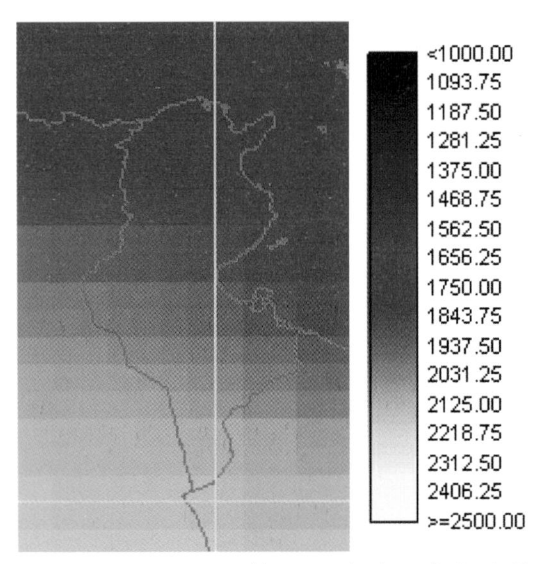

Figure 2.5 Average annual horizontal solar radiation in Tunisia in kWh/m^2 [13]

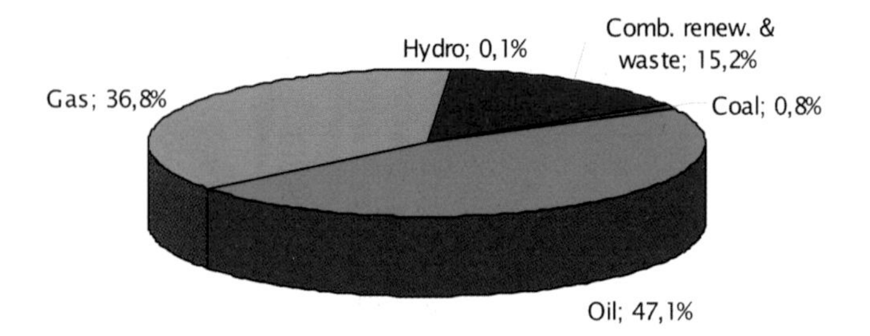

Figure 2.6 Primary energy supply in Tunisia in 2002[14]

coast in the north. The most important urban area is the capital region of Tunis, with 1,8 million inhabitants. The total area of Tunisia is 163 610 km^2, almost twice the size of Portugal. It has a population density of 61 persons/km^2. The income per capita was 2 200 EUR in 2002, the highest of all countries in North Africa apart from Libya.

The average annual horizontal global radiation in Tunisia increases from about 1 700 kWh/m^2/year in the north at Tunis to intensities of more than 2 100 kWh/m^2/year in the south in the Sahara Desert, as shown in Figure 2.5.[13]

Located between two major producer countries, Tunisia only has limited energy resources of gas and petroleum. The primary energy supply is based on oil (47,1 %) and gas (36,8 %). Renewable energy resources make up a remarkable 15,3 %.

The demand for energy, notably electricity, is rising sharply.[15] In 2002, 11 846 GWh of electricity were produced in Tunisia and supplied to end customers, who numbered some 2,3 million. Electricity generation is based almost exclusively on oil (55 %) and natural gas (44,6 %). Hydropower and wind accounted for a share of less than 0,5 %. Due to economic growth and continuous improvement in living standards, the demand for electricity has constantly risen in recent years. Electricity consumption has increased every year with the growth attributable to consumption by private households. According to projections in the tenth development plan, which sets out infrastructure measures for the period from 2003 to 2007, demand will increase to 14 140 GWh by 2007. The intention is to satisfy this growth in consumption by building an additional 770 MW of power station capacity. The total required capacity by 2011 is expected to be 4 400 MW.[16]

The electricity market in Tunisia has been liberalized since the 1990s, when the market had already become fully controlled by the Société Tunisienne d'Electricité et du Gaz (STEG). STEG is a state-owned enterprise reporting to the Ministry of Industry and Energy. It is responsible for the supply of gas and electricity, electricity transmission and supply to the end customers. There is no regulatory presence in the energy sector. Since 1996, STEG has no longer enjoyed a monopoly over electricity generation because private producers were allowed to enter the market. The door to private-sector involvement in the electricity sector was opened with the adoption of the 1996 law on the de-monopolization of the state-owned power utility STEG. Fundamentally, this allows private companies to generate electricity presuming that they have been successful in a bidding process, and to sell that electricity to STEG as a single buyer. The feed-in tariff for electricity generation based on renewable energies in the STEG net is – as for conventional power plants – only possible on the basis of individual contracts between private producers and STEG. Despite these initial steps in the direction of opening the market, the electricity sector in Tunisia is still largely subject to state control and is only geared to limited competition. To date, as is the case with conventional power stations, electricity from renewable energy sources can only be fed into the STEG grid. Even in 2002, 73 % of the national electricity-generating capacity belonged to STEG, while only 19 % was operated by independent producers and 8 % by self-generators. One result of liberalization was a gas-fired power plant brought into service in 2002 by a consortium of foreign corporations and STEG. In the third quarter of 2004, British Gas intends to begin the construction of a 500 MW power station. Commissioning is scheduled for 2006.[15,16]

The electricity tariffs for end customers are state regulated and set by the Ministry of Industry and Energy, based on proposals from STEG. By international standards, the electricity tariffs are extremely low. The tariffs are broken down in two ways: according to voltage level and according to the time of use, as shown in Table 2.6. In the low-voltage sector, the tariff structure has a progressive component.[15]

In Tunisia, it is the Ministry of Industry and Energy and its General Directorate for Energy that draws up plans for expanding the energy infrastructure and

Table 2.6 Electricity tariffs in Tunisia in 2002 (tax not included)[16]

	Daytime tariff (EUR cents/kWh)	Night tariff (EUR cents/kWh)
High voltage	2,53	1,72
Medium voltage	3,95	3,95
Low voltage	–	–
0–50 kWh/month	3,82	3,82
over 50 kWh/month	5,37	5,37

implements the energy policy adopted by the government. Most of the state actors in the energy sector are answerable to the ministry, including the Commission Supérieure de la Production Indépendante d'Electricité (CSPIE) and the Commission Interdépartementale de la Production Indépendante d'Electricité (CIPIE), which were both set up in 1996. The CSPIE decides on the procedures and selection criteria for public tender processes, awards contracts to independent power producers and passes rulings on granting tax incentives to investors. The CIPIE carries out preliminary work for the CSPIE by selecting projects for tendering, preparing bidding procedures, evaluating offers, flanking the contractual negotiations between the independent producers and the Energy Ministry, and securing the granting of public subsidies on a case-by-case basis. In 1985, the Tunisian National Agency for Renewable Energies (ANER) was established and subordinated to the Ministry of Industry and Energy. ANER is responsible for promoting renewable energy sources and raising energy efficiency, including all training matters, awareness-raising and information campaigns, as well as research and development. Its main task, however, is to manage and provide back-up support for relevant projects. In 2004, the Mediterranean Renewable Energy Centre (MEDREC) was opened in Tunis. The centre, in the beginning financed by the Italian Territorial and Environment Ministry (IMET), offers information, education, networking, technology transfer, energy efficiency and the development of pilot projects in the renewable energy sector. In addition, MEDREC is responsible for developing legal and financial frameworks and instruments. Their first project is SOLdinars, which supports distributing solar cooking stoves.[16] Some of the most important institutions regarding renewable energy and their functions are summarized in Table 2.7.

Tunisia ratified the Framework Convention on Climate Change in July 1993, and the first national climate report was presented in October 2001. The Kyoto Protocol was signed in June 2002. ANER has been given the responsibility for balancing Tunisia's greenhouse gas emissions. A first portfolio of potential projects has been prepared by ANER; alongside schemes in energy efficiency and transport, it also includes investment opportunities in wind energy, biogas exploitation and solar thermal energy. Since oil and gas are limited and will contribute only a small part of future national energy supply, the main focus for the energy policy is to promote and increase the utilization of renewable energy, especially from wind and solar energy.[16] Table 2.8 shows the share of the different energy sources for electricity generation as planned for 2010 and 2020.

In 1977, the Tunisian government launched a national research programme, PNR, to bring all actors in the energy field around the table. One area of the programme covers renewable forms of energy – specifically solar energy – and their use for generating electricity. In 2001, about 20 presidential decrees were issued on using renewable energy. One was a resolution on the obligatory use of solar thermal energy in all newly constructed public buildings. In March 2003, two commissions were called into being to work out detailed proposals for institutional capacity-building and changing the regulatory framework. However, as yet there are no specific systems in place in Tunisia to promote electricity production from renewable energy sources.[16]

Table 2.7 Tunisian institutions on energy, renewable energy and PV[15]

	Function	Task
CSPIE	Governmental commission for law framework	Develop procedures, criteria for public tender processes, rules of tax incentives for investors and contracts for independent power producers
CIPIE	Governmental commission for law framework	Execute tendering: selection of projects; preparing bidding procedures; evaluating offers; flanking contractual negotiations
STEG	National gas and electricity supplier	Generate gas and electricity; transmit to customers; support renewable energies
ANER	National agency for renewable energies	Campaign for, promote and administer renewable energy projects; increase energy efficiency, qualification, information and sensibility
Centre International des Technologies de l'Environnement (CITET)	International centre for environmental technologies	Promote and support environmental technologies and qualification

Table 2.8 Expected share of different energy sources for electricity generation in 2010 and 2020[17]

	Capacity installed in 2010	Capacity installed in 2020
Conventional energy	4,63 GW	7.85 GW
Hydropower	70 MW	n.a.
Wind power	40 MW	80 MW
Solar	0,40 Mtoe	0,45 Mtoe
Biomass	0,10 Mtoe	0,15 Mtoe

Note: n.a. = data not available. Mtoe = million tonne oil equivalent

At the end of 2002, the installed power station capacity in Tunisia totalled roughly 2 900 GW. Only 60 MW of this was provided by hydroelectric power plants and 10 MW by wind power, while the remaining capacity was entirely accounted for by thermal power stations. The petroleum and natural gas used in the thermal power stations has until now mostly originated from Tunisian sources. Until now, solar power plants have played no role in the primary energy supply. Despite excellent conditions, there is currently little use of solar energy for generating electricity. PV systems are mostly used for decentralized power supply for border posts, lighthouses, water pumps, water desalination plants and telecommunication facilities. In 2002, solar home systems were installed in 1 200 remote households and 15 schools, which were not planned to connect to the STEG grid in the foreseeable future. Electricity generation from solar energy is supposed to account for 3 % of the further electrification of rural regions in the course of this decade.[16]

Regarding financial support for PV-generating companies, investors can generally benefit from tax and customs duty relief. Customs duties can be reduced to the minimum 10 %. Value-added tax can be reclaimed entirely in the case of imported goods, if they cannot be manufactured in Tunisia, and for locally produced capital goods. In specific cases income tax can be remitted for up to five years and investment subsidies can be granted. The Tunisian state can also contribute to the costs of expanding the infrastructure. If a project is considered particularly important on account of the magnitude of the investment or the number of jobs created in Tunisia, the state can also make the required land available at a nominal price. Decisions on concessions of this nature are made by CSPIE.[16]

Currently, there is no PV market in Tunisia. Only a few companies import and distribute photovoltaic systems, such as the Société Internationale de l'Energie et des Sciences (SINES) and Energy Engineering Maintenance Service, which is a Shell PV product dealer.[18]

Portugal

Portugal, as shown in Figure 2.7, is located in South-Western Europe on the western side of the Iberian Peninsula and has a population of 10,2 million concentrated in the plains along the Atlantic Coast. The most important urban areas are the capital regions of Lisbon, with 1,9 million inhabitants, and the region of Porto. The total area of Portugal, including the Azores (2 247 km^2) and the Madeira Islands (794 km^2), is 92 040 km^2. Its population density is 111 persons/km^2. The income per capita was 12 500 EUR in 2002.[2]

The average annual horizontal global radiation increases from about 1 500 kWh/m^2/year in the north-west to over 1 760 kWh/m^2/year in Lisbon, to Europe's highest at 1 900 kWh/m^2/year in the Algarve and Alentejo regions in the south.[19]

The primary energy demand in Portugal is based on oil and coal. Portugal has no remarkable fossil energy resources so it depends upon energy imports, which make up about 87 % of all energy resources supplied. In 2002, oil accounted for 62,4 % and coal for 13,3 % of the total primary energy consumption, as shown in Figure 2.9.

The entire demand for coal is satisfied by imported coal. Gas comes from Algeria. Nuclear energy plays no role in Portugal. In 2002, a total of 46,107 GWh of electricity were produced and 41,473 GWh consumed. The country's largest domestic energy resource is hydropower, which was around 30,2 % in 2001. However, its contribution fluctuates depending upon rainfall, making it, to a great extent, unreliable. Photovoltaic, wind and tide power together deliver approximately 0,5 % of the total energy production.

Traditionally, hydropower is the most important energy source in Portugal. Portugal is also experienced in biomass. Wind power has shown the highest growth rate in recent years, but the level is still low. The

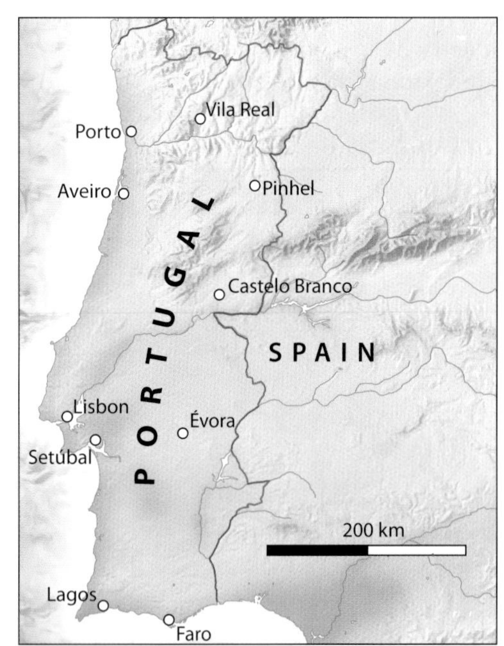

Figure 2.7 Physical map of Portugal[3]

Figure 2.8 Average annual horizontal solar radiation in Portugal and Spain in kWh/m² [19]

installed electrical capacity of renewable energy is shown in Table 2.9.

The level of liberalization of the Portuguese electricity market was around 45 % in 2002. It is still open as to when a completely liberalized and deregulated market will be achieved. Until now, the electricity market has been dominated by the company Eletricidade de Portugal (EDP). The former public-owned monopoly is already privatized. In addition, some foreign electricity corporations, such as EDF (Electricité de France), ENDESA (Empresa Energetica Espanola, in Spain) and ENI (Ente Nazionale Idrocarburi, in Italy) operate in the Portuguese energy market. In 1995, the energy sector was newly structured and the national electricity distribution system, Sistema Eléctrico Nacional (SEN), was established. It is divided into the existing public system, the Sistema Eléctrico Publico (SEP), and the new independent system, the Sistema Eléctrico Independente (SEI). Independent operators of renewable energy plants integrate with SEI. Since 2002, the independent public authority, Entidade Reguladora dos Serviços Energéticos (ERSE), is responsible for regulating the national sectors of gas and electricity. ERSE ensures a free entry to the grid, acceptable transfer tariffs and no misuse. During 2003, the Mercado Ibérico de Electricidade (MIBEL) was announced with the aim of creating an electricity market union for the Iberian Peninsula.

In Portugal, the price of electricity for end users or private households is approximately 0,12 EUR/kWh, including all duties, but excluding value-added taxes, and the yearly consumption is about 3 500 kWh. Operators of PV installations receive an allowance of 0,31 EUR/kWh for large systems of more than 5 kW and 0,55 EUR/kWh for small systems of less than 5 kW, according to the Portuguese Renewable Energy Act. Table 2.10 shows the end prices without value-added taxes for PV applications in Portugal.

Portugal, as a member of the European Union, has to conform with the Kyoto Protocol. All relevant greenhouse gases should not increase more than 27 %

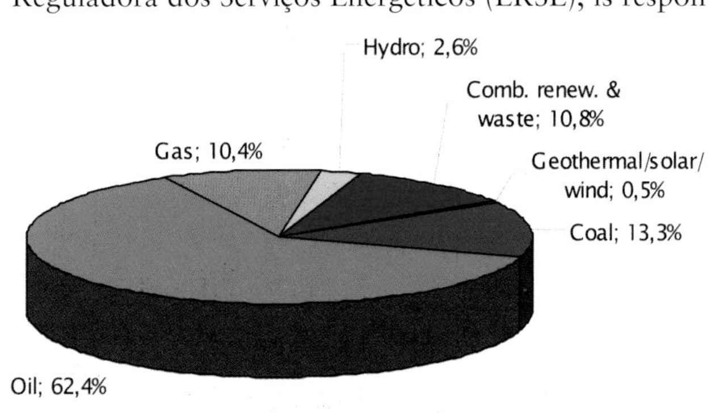

Figure 2.9 Primary energy supply in Portugal in 2002[20]

Table 2.9 Installed capacity of renewable energy in Portugal (not including the Azores and Madeira)[20]

	2001 (in MW)	2002 (in MW)	2003 (in MW)
Wind	125	190	268
Maxi hydropower (over 10 MW)	4 049	4 061	4 062
Mini hydropower (less than 10 MW)	281	293	297
Photovoltaic	1,34	1,51	2,07
Biomass	441	478	458
Waste	88	97	97
Geothermal	18	18	18
Total	4 915,34	5 041,51	5 105,07

Table 2.10 Electricity tariff in Portugal in 2002

Consumption per month	Electricity tariff in EUR/kWh
Price for private households for conventional electricity	0,12
PV systems less than 5 kW	0,31
PV systems over 5 kW	0,55

Table 2.11 Expected installed capacity of renewable energy in Portugal in 2010

	Original goal until 2010 in MW*	Revised goal until 2010 in MW**
Maxi hydropower	4 995	5 000
Wind power	2 930	3 750
Mini hydropower	500	400
Waste	130	130
Biomass plus biogas	150	200
Photovoltaic	50	150
Tide power	50	50
Total	8 805	9 680

Source: * DGE; ** Resolution No 63 2003

compared to 1990. At the end of 2000, that number was about 30,1 %, making a change in energy policy necessary. Since 1999, the Portuguese energy policy has focused on modifying the energy sector, supporting gas use, and implementing measures to increase energy efficiency and renewable energies. According to a European directive from 2001, renewable energy should make up 38 % of the primary energy consumption in 2010. In order to reach this goal, the Forum for Renewable Energies in Portugal was created and the E4 Programme introduced. The latter formulates the national energy policy and strategy until 2010. In 2003, the Portuguese government increased the goal of 50 MW installed PV capacity to 150 MW by 2010. The E4 Programme formulates the political goal of 50 MW electricity from PV by 2010, which correlates with a yearly growth rate of about 40 %. Experts do not believe these goals to be realistic. The actual growth rate is about 25 % and cannot meet the expectations and prognosis of 10 MW by 2010. However, in 2003 the Portuguese government fixed a new goal for the PV industry of 150 MW installed PV capacity by 2010. The actual goals for renewable energies are shown in Table 2.11.

The Portuguese energy policy, including the support for renewable energies, is directly subordinate to the Ministry of Economy. The ministry is supported by three institutions: the General Directorate for Energy (DGE), the regulation authority Entidade Reguladora dos Serviços Energéticos (ERSE) and the national energy agency (ADENE). The DGE is responsible for the conception, implementation and control of the national energy policy. Since 2001, ERSE is the central authority for the approval of grid connection, installation and operation of energy projects. ADENE implements the energy policy made by the government. The most important associations for the national PV industry are the Associação Portuguesa da Indústria Solar (APISOLAR) and the Sociedade Portuguesa de Energia Solar (SPES). APISOLAR was established in 1998 to support the solar thermal and photovoltaic market. SPES promotes the advantages of renewable energies, especially photovoltaic energy. Table 2.12 summarizes the relevant institutions for renewable and photovoltaic energy.

In 1988, the Portuguese renewable energy law obliged the grid operator to purchase electricity from renewable resources and to pay an allowance to the generating company. In 1995 and 1999, the feed-in law was adopted for the new status of liberalization and climate protection. In 2001, a new allowance system was created to determine the electricity prices for renewable energy, considering technology and location. The law is a good incentive to investment in renewable energy; but a defined valid period is missing. Only a subordinated guideline defines a guarantee of 12 years. Operators of renewable energy plants are, automatically, attendants of the SEI electricity system. The public grid operator Rede Eléctrica National (REN) is obliged to buy the produced electricity. Additionally, a special regulation, Decree 68/2002, applies for operators of small plants for self-supply.[21]

Besides the feed-in law there is no other national subsidy programme for photovoltaics in Portugal. However, there is a superior programme for renewable energy, the Programa Operacional de Economia (POE), which is embedded in the EU structural fund programme. It supports projects for energy efficiency

Table 2.12 Portuguese institutions on energy, renewable energy and PV[21]

	Function	Task
DGE	General authority for energy	Conception, implementation and control of national energy policy
ERSE	Regulatory authority	Approval of grid connection, installation and operation of energy projects
ADENE	National energy agency	Implementation of political decisions on energy; support of renewable energies
APISOLAR	Association of solar industry	Promotion and support of solar thermal and PV technology and PV market development
SPES	Portuguese society of renewable energies	Promotion of the advantages of renewable energy and PV

and renewable energies. Medida de Apoio ao Aproveitamento do Potencial Energético e Racionalização de Consumos (MAPE) offers grants and interest-free loans for investments in energy efficiency and renewable energy. The non-repayable grants are up to a maximum of 20 % of the investment and 300 000 EUR per project. In addition, interest-free loans up to 50 % of total investment and for a valid period of 12 years, as well as two repayment-free years, are offered. From 2000 till 2006, the European QCA programme offers 5,1 million EUR, of which 1,5 million EUR are for the energy sector, including a budget for the grid operator for improving the grid. The General Directorate for Energy (DGE), as well as local directorates from the Ministry of Economy (DER), is responsible for the administration of MAPE grants.

In 2003, the Portuguese Mutual Savings Bank, Caixa Geral dos Depositos, started the first credit programme, Credicaixa Ambiente, for environmental projects. This programme supports investors in financing and purchasing small units, with credits of about 1 000 to 25 000 EUR, low interest rates and a duration of about five years. The Portuguese Ministry of Finance grants tax benefits for investors and operators of renewable energy plants. Private persons can obtain a credit note of up to 30 % of the total investment of up to 700 EUR for investments in renewable energies. The actual value-added tax is about 12 %.

Despite excellent solar conditions – Portugal has the third greatest solar potential in Europe, after Spain and Greece – there is currently little market development. Portugal has an installed PV capacity of 1,66 MW in 2002, ranking 12 in the list of European countries. 76 % of the PV systems are stand-alone installations, including about 29 % very small installations, such as SOS-telephones and telecommunications. At the end of 2002, the greatest installations were stand-alone systems in Ourique (wind/PV hybrid plant with a capacity of 42 kW), Castro d'Aire (19,3 kW) and Berlengas (13 kW). The market for grid-connected applications grows slowly, with only about 24 % of the total installed capacity. In 1998, British Petroleum (BP) Solar started the Sunflower project in Portugal and installed a total of about 250 kW grid-connected solar modules on the top of BP petrol stations across Portugal.

The market for the PV industry in Portugal is slightly more developed than in Morocco and Tunisia. In 2001, Shell Solar opened a factory in Évora for producing mono- and multi-crystalline modules for export. One year later, the factory produced about 8 000 mono-crystalline modules with a total capacity of 10,8 MW. The total production capacity in Évora is about 15 MW. Shell Solar plans to expand the capacity up to 40 MW by 2010. Shell Solar is currently the only cell producer in Portugal, while a dozen companies distribute and import modules and components from the EU, US and Japan, and offer project planning, installation and services. In 2002, there were no cell producers in Portugal. The solar cells for Shell Solar came from the US and Germany. In addition, there is no consistent distribution structure for trading PV systems. There are some companies that cooperate with one partner, such as F. F. Sistemas de Energias Alternativas, which cooperates with BP Solar, and Lobosolar, which keeps close business relations with Shell Solar. Other companies have only informal contacts and source modules from different suppliers. All actors supply the end users, especially for larger projects. The most important PV companies in Portugal are shown in Table 2.13.

Table 2.13 PV companies in Portugal[12]

Company	Activity
Shell Solar	Production of mono- and multi-crystalline modules
F. F. Sistemas de Energias Alternativas Lda	Sales of BP Solar modules
Lobosolar	Sales of Shell Solar modules
J. A. Revéz & Filhos, Lda	Sales of Kyocera and BP Solar modules
Power5 – Engenharia, Lda	Sales of Eurosolare and BP Solar modules
Solar Blaser	Sales of Eurosolare and Solar-Fabrik modules

Spain

Spain occupies the greater part of the Iberian Peninsula, as shown in Figure 2.10. With about 40.8 million people, it is the third biggest of the northern Mediterranean countries, with a high 80 % urbanization rate. Apart from the capital, Madrid (3 million people), large areas of the centre of Spain are sparsely populated. The area of Spain, including the enclaves Melilla and Ceuta in Northern Africa and the Canaries and Balearics, is 505 960 km^2. The population density is 81 persons/km^2. The average income per capita was 17 100 EUR in 2002.[2] In Spain, as shown in Figure 2.8, the average annual horizontal global radiation increases from about 1 200 kWh/m^2/year in the north towards the best solar conditions along the southern Costa del Sol, with a radiation of about 1 800 kWh/m^2/year – for example, in Almería.[19]

Economic growth and accelerated industrialization associated with EU membership have led to increased Spanish energy demand since the mid 1970s. The primary energy supply is based on oil (51,3 %), coal (16,5 %), nuclear power (12,5 %) and gas (14,3 %), as shown in Figure 2.11.

Although its share of primary energy consumption in Spain is decreasing, oil continues to play a major role in the country's energy supply. Spain imports almost all

Figure 2.10 Physical map of Spain[3]

of its oil, mainly from Russia, Libya, Mexico, Saudi Arabia and Nigeria. About 75 % of gas is imported from Algeria. The country's largest domestic energy resource is coal, making the import rate for coal comparatively low at approximately 60 %. However, Spain has the highest import rate for fossil energy resources in the EU, followed by Luxembourg, Portugal, Ireland and Italy. The renewable energy resource share of the total primary energy supply is only about 5,4 %. In 2002, Spain had the fifth largest electricity market in the EU behind Germany, France, the UK and Italy. Total electricity generation was approximately 246,1 TWh. Spain's generation mix comprised oil, gas, coal, hydropower, nuclear and renewable energy. Coal delivers the biggest share, followed by nuclear power. The renewable energy resources of hydropower, wind and biomass supplied 20 % of Spanish electricity generation.[22]

In Spain, economic growth and accelerated industrialization associated with EU membership have led to increased energy demand of over 100 % since the mid 1970s. In September 2002, the Spanish Cabinet approved a ten-year energy plan focusing on meeting the country's future energy requirements. The plan

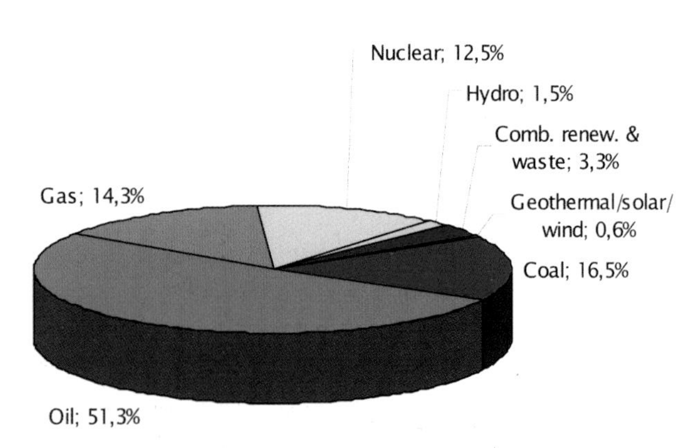

Figure 2.11 Primary energy supply in Spain 2002[22]

Table 2.14 Installed capacity of renewable energy in Spain[22]

	2001 (in MW)	2002 (in MW)
Wind	3 244	4 144
Maxi hydropower (over 10 MW)	16 399,3	n.a.
Mini hydropower (less than 10 MW)	1 618,7	n.a.
Photovoltaic	15,7	20,8
Biomass	167	208
Waste	51,3	65,6
Geothermal	94,1	n.a.
Total	21 590,1	n.a.

Note: n.a. = data not available.

reflects an increased reliance on making gas and renewable energy Spain's primary energy consumption by 2011. It also encourages revamping the country's basic energy infrastructure, investing heavily in the electricity sector and expanding gas distribution networks. The current installed capacity of renewable energies is shown in Table 2.14.

The stage-to-stage liberalization of the Spanish electricity market started in 1998 and was already finished in 2003, making Spain one of the most progressive countries in Europe. The regulation authority, Comision Nationál de Energía – Spanish Regulation Authority (CNE), ensures consistent competition conditions for grid connection, feed-in and transfer of electricity under the special guidelines for conventional operators (Regimen Ordinario) and renewable operators (Regimen Especial). The Spanish Energy Exchange is run by the Companía Operadora del Mercado Espanol de Electricidad (OMEL). A consistent pool price is calculated daily and shown on the OMEL homepage (see www.omel.es). The liberalization of the market attracts domestic as well as foreign companies to invest in the Spanish market. Currently, the Spanish electricity market is dominated by four great corporate groups: ENDESA, Iberdrola, Union Fenosa and Hidrocantábrico.

Spain has signed the Kyoto Protocol and fixed the value for greenhouse emissions. It has already passed the limit of greenhouse emissions by 15 %. In response, the Spanish Cabinet formulated a new national energy plan in 2002 for up to 2011. Gas as a primary energy resource plays an important role and is expected to increase from 12,3 % in 2000 to 22,5 % in 2011. Renewable energy is projected to increase its share of electricity generation considerably, while nuclear, coal and oil are expected to account for smaller percentages of the total generation. The planned capacity of electricity generated from renewable sources for 2011 is shown in Table 2.15.

In 1999, the Spanish government fixed the objectives in the *Plan de Fomento de las Energías Renovables* (PFER). By 2010, an increase of the installed PV capacity to 144 MW (61,2 MW by 2006) is expected, of

Table 2.15 Planned electricity capacity from renewable energy in Spain in 2011[23]

	Planned capacity installed in 2011 (in MW)
Wind	13 000
Maxi hydropower (over 10 MW)	16 571
Mini hydropower (less than 10 MW)	2 380
Photovoltaic	144
Solar thermal	200
Biomass	3 098
Biogas	78
Waste	262

which 135 MW are from grid-connected PV systems. In 2004, the limitation for PV was raised from 50 MW to 150 MW for PV installations of up to 100 kW. This is already more than the fixed value of 135 MW in the PFER for 2010. The national energy agency, Instituto para la Diversificación y Ahorro de la Energía (IDAE), also declared that the new socialist government is willing to further increase this value and formulate a binding development goal for 2020. The national energy plan also prescribes the *Plan de Ahorro y Eficiencia Energético* (PAEE) and the *Plan de Fomento para Energías Renovables* (PER), further specified in royal decrees. PER is carried out by IDAE and formulates an increase of the rate of renewable energy of the total primary energy consumption from 5,2 % to 12 % by 2010, when 0,11 % will be from PV.

The institution implementing the national energy plan is IDAE, which is subordinate to the Ministry for Science and Technology. As shown in Table 2.16, the most important associations for the PV industry are the Asociación de Productores de Energías Renovables (APPA) for producers, the Asociación para la Indústria Fotovoltáica (ASIF) for the PV industry and the Confederación Nacional de Instaladores (CNI) for installers.

During recent years, several acts have been passed in order to support energy from renewable sources. In March 2004, there was the still-valid royal Decree 436/2004. This includes a guarantee for feed-in tariffs over the total operation time of PV installations. PV plants with a capacity of up to 100 kW profit from a higher feed-in payment, which is linked to a consistent average tariff. In addition, the operator has a choice between a fixed price and a premium price (bonus) on top of the market price. Table 2.17 provides an overview of the 2004 tariffs.

Table 2.17 Spanish feed-in prices: Fixed prices in 2004 and bonus system in comparison[21]

	Fixed price, 2004	Bonus price (market price about 3–4 EUR-cents/kWh), 2004
PV plants under 100 kW	575 % of reference price (RP) in the next 25 years, later 460 % of RP	Not possible
	41,44 (33,15) EUR cents/kWh	Not possible
PV plants over 100 kW plus solar thermal high temperature plants	300 % of RP in the first 25 years, later 240 % of RP	250 % of RP in the first 25 years, later 200 % of RP (+ 10 % bonus for market participation)
	21,62 (17,30) EUR cents/kWh	18,74 (15,14) EUR cents/kWh

Note: Reference price (RP) 2004 = 7.2072 EUR cents/kWh.

In Spain, the electricity price for end users for private households is 0,087 EUR/kWh, including all duties, but excluding value-added taxes; yearly consumption is slightly over 3 500 kWh.

Based on the new law, market observers expected a PV boom. However, the approval procedure for installation and grid connection is still inefficient, inhibiting the Spanish solar market. In addition, there is a lack of efficient implementation by national and regional authorities. In conclusion, IDAE, the responsible national and regional authorities, and the electricity industry are planning a new national regulation for 2005.

Apart from the national programmes and decrees, decision-making power for the energy policy was delegated to the autonomous regions in 2003. As a result, energy policy in Spain differs from region to region. Energy agencies exist in Andalusia, the Balearics, Castilla la Mancha, Castilla y León, Catalonia, Valenciana, Madrid, Pais Vasco and Galicia. Very active autonomous regions regarding PV are Navarra and Catalonia, which aspire to a yearly PV growth of approximately 1 MW. In 2002, Navarra (1,37 MW) and Catalonia (1,02 MW) had approximately 50 % of the total market growth. Andalusía, in spite of being one of the sunniest regions in Spain, only installed about 684 kW, while the Basque region installed 456 kW and Valencia 378 kW.

The Spanish solar market is strongly affected by national incentive programmes. Investment in PV is

Table 2.16 Spanish institutions on energy, renewable energy and PV[21]

	Function	Task
CNE	Regulatory authority	Regulation of energy market, feed-in, etc.
IDAE	National energy agency	Support of diverse electricity sources and efficient use of energy
APPA	Association of producers	Support of PV producers
ASIF	Association of solar industry	Promotion, support and development of PV technology and PV market
CNI	Association of installers	Support of installers of PV modules and projects

based on attractive financing, such as non-repayable grants, low interest loans and tax benefits. In the past, the PV programme was inhibited by bureaucracy. Based on IDAE decisions, investment grants and low interest loans were merged into one financing instrument. PV projects of less than 100 kW can be financially supported up to 90 % of the total investment costs, and for projects of more than 100 kW, up to 70 %. As a result of the modified feed-in law and increasing demand, the fund for PV systems up to 100 kW was exhausted in July 2004. IDAE announced an approval stop. In 2003, there were 2 438 applications, and 2 169 projects with a financial volume of about 130,02 million EUR were co-financed, of which 20,22 million EUR were IDAE grants, while the rest was given to investors as low interest loans.

Today, about 10 % of the total investment costs for a PV plant are tax deductible from income tax and corporate income tax. Since the beginning of 2003, Spanish communities are authorized to grant credit notes for up to 50 % of the cost from the excise tax to companies using or generating electricity from renewable energies. Private investors who use PV for self-supply can reduce their construction tax by up to 95 %. So far, the generated electricity is exclusively for self-supply without any feed-in to the grid, and private investors can reduce their property taxes.

In 2003, IDAE announced an installed PV capacity of about 27,3 MW (ASIF announced a capacity of 28,8 MW). In 2002, Spain ranked third in Europe regarding PV installation, following Germany and The Netherlands. In 2004, the market growth went along with the national support programme for PV. In order to reach the national development goal of 144 MW in 2010, an average annual growth of 16,7 % and 95 % market growth are necessary. In 2001, the majority of PV installations were small and stand-alone systems (two-thirds of total installation). Recently, market growth has been based on grid-connected PV plants, which profit directly from the new feed-in tariff. In 2003, there were only 5 PV plants with a capacity of more than 100 kW, 3 MW in total, and 65 PV plants with 5 to 100 kW, 2 MW in total. Small and stand-alone systems for private houses and rural installations dominate the market. For a short time, the number of large PV plants was growing. First, solar parks were installed in Navarra (1,2 MW) and Toledo (1 MW), and another project in Sevilla (about 1,2 MW) is under construction. The energy supplier Iberdrola is planning a PV project with a capacity of about 12 MW in Murcia. The major projects are for demonstration technology testing purposes. They are erected under powerful public involvement, financed by IDAE and Spanish and European research programmes. This leads to the conclusion that only a guaranteed feed-in tariff over a fixed duration makes major commercial projects possible and financially attractive.

Table 2.18 PV companies in Spain[18]

Company	Activity
Isofotón	Production of mono-crystalline cells and modules
BP Solar	Production of crystalline cells and modules
Atersa	Production of crystalline cells and modules

The Spanish PV industry is one of the market leaders in Europe. According to ASIF, the total production of mono-crystalline PV cells in Spain is about 56,2 MW. With a market volume of 6,5 MW, around 90 % of the total production was exported. The Spanish share of the world market is about 7 %. The Spanish producers have a production capacity of about 150 MW, which can reach up to 270 MW without great investment. Like other countries, Spain suffers from a shortage of silicon and wafer, slowing down the market growth. Basically, there are three PV companies operating in the Spanish market (see Table 2.18).

In 2003, BP Solar produced approximately 15,5 MW of mono-crystalline cells with its 40 MW manufacture line. With a production of 35,2 MW, Isofotón became the second largest cell manufacturer. The rest is produced by the Spanish manufacturer Atersa. So far, a functional mass market is still missing in Spain, so there is not a widespread distribution channel for PV. Cell and module producers often undertake turnkey development, such as planning, financing, installation and implementing for sophisticated major projects with high prestige.

2.2.2 Overview of most important country data

Table 2.19 presents further key data for Mediterranean countries.

2.2.3 Evaluation with respect to VLS-PV

Based on geographic and climatic data, all countries under consideration offer favourable conditions for implementing VLS-PV. The population density is comparatively low, and there are ample areas with suitable geographic conditions that feature small populations and vegetation, combined with smooth slopes. We have visited several possible sites of 200 to 300 ha land size required for a 100 MW-scale PV system. The irradiation is highest with 1 700 to 2 100 kWh/m^2/year in Morocco and Tunisia, but southern Portugal and Spain also offer regions with up to 1 900 kWh/m^2/year annual global irradiation.

The economic key data shows stronger differences between the two groups of Northern African and Southern European countries (see Figure 2.12).

Southern European countries exhibit distinctly

Table 2.19 Key data of the Mediterranean countries

	Morocco (2002)	Tunisia (2003)	Portugal (2002)	Spain (2002)
Population (millions)	29,7	9,9	10,2	40,8
Area (km^2)	446 550	163 610	92 040	505 960
Population density (1/km^2)	67	61	111	81
GDP per capita (EUR)	1 532	2 200	12 500	17 100
Energy consumption per capita (toe)*	0,48	0,83	2,52	3,24
Total energy consumption (Mtoe)	14,3	8,2	25,7	132,2
Total electricity production (TWh)	17,2	11,8	46,1	246,1
Electricity price level (EUR cents/kWh)	~8–12	~2–5	~12	~9
Annual solar radiation (kWh/m^2/year)	1 700–2 100	1 700–2 100	1 500–1 900	1 200–1 800
Feed-in tariff for renewable electricity	No	No	Yes	Yes
Cap for PV (megawatt peak, MWp)	–	–	150	150
PV industry	No	No	Small	Large

Note: *1 toe = 1 tonne oil equivalent = 11 630 kWh = 11,63 MWh

higher gross domestic product (GDP) and energy consumption per capita, leading to the conclusion that the economic power of Portugal and Spain might be favourable for the economic support required for the market introduction of renewable energy sources (RES), in general, and PV, in particular, on a large scale. Furthermore, the pressure to introduce RES might be higher with higher energy and electricity consumption from an environmental responsibility point of view. The existence of a traditional and strong PV industry in Spain, and on a smaller scale in Portugal, provides an additional factor contributing to the economic and political environment favouring the application of PV.

Interestingly, these assumptions are reflected by both Portugal and Spain, having passed long-term national plans for developing renewable energies and PV, in particular, which is not the case in Morocco and Tunisia. The latter countries also have an increasing application of PV, but more in the way of small off-grid rural solar home and village PV systems. The existence of feed-in tariff legislation for grid-connected PV, as well as other RES, in Portugal and Spain is consistent with these general remarks and provides favourable framework conditions for VLS-PV, in principle, based on the currently valid assumption that unsupported generation cost for PV electricity is not yet low enough to compete with conventional grid electricity prices. The actual 150 MW cap for PV in both Portugal and Spain leaves capacity for large scale systems, which renders it worthwhile to perform closer calculations of economic feasibility for large PV systems.

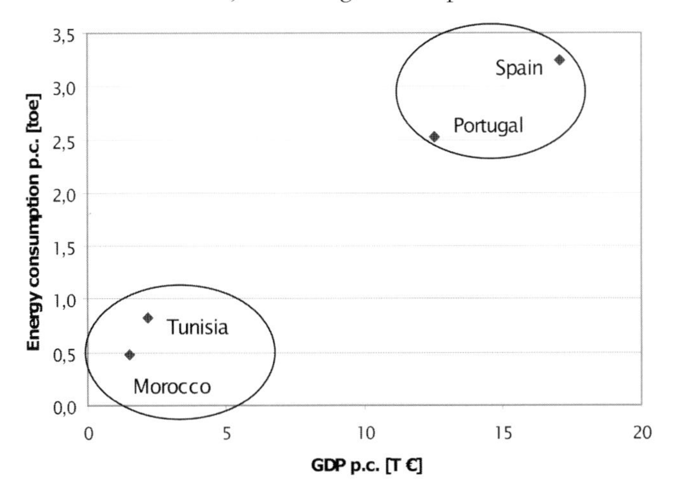

Figure 2.12 Relation of GDP per capita to energy consumption per capita

2.3 ECONOMIC FEASIBILITY OF VLS-PV SYSTEMS

2.3.1 PV electricity-generation cost

Experience with already realized MW systems shows that a stationary (non-tracking, flat-plate) large scale PV installation can now be realized at around 4 010 EUR/kW as represented in Table 2.20. 4 000 EUR/kW therefore serves as a fair approximation for the following calculations, including a limited overhead cost of 8 %. Note that this overhead does not yet include a further 6 to 8 % capital acquisition cost, which is typically required if the project is sold to private or fund investors, a frequently encountered way of project financing today. Three-quarters of the system cost is for the PV modules, the module prices thus being the main parameter for future cost reduction.

The annual cost is represented in Table 2.21. The model parameters are 20 years' linear depreciation and 100 % loan financing at 5 % interest, which, of course, need to be adapted for a concrete project proposal. In our model, financing forms a major contribution to annual cost. No investment for land was considered here, and the estimated land renting cost is instead included in the 2 % annual operation and maintenance cost.

The annual global irradiation and annual energy yield were calculated using a PV system[24] for two

Table 2.20 Cost structure for stationary VLS-PV systems in the MW range

PV modules	2 950 EUR/kW
Mounting racks (material cost)	210 EUR/kW
Inverter (assumed N * 300–400 kVA)	280 EUR/kW
Cables (including both DC and AC)	45 EUR/kW
Transformer (to 10–20 kW grid level)	58 EUR/kW
Installation (labour and auxiliary material)	120 EUR/kW
Permit fees, infrastructure, transportation	50 EUR/kW
Installed system cost (net)	3 713 EUR/kW
8 % overhead (project development, attorney and bank fees, etc.)	297 EUR/kW
Total system cost (net investment)	4 010 EUR/kW

Table 2.21 Annual cost per kW for a total system cost of 4 000 EUR/kW

Depreciation (linear over 20 years)	200 EUR/kWyear
Interest (100 % financing @ 5 % p.a.)	100 EUR/kWyear
Operation and maintenance (2 % p.a.)	80 EUR/kWyear
Total annual cost (net value)	380 EUR/kWyear

Note: p.a. = per annum.

locations per country (see Table 2.22). A virtual stationary PV system was used as input for the programme. The irradiation represented is the annual global radiation onto a horizontal surface for the given site. The specific energy yield per kW represented is the output of the PV system computer programme for optimum orientation (south) and inclination (in most cases, 30°). This is multiplied by a factor of 0,93 – that is, including a 7 % safety margin. The generation cost for PV shown was determined by the total annual cost per kW (assumed to be equal to 380 EUR/kW at all locations) and the annual energy yield.

The investment for tracking systems is higher than for stationary systems because of the high cost of the electromechanical structure compared with stationary mounting racks. The operation and maintenance costs are also higher because of moving parts, resulting in a total annual cost per kW of 25 to 30 % (depending upon the exact type) higher than for stationary systems. The energy yield gain of a single-axis azimuthal tracking system amounts to 30 to 35 % over a south-oriented fixed system in the Mediterranean region according to experience within the group. This effectively gives a slight economic advantage to tracking systems. We restrict the following discussion to the stationary system results because this difference does not substantially influence the conclusions of the study.

2.3.2 Main results and discussion

PV electricity-generation costs in the analysed Mediterranean countries are between 23,9 and 37,7 EUR cents/kWh. As a result of the method of determination, the generation cost inversely follows the irradiation conditions.

As expected, the generation cost for PV calculated with the described parameters is higher than the price level of conventional electricity drawn from the grid in all places. In this context, it is important to note that the assumed 100 % loan financing makes up a substantial proportion of the generation cost. Generation costs well below 20 EUR cents/kWh result for almost all sites (data not shown) without including the financing cost and the 7 % safety margin in the annual energy yield. This confirms that PV generation cost is not too far above the conventional price line, and the cost could reach or even fall below this line after a price decrease of PV modules. This decrease is already anticipated by foreseeable advances in technology and economies of scale in increasing mass production.[25]

If the generation costs are compared to local feed-in tariffs, the results indicate a clear ranking of the economic feasibility.

Although the lowest generation cost of 23,9 EUR cents/kWh is reached in Quarzazate, this is not low enough to be attractive for a buy-back scheme in Morocco, even considering that the general electricity price level is comparatively high there. Tunisia has a centralized electricity industry with a low price level, making the situation for PV even more difficult. Morocco and Tunisia have no specific legal frameworks

Table 2.22 Solar irradiation, energy yield and PV electricity-generation cost data compared with the conventional electricity price level and local feed-in tariff rates for stationary systems at two representative sites in four Mediterranean countries

Country	Site	Annual global irradiation (kWh/m²/year)	Annual energy yield (kWh/kW/year)	Generation cost for PV (EUR cents/kWh)	Conventional grid electricity price level (EUR cents/kWh)	Feed-in tariff rate (EUR cents/kWh)
Morocco	Casablanca	1 772	1 337	28,4	~8–12	None
	Quarzazate	2 144	1 589	23,9		
Tunisia	Tunis	1 646	1 219	31,2	~2–5	None
	Gafsa	1 793	1 339	28,4		
Portugal	Porto	1 644	1 312	29,0	~12	~55 <5 kW
	Faro	1 807	1 360	27,9		~31–37 >5 kW
Spain	Oviedo	1 214	1 008	37,7	~9	41,44 <100 kW
	Almería	1 787	1 372	27,7		21,62 >100 kW

to support electricity generation by PV and no existing feed-in tariff. Therefore, the economic feasibility for VLS-PV is low in these Northern African countries if it is based on achieving income from electricity sales to consumers or the grid alone, without considering any investment subsidies.

Portugal and Spain also have conventional electricity prices much lower than the calculated PV electricity-generation cost. In these countries, however, smaller systems appear to be economically feasible with the available feed-in tariffs in higher irradiation sites. The exciting question for VLS-PV is if large systems could also be economically operated under special circumstances. To answer this question requires a closer look at the conditions in these Southern European countries.

2.4 A CLOSER LOOK AT PORTUGAL AND SPAIN

2.4.1 VLS-PV in Portugal

The main problem in Portugal related to grid-connected PV is the low transparency of both legislation and administration related to PV electricity feeding into the grid. This refers to the per kWh price and total capacity of the feed-in tariff, as well as the approval procedure for applicable systems. The original values of 41 EUR cents/kWh (for systems less than 5 kW) and 22,4 EUR cents/kWh (greater than 5 kW) were subject to a complicated inflation correction formula. In 2004, the respective values were –55 and –32 EUR cents/kWh. Only recently, in February 2005, the feed-in tariff for large systems (greater than 5 kW) was increased to 37 EUR cents/kWh, together with raising the cap from 50 to 150 MW. This was exciting because large scale systems could be economically feasible with these values. A few weeks later, however, the rate was apparently decreased again to 31 EUR cents/kWh, rendering an economic operation impossible for our parameters. The political attitude towards PV is generally positive in Portugal; but this is not clearly valid for large scale PV, possibly because the presence of the PV industry is small there.

Interestingly, the world's largest PV system proposals, which are prototypes of real VLS-PV, are waiting for approval under the described uncertain circumstances in Portugal. The first one is the 64 MW Amper Central Solar S.A. project proposed by BP Solar for a site close to the community of Moura in the Alentejo region of southern Portugal.[26] The second one is a 116 MW project. It appears that there is no clear time frame for either of these projects. The realization of more than one VLS-PV project would furthermore require additional raising of the 150 MW cap, which cannot be realistically expected.

2.4.2 VLS-PV in Spain

The stated feed-in tariffs for PV have been valid since 2004. They represent an important improvement because now the higher tariff rate is available for systems up to 100 kW. Before, it was only up to 5 kW. We do not comment here on further details, such as the duration of the feed-in tariff payment, possible alternative rates and/or accumulation with additional investment subsidies.[21,27] It is important to note that practical experience makes it clear that the application of feed-in tariffs is a well-organized procedure in Spain, with only limited administrative hurdles in many cases. The attitude is clearly positive towards PV, possibly because of the stronger presence of the PV industry in Spain.

This has already led to a number of MW-scale PV systems under the former tariff conditions by combining many systems with less than 5 kW into one MW system. One example is the Huertaesolar system concept realized by AeSol in several locations of Arguedas, Navarra. Another prominent project under the increased tariff legislation is the SOLTEN (Sol Tenerife) proposal for Tenerife, with plans to combine 150 lots of 100 kW systems into one 15 MW system, with even larger systems under consideration.[28] Also, for Spain, the implementation of VLS-PV on a larger scale would require a further lifting of the present 150 MW cap.

In addition to Portugal, we expect that Spain will follow the German example and improve legislation by steps, further raising and eventually removing both the permitted maximum system size as well as the cap. Such a development could lead to very attractive conditions for VLS-PV in Spain in the future, thus winning back European leadership in this field.

2.5 SUMMARY AND CONCLUSIONS

The economic boundary conditions for VLS-PV are still unsatisfactory at current PV system prices without supporting schemes such as feed-in tariff legislation. The lack of such supporting schemes renders VLS-PV currently improbable in Morocco and Tunisia. The boundary conditions in Portugal and Spain are generally more favourable because of the existence of feed-in tariff schemes with sufficiently large caps. However, these countries also have generation costs of PV electricity exceeding the conventional electricity price level and the feed-in tariffs applicable for larger systems in nearly all of the cases studied, with the possible but uncertain exemption of Portugal. The example of Spain shows that applying intelligent financing schemes will most probably lead to VLS-PV in the very near future there.

In summary, we expect the best conditions for VLS-PV to develop in Spain on an intermediate time scale of two to five years, even though there are now some larger projects proposed in Portugal. Concrete realization of VLS-PV projects depends upon successful negotiation between project developers, PV and electricity industries and politics (acceptance, sustainability and incentives) for every single project.

Generally, VLS-PV as a centralized electricity source needs to compete with conventional electricity sources and other centralized RES, such as solar thermal[4] and wind energy, which are also proposed and implemented strongly in the studied region, in addition to decentralized small scale PV. Lowering investment requirement, additional support, and/or higher feed-in tariffs for large systems are required, in addition to intelligent financing schemes in order to make VLS-PV economically feasible on a larger scale in the Mediterranean region considered.

ACKNOWLEDGEMENTS

Contributions from Deutsche Energie Agentur (DENA), assistance from B. Rekow and financial support from the German Federal Ministry of Environment (BMU), granted through Projektträger Jülich (PTJ), are gratefully acknowledged.

REFERENCES

1. Kurokawa, K. (ed) (2003) *Energy from the Desert: Feasibility of Very Large Scale Photovoltaic Power Generation (VLS-PV) Systems*, James and James, London
2. Auswärtiges Amt (2004) Länderinformationen, www.auswaertiges-amt.de/www/de/laenderinfos/, accessed 10 December 2004
3. MapSet Ltd using map data from Mountain High Maps.
4. Deutsches Zentrum für Luft und Raumfahrt (DLR) (2005) *Concentrating Solar Power for the Mediterranean Region*, MED-CSP, Stuttgart
5. IEA (2004) *IEA Energy Statistics*, Morocco, www.iea.org/dbtw-wpd/Textbase/stats/noncountryresults.asp?nonoecd=Morocco, accessed 23 December 2004
6. Czisch, G. (1999) *Potentiale der regenerativen Stromerzeugung in Nordafrika: Perspektiven ihrer Nutzung zur lokalen und großräumigen Stromversorgung*, www.iset.uni-kassel. de/abt/w3-w/folien/Allgemein/overview_allgemein. html and www.iset.unikassel.de/abt/w3w/projekte/Pot_Strom_Nordafrika.pdf, accessed 1 November 2004
7. GTZ (2004) *Energiepolitische Rahmenbedingungen für Strommärkte und erneuerbare Energien. 21 Länderanalysen: Teilstudie Marokko*, Eschborn, www2.gtz.de/wind/deutsch/studien_download_2004.html, accessed 25 October 2004
8. Commission of the European Communities (2004) *European Neighbourhood Policy. Country Report: Morocco*, Commission Staff Working Paper, Brussels, www.europa.eu.int/comm/world/ enp/document_en.htm, accessed 1 December 2004
9. ONE (2004) Office National de l'Electricité du Maroc, www.one.org.ma
10. EIA (2004) *Arab Maghreb Union Country Analysis Brief*, EIA, www.eia.doe.gov/emeu/cabs/ maghreb.html, accessed 6 December 2004
11. MEDNET (2005) *Country Profiles: Morocco*, www.cres.gr/mednet/cprofiles/morocco1.htm, accessed 29 May 2005
12. Solarbuzz (2004) *Moroccan Solar Energy Organizations*, www.solarbuzz.com/CompanyListings/Morocco.htm, accessed 7 December 2004
13. Deutsches Zentrum für Luft und Raumfahrt (DLR), Institut für Technische Thermodynamik, 23 December 2004
14. IEA (2004) *Information Center: Tunisia*, www.iea.org/Textbase/stats/noncountryresults.asp?nonoecd=Tunisia
15. Commission of the European Communities (2004) *European Neighbourhood Policy. Country Report. Tunisia*, Commission Staff Working Paper, Brussels, www.europa.eu.int/comm/world/ enp/document_en.htm, accessed 1 November 2004
16. GTZ (2004) *Energiepolitische Rahmenbedingungen für Strommärkte und erneuerbare Energien. 21 Länderanalysen. Teilstudie Tunesien*, GTZ, Eschborn, www2.gtz.de/wind/deutsch/studien download_2004.html, accessed 6 December 2004
17. MEDNET (2004) *Country Profiles: Tunisia*, www.cres.gr/ mednet/cprofiles/tunisia3.htm, accessed 3 December 2004
18. Solarbuzz (2004) *Tunisian Solar Energy Organizations*, www.solarbuzz.com/CompanyListings/Tunisia.htm, accessed 7 December 2004
19. Huld T., Šúri M., Dunlop E., Albuisson M, Wald L (2005) 'Integration of HelioClim-1 database into PVGIS to estimate solar electricity potential in Africa', *Proceedings from 20th European Photovoltaic Solar Energy Conference and Exhibition*, 6–10 June 2005, Barcelona, Spain
20. IEA (2005) *Information Center: Portugal*, www.iea.org/Textbase/subjectqueries/countryresult.asp?country=PT&SubmitA=Submit, accessed 29 June 2005
21. DENA (2003) *Exporthandbuch Photovoltaik 2003/04. Die europäischen Märkte im Vergleich*, DENA, Berlin, pp182–193, 224–241
22. IEA (2005) *Information Center: Spain*, www.iea.org/Textbase/subjectqueries/countryresult.asp?country=ES&SubmitA=Submit, accessed 29 June 2005
23. MINECO (2003) *Bekanntmachung der Generaldirektion für Energiepolitik und Bergbau*, www.mineco.es/ energia, accessed 20 August 2003
24. *FhG-ISE/econzept: Energieplanung* (2000) Freiburg
25. Hoffmann, W. (2005) 'A clear vision: EPIA sets its sights on PV technology to 2030 and beyond', *Renewable Energy World*, May–June, p56
26. Hirshman, W. P. (2004) 'Sexy project to go all the way?

BP Solar will build Portuguese module factory only if huge 64 MW system approved', *Photon International*, January, 2004, p12
27. DENA (2004) *Exportinitiative Erneuerbare Energien*, www.exportinitiative.de/index2.cfm?cid=838&tid=103, accessed 21 October 2004
28. Hirshman, W. P. (2005) 'Island in the sun: Tenerife eyes feed-in tariff for huge PV plants', *Photon International*, January, p18

FLA. SOLAR ENERGY CENTER LIBRARY

The Middle East region: A top-down approach for introducing VLS-PV plants

3.1 INTRODUCTION

A top-down approach to providing solar electricity to any given region must address the following five questions:

1. How much land area is available for the harvest of sunshine, and how much electricity could this resource provide?
2. How much electricity is required?
3. What kind of technology should be used, and how much of it would be needed for the task?
4. At what rate should the technology be introduced?
5. What monetary resources would be required and how could these resources be provided?

This chapter will address these questions and provide answers for the principal electricity-consuming countries in the Middle East. These results can also be easily modified for application in other countries, both within the region and elsewhere.

3.2 SOLAR ELECTRICITY POTENTIAL BASED ON LAND AREA

Table 3.1 lists geographic, demographic and electricity-generation data for the countries in the Middle East, which will form the basis for our discussion. The specific data in Table 3.1 is for 2002 because this is the latest year for which comprehensive electricity production data are currently available. However, the latest edition of the *World Factbook*[1] provides population estimates

Table 3.1 Middle East statistics for 2002

Country	Land area (km^2)[1]	Population in 2002[1]	Growth rate in 2003 (%/year)[1]	Electricity (2002)[2] Capacity (GW)	Electricity (2002)[2] Production (TWh)	Electricity production (kWh/capita/year)
Bahrain	665	656 666	1,6	1,4	6,9	10 446
Cyprus	9 240	724 153	6,6	1,0	3,6	4 913
Egypt	995 450	73 340 005	1,9	17,6	81,3	1 108
Iran	1 636 000	67 549 294	1,1	28,0	129,0	1 909
Iraq	432 162	24 015 677	2,8	9,5	34,0	1 415
Israel	20 330	6 116 533	1,4	9,7	42,7	6 976
Jordan	91 971	5 312 575	2,8	1,7	7,3	1 375
Kuwait	17 820	2 112 600	3,3	9,4	32,4	15 353
Lebanon	10 230	3 678 412	1,3	2,3	8,1	2 193
Oman	212 460	2 715 346	3,4	2,4	9,8	3 603
Qatar	11 437	794 257	2,9	1,9	9,7	12 247
Saudi Arabia	1 960 582	23 524 590	3,3	23,8	138,2	5 874
Syria	184 050	17 585 540	2,5	7,6	26,1	1 487
Turkey	770 760	67 328 459	1,2	28,3	123,3	1 832
United Arab Emirates	82,880	2 446 492	1,6	5,8	39,3	16 064
Yemen	527,970	18 709 999	3,4	0,8	3,0	162

Table 3.2 Potential electrical yield of VLS-PV technologies in the Negev[3]

VLS-PV yield	System type			
	Static, 30° tilt	One-axis tracking	Two-axis tracking	Concentrator PV (CPV)
GWh/year GWp^{-1}	1 644	2 071	2 279	1 754
kWh/year m^2 aperture	189	238	262	385
GWh/year km^2 land area	65,4	69,6	33,2	48,7

for the year 2003, together with their associated growth rate estimates. These were therefore used to calculate the 2002 population figures shown in Table 3.1. Table 3.1 does not include 3 512 062 Palestinians who reside in the Gaza Strip and West Bank territories (6 000 km^2), and who purchase most of their electricity from Israel.

In our previous report,[3] we estimated the land area that would be needed in the Negev Desert for potential very large scale photovoltaic power generation (VLS-PV) technologies. The results are reproduced in Table 3.2. Since the Negev is representative of many much larger deserts in the region, the data provide a useful starting point for our current discussion of the entire Middle East.

If we combine the last line of Table 3.2 with the land areas in Table 3.1, we can obtain our first estimate of the solar electricity potential of each country. The results are shown in Table 3.3.

From Table 3.3, one can see that, in most cases, the solar electricity-generating potential of a country is *at least one order of magnitude* larger than that country's present electrical production. The singular exception is Bahrain, where solar could generate only *seven* times its present electricity production. At the other extreme, Yemen has a solar generating potential that is *four orders of magnitude* larger than its present electricity production.

3.3 ELECTRICITY REQUIREMENT

The question of electricity requirement is a highly country-specific matter, which depends upon factors not relevant to this chapter. Table 3.1 indicates that the Middle East includes countries such as Bahrain, Kuwait, Qatar and the United Arab Emirates, whose electricity generation per capita is comparable to the most developed Western societies. There are also countries that fall far short of this norm.

Another important issue is that even supposing that each country currently generates enough electricity for its needs, we do not know how future needs will grow with increasing populations, or even how the population of each country will increase with time.

For these reasons, we have adopted a transparent, albeit slightly simplistic, approach to future electricity needs:

- We will discuss the timetable and costs involved in establishing a manufacturing facility capable of producing solar collectors and storage units with annual throughput on the GW scale.
- We will discuss the capital investment needed to install VLS-PV plants at a linear rate of one plant per year for 30 years.
- For each country in the Middle East, we will tailor its VLS-PV production rate to enable the solar

Table 3.3 Estimated annual solar electricity potential of countries in the Middle East

Country	Technology				Electricity Production
	Static, 30° tilt (TWh/year)	1-axis tracking (TWh/year)	2-axis tracking (TWh/year)	CPV (TWh/year)	in 2002 (TWh) [2]
Bahrain	43,5	46,3	22,1	32,4	6,9
Cyprus	604,3	643,1	306,8	450,0	3,6
Egypt	65 102,4	69 283,3	33 048,9	48 478,4	81,3
Iran	106 994,4	113 865,6	54 315,2	79 673,2	129,0
Iraq	28 263,4	30 078,5	14 347,8	21 046,3	34,0
Israel	1 329,6	1 415,0	675,0	990,1	42,7
Jordan	6 014,9	6 401,2	3 053,4	4 479,0	7,3
Kuwait	1 165,4	1 240,3	591,6	867,8	32,4
Lebanon	669,0	712,0	339,6	498,2	8,1
Oman	13 894,9	14 787,2	7 053,7	10 346,8	9,8
Qatar	748,0	796,0	379,7	557,0	9,7
Saudi Arabia	128 222,1	136 456,5	65 091,3	95 480,3	138,2
Syria	12 036,9	12 809,9	6 110,5	8 963,2	26,1
Turkey	50 407,7	53 644,9	25 589,2	37 536,0	123,3
United Arab Emirates	5 420,4	5 768,4	2 751,6	4 036,3	39,3
Yemen	34 529,2	36 746,7	17 528,6	25 712,1	3,0

Figure 3.1 A typical large dense array CPV collector, in which an array of cells is illuminated by a single light-gathering element (in this case, a parabolic reflecting dish)
Source: Ben-Gurion University

Figure 3.2 A typical large individual cell CPV collector, in which each individual cell in the array is illuminated by its own light-gathering element (in this case, a Fresnel lens)
Source: Amonix Corp

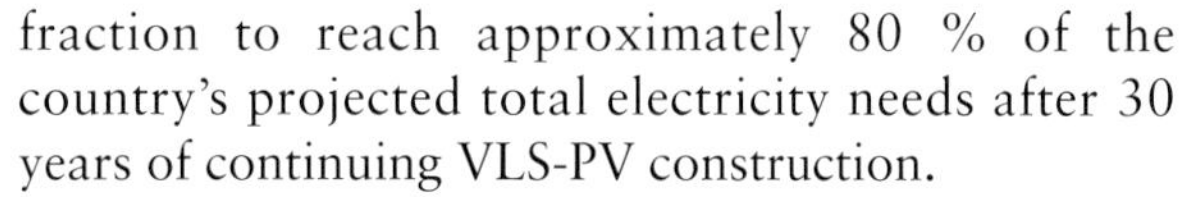

fraction to reach approximately 80 % of the country's projected total electricity needs after 30 years of continuing VLS-PV construction.

- We will project each country's electricity needs into the future by assuming that their current production will increase linearly according to the present population growth.

Under such a scenario, typical Middle East countries would find themselves with most of their total electricity-generation capacity coming from solar energy within approximately one generation. Furthermore, however simplified this quantitative picture may be, having such a manufacturing facility would enable individual governments to plan their own rates of VLS-PV introduction according to the availability of investment capital and the changing electrical needs of their own society.

3.4 SUGGESTED TECHNOLOGY

3.4.1 Concentrator photovoltaic (CPV) technology

In our previous report,[3] we discovered that, in the long term, the lowest-cost VLS-PV systems are likely to involve concentrator photovoltaic (CPV) technology. The basic reason for this is that very *large areas* are necessary for solar energy collection, regardless of the conversion technology, because solar is a very *dilute* form of energy (desert regions typically receive the energy equivalent of one barrel of oil per m^2 per year). However, a fundamental economic problem with PV technology is that its high cost is due to employing the same material both for energy collection and conversion. This can be overcome by separating these two functions and employing a large area of low-cost material such as plastic, aluminium and steel to collect and concentrate the energy onto a small area of high-cost PV material (that is, concentrator solar cells) where the conversion is performed.

In *Energy from the Desert: Feasibility of Very Large Scale Photovoltaic Power Generation (VLS-PV) Systems*[3] we emphasized one specific type of CPV technology that employs a dense array of CPV cells illuminated by a single reflecting dish concentrator, as shown in Figure 3.1. Alternatively, one can think in terms of a large array of individual CPV cells, each illuminated by its own concentrator. The relative advantages and disadvantages of these two approaches have been discussed.[4] However, for our current purposes, the single greatest advantage of the latter type of technology is that it is already commercial. Therefore, there are fewer uncertainties in estimating the costs involved in up-scaling the existing technology to VLS-PV scales of size. Figure 3.2 shows a 25 kW CPV collector of a kind demonstrated by the Amonix Corporation in the US. One of the present authors has extensively considered how such units could be mass produced and what the resulting energy costs would be.[5] The following sections extend the results of this study[5] to the entire Middle East.

3.4.2 The individual CPV units and their expected performance

The basic collector will have an area of 200 m^2 and will contain an array of 3 200 individual CPV cells, each with an area of 1 cm^2. The geometric concentration ratio of these collectors is accordingly 625X.

Existing multi-junction CPV cells already have efficiencies in excess of 32 %; but this is the nominal efficiency that we will take for our present calculations. After allowing for the loss mechanisms at the system level discussed in *Energy from the Desert: Feasibility of Very Large Scale Photovoltaic Power Generation (VLS-PV) Systems*,[3] the effective operating efficiency of the collectors will be 25 %. Hence, each collector will have a true peak power rating of 50 kW.

Our reference data are the *Typical Meteorological Year TMY v.3* data set for Sede Boqer.[6] This is an updated version of the data set used in *Energy from the Desert*[3] and differs only slightly from the latter. At 25 % efficiency, a 50 kW collector would generate 119 700 kWh of electrical energy in a typical year at Sede Boqer.

Regarding land usage, Table 3.2 assumes a land area/collector area ratio of 8:1 for our previous CPV calculations in order to minimize mutual shading losses. However, if one wants to supply most of an entire country's electrical needs from solar energy, a trade-off is required between shading losses and gained land area. By reducing the ratio to 3:1 we lose 10 % of the energy, but decrease the required land area to one third. Hence, although a single un-shaded collector would produce 119 700 kWh per year, a more closely packed field of collectors of the density described here would produce only 107 700 kWh per collector per year. Note that we have assumed the use of bypass diodes so that shading losses in the field are directly proportional to the area of collectors that is shaded.

With these considerations, a VLS-PV plant of 1 GW rated capacity would consist of 20 000 collectors and occupy 12 km^2 of land. Its annual electricity production would amount to 2,15 TWh in the Negev. However, in this study we have adopted a downward-rounded annual figure of 2,0 TWh/GW for all Middle East calculations, including Israel. For comparison, Table 3.1 reveals that all of Bahrain's current electricity requirements could be provided by four such 1 GW VLS-PV plants. At the other extreme, it would take 65 such plants to provide the electricity requirements of Saudi Arabia.

3.4.3 Storage

Because PV plants can only generate electricity during the day and CPV plants only during periods of direct sunlight, some form of storage will be necessary for no sunshine or partial sunshine conditions. This is not initially a big issue if a small number of plants are to be integrated within an existing fossil-fuelled grid system. However, the issue gains increasing importance as the solar-to-fossil ratio increases. One cannot disregard the influence of a fluctuating solar input on grid stability, and the energy supplied must respect the standards of utility power plants. The entire complement of energy sources (solar plants, storage and base-load fossil plants) must be able to generate and supply the entire energy demand and peak load. The primary function of the storage plant is to accumulate the excess solar energy or base-load generated energy and deliver it when required. A plant configuration that includes only solar plants and limited storage batteries is not optimal for the sole supply of energy because there are periods during the year (mainly in winter) when sun is not available for some days. This configuration is not acceptable. In practice, storage power of about one third of the solar power installed is needed when the grid includes fossil or nuclear plants as power back-up, in addition to the storage capacity.

3.4.4 Transmission

Even though most of the Middle East states are largely, if not entirely, desert areas, VLS-PV plants do not need to be located close to population centres. It is far more likely that they will be distant and that transmission losses will need to be considered. If a suitable grid line exists, then the economic cost of transmission losses from solar plants will be comparable per kW with losses from base-load plants. If new transmission infrastructure must be introduced, it is desirable to minimize losses. This issue will be touched upon after we have discussed the economics of VLS-PV plants.

3.5 THE RATE OF VLS-PV INTRODUCTION

Each specific country will naturally decide its appropriate rate of VLS-PV plant introduction. Such decisions would probably rest principally upon economic considerations, the rate at which population actually grows, and the change in energy consumption per capita. Since these factors require the ongoing attention of local governments, they cannot be guessed effectively from the outside. Therefore, we have adopted a simple energy growth model for each country, based on its current energy production per capita, and a constant population growth rate equal to its present value.

We assume that each country constructs a manufacturing facility capable of producing enough components for one VLS-PV plant per year. The plant rating required for each country is chosen to enable its electrical solar power economy to reach approximately 80 % of its total projected electricity needs after a 30-year VLS-PV construction period.

Of course, individual countries might decide to construct a larger manufacturing facility to attain such a goal more rapidly, produce hydrogen as substitute transportation fuel, or export to other countries without the possibility of producing their own solar plants. Others might elect to introduce VLS-PV plants at a slower, and possibly non-linear, rate according to local requirements. Furthermore, because the basic collectors are 50 kW units, there is no requirement for any country to introduce VLS-PV plants of any standard capacity, such as 1 GW. Whatever implementation strategy is adopted, though, the results of this study will facilitate its formulation.

3.6 THE COST OF VLS-PV PLANTS

In this section we address the cost of constructing a 1 GW VLS-PV plant, the cost of an appropriately sized storage plant, and the cost of constructing a manufacturing facility capable of producing one such plant per year.

3.6.1 The cost of a 1 GW plant

Summing the following four items together gives a total cost of 850 MUSD for a VLS-PV plant of 1 GW rating, or 850 USD/kW:

1. *Cost of solar cells:* each VLS-PV plant requires 64 million CPV cells. For this study MST Ltd discussed the production logistics and received price quotations from two well-known cell manufacturers, as well as a third that has the technology to build such devices. All agreed to provide 1 cm^2 multi-junction cells at 32 % efficiency, at a price of less than 2,5 USD per cell. Thus, the total cell cost per plant = 160 MUSD.
2. *Cost of Fresnel lenses:* the cost of these elements is estimated by MST Ltd at 1 USD per cell. Thus, the total optics cost per plant = 64 MUSD.
3. *Cost of DC to AC inverters:* in *Energy from the Desert*[3], it was estimated that inverters in the size range of a few hundred kW would cost 300 USD to 450 USD/kW. However, for this report, it has become clear that inverters can be manufactured for very much less by using larger sizes. One well-known manufacturer gave MST Ltd a price quotation of 100 000 USD for a 2,5 MW inverter (that is, 40 USD/kW), even if purchased at low quantities. Each such inverter would, of course, be fed by 50 solar collectors. Thus, the total inverter cost per plant = 40 MUSD.
4. *Cost of balance of system (BOS):* in addition to the specific component costs enumerated above, MST Ltd estimates that the production of 20 000 collectors and trackers, as well as transportation to and physical erection at the VLS-PV plant site, will cost 586 MUSD. This works out to 146,5 USD/m^2 of collector – a figure that is well within recent cost estimates for solar concentrator technology.[7]

3.6.2 The cost of storage batteries

The precise amount of storage is a complicated economic issue related to the entire electrical grid structure. Large vanadium redox flow batteries now appear to be a promising technology. Under mass production they should cost 850 USD/kW for an operationally effective storage capacity of six hours. The amount of storage power and capacity to be installed depends upon the amount of fossil power available for back-up and is specific for each economy. As a rule of thumb, the storage power has been found to satisfy normal system constraints when its level is approximately one third of the installed solar power.

At first glance, this may appear to increase the cost of a 1 GW VLS-PV plant to the seemingly undesirably high figure of 1 150 USD/kW. This is incorrect due to battery storage, which can store solar- or fossil-generated energy. This enables VLS-PV plant operations at a large capacity factor – well beyond that of the available solar power or of conventional fossil-fuelled power plants. Thus, since a solar plant must include storage to regulate its power as part of a more complex system, its cost per kW is not the simple sum of its parts. What matters, of course, is not the cost per kW, but whether the generated energy is economically viable. This, as will be seen, is the case presented in this study.

3.6.3 The cost of a manufacturing facility

MST Ltd estimates that a manufacturing facility capable of producing 20 000 collectors per year with a total rating of 1,0 GW will cost 530 MUSD. The first facility will also require an estimated cost of 220 MUSD for engineering designs. This cost will be included in the investment algorithm for each country.

3.7 ISRAEL AS A CASE STUDY, AND A SUGGESTED INVESTMENT ALGORITHM

3.7.1 Projected revenue

Considering the prevailing electricity tariffs (regulated by the government of Israel) – which vary according to the time of the day, day of the week, season of the year – and hourly output of a 1 GW VLS-PV plant in the Negev, the yearly value of the electricity produced amounts to 180 MUSD. This is equivalent to an average electricity price of 9 US cents/kWh.

The operation and maintenance (O&M) cost per kWh is estimated at 0,5 US cents. Thus, the net income from solar power sales of a 1 GW plant in the Negev will be 170 MUSD. This figure includes the plant amortization and return on investment. However, these issues will be addressed when presenting the general investment algorithm. The result, as will be shown, is that solar electricity with or without storage is actually *cheaper* than fossil-based electricity, thus representing a profitable investment on a national-size scale.

It is expected that CPV cell efficiencies will rise above the 32 % level assumed for our current calculations. Indeed, cells with efficiencies greater than 35 % have already been demonstrated[8] and might even reach figures close to 50 %. The electrical output from the collectors is directly proportional to the cell efficiency, and the resultant electricity investment cost falls in inverse proportion. With the widespread adoption of VLS-PV systems, cell costs would probably fall below our assumed figure of 2,5 USD per cm^2. The lower boundary will probably be at approximately 1,5 USD per cm^2. As solar cells constitute only 20 % of the plant cost, such cost reduction will not impact upon the plant cost as much as the increase in cell efficiency.

3.7.2 A suggested investment algorithm

We have indicated that solar electricity from a VLS-PV plant could be profitable if the erection cost is sufficiently low (below 1 000 USD/kW – the typical cost of fossil-burning plants). Even at such low prices, though, the infrastructure cost for VLS-PV is large and we must therefore find a method compatible with banking financing practices that will allow erecting a broad base of these systems. Specifically, we shall require that within 36 years, which is 30 years from completion of the first VLS-PV plant, approximately 80 % of the country's projected electricity requirements will be met by solar.

We started with a number of recent statistical facts published by the Israel Electric Corporation:[9]

- Installed generating capacity at the start of 2004 was 10 GW.
- Expected electricity production for 2004 is 44 TWh.
- Yearly expected increase in demand is 2,5 %.

For Israel's VLS-PV plants we made the following assumptions:

- Yearly output of a 1 GW plant is 2,0 TWh.
- Sale price of solar electricity is 9 US cents/kWh.
- O&M cost of solar electricity is 0,5 US cents/kWh.
- O&M cost of solar electricity is 0,5 US cents/kWh.
- Plant engineering design costs equal 220 MUSD.
- Investment required to construct a solar production line of 1,5 GW per year is 800 MUSD.
- Investment required to construct a storage production line of 0,5 GW per year is 150 MUSD.
- Cost of erecting a 1,5 GW generating plant is 1 275 MUSD.
- Cost of erecting a 0,5 GW storage plant is 425 MUSD.
- G&A overhead costs are assumed to be 5 % of sales, for a ceiling of 100 MUSD per year.
- A credit line to cover all expenses and cash flow requirements is available at 3 % interest.

All calculations of investment costs, interest and electricity revenue are in constant dollars (that is, without inflation). Given the above assumptions, our suggested investment algorithm is as follows:

1. The government or other investors make available a credit line that reaches its peak at 9 781 MUSD in year 13.
2. The credit line during the first four years will cover the following cost items:
   - 800 MUSD investment in collector production line of 1,5 GW;
   - 150 MUSD investment in storage production line of 0,5 GW;
   - 220 MUSD engineering costs for plant design.

   This will be made available in successive yearly amounts of 234 MUSD, 292 MUSD, 234 MUSD and 410 MUSD.
3. From year five and on, an additional yearly investment of 1 700 MUSD will be made to erect one 1,5 GW solar generating plant and one 0,5 GWe storage plant each year.
4. From year six and on, a yearly net income of 255 MUSD per 1,5 GW plant will be realized from the 3 TWh of electricity that each plant sells.
5. Electricity will be sold at the price of 9 US cents/kWh from years 6 to 21 to pay off *all the accrued costs plus interest.*
6. The price of electricity can be lowered to 5,5 US cents/kWh thereafter.
7. A yearly (compound) interest of 3 % will be charged on all credit owing at the end of each year.

The credit line will be as follows:

- During the first four years the credit will accumulate to 1 170 MUSD for the VLS-PV production line, plus an additional 83 MUSD to cover the interest.
- Thereafter, each year will require an additional investment of 1 700 MUSD for each successive VLS-PV plant, plus additional amounts to cover the accrued interest.
- From year six begins a steadily increasing annual income of 255 MUSD for each plant as each successive VLS-PV plant goes on line.

Under these conditions it turns out that at year 13 the annual income is sufficiently large that the credit line reaches its maximum of 9 781 MUSD. After that, the still increasing annual income causes the credit line to decrease, reaching zero in year 22.

No further investment will be needed, all debts having been paid off, and *all further VLS-PV plants are being paid for out of profit from electricity sales.* The

Table 3.4 Expected economic benefits to Israel of VLS-PV plant introduction during the first 36 years

Interest rate	3	%/year
Yearly added solar power	1,5	GW
Yearly added six-hour storage power	0, 5	GW
Credit line capacity required for the entire project	9 781	MUSD
Interest paid	3 397	MUSD
Loan repaid after	21	Years
Total solar power installed	46, 5	GW
Total storage power installed	15, 5	GW
Electricity price after five years, when solar electricity sales start	9	US cents/kWh
Electricity price after 22 years, when all debts are paid off	5, 5	US cents/kWh
Land area required for installation	558	km^2
Fraction of total national land area	2, 7	%
Yearly manpower requirements for solar production	4 500	Jobs
Yearly manpower requirements for solar operation	11 625	Jobs
Yearly manpower requirements for storage production	1 500	Jobs
Yearly manpower requirements for storage operation	3 875	Jobs
Headquarters and engineering	1 395	Jobs
Total number of jobs after 36 years	22 895	Jobs

algorithm is thus based on erecting a large enough production base that will both repay the initial investment and produce the funds needed to erect future plants for an indefinite period of time into the future. Naturally, after 30 years of service we have to decommission one plant per year; but the process turns out to be self-sustaining, without needing to borrow money for replacing the decommissioned plants. The revenue from electricity sales will be so large by the end of year 34, when the first 'old' VLS-PV plant has to be decommissioned, that two new plants can be built annually from then onwards.

As a large amount of cash will accumulate after repaying the debt/credit line, the sale price of electricity can be reduced from 0,09 USD/kWh to 0,055 USD/kWh in year 22. The results for the Israel example are summarized in Table 3.4 and may be extended, *mutatis mutandis*, to the other countries in the Middle East, as the following sections will indicate.

3.7.3 Results for the Israel example

Under the above assumptions, Table 3.4 lists the expected economic benefits to Israel during the first 36 years of a top-down VLS-PV programme. Our estimated algorithm for the number of jobs generated by such a programme is as follows:

- 3 000 people per 1 GW of solar plant production line;
- 250 people per 1 GW of solar plant operation;
- 3 000 people per 1 GW of storage production line;
- 250 people per 1 GW of storage plant operation;
- 30 people per 1 GW at headquarters and engineering.

Since Israel's current fossil-based generating capacity is 10 GW, if none of this infrastructure is retired or expanded, the solar part of the country's total generating capacity at the end of 36 years would be 82 % and the solar part of the total electrical energy production at that time would be 87 %. The project would cover 560 km^2 of land and generate approximately 23 000 jobs.

The graphs in Figures 3.3 (a) to (d) illustrate in greater detail the economic results summarized in Table 3.4.

3.7.4 Discussion of the Israel results

A number of interesting results have emerged from this Israel case study that are worthy of attention before moving on to other countries in the Middle East. First is the results of MST Ltd's previous study[5] that large CPV systems of the individual cell type[4] can probably be mass produced for under 1 000 USD/kW, a finding that we had previously concluded[3] only for large dish-based systems of the dense array variety.[4] The importance of this is that solar collectors of the kind considered here have already been successfully tested at the Arizona Public Services (APS) test facility.[10] Thus, this technology is ripe for mass production.

A second important point is that the efficiency of such collectors is directly proportional to the efficiency of the CPV cells employed. The Amonix Corp collectors being tested at APS employ silicon cells with efficiency slightly above 20 %. Our current study assumes the use of multi-junction GaAs cells with a nominal efficiency of 32 %. Such cells already exist and this figure is by no means the highest efficiency that has been demonstrated.[8] Accordingly, the power production figures

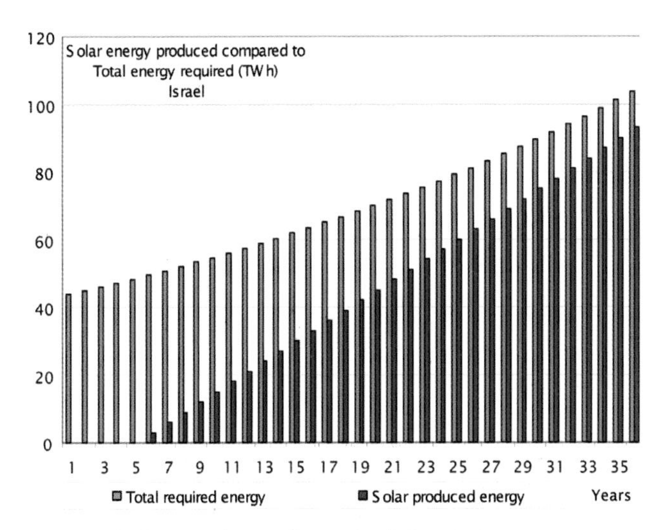

Figure 3.3 (a) Projected annual growth of electrical energy generation in Israel for the first 36 years of a top-down VLS-PV programme, starting from an all-fossil 44 TWh per year in 2004 and reaching 107 TWh, of which 87% is solar

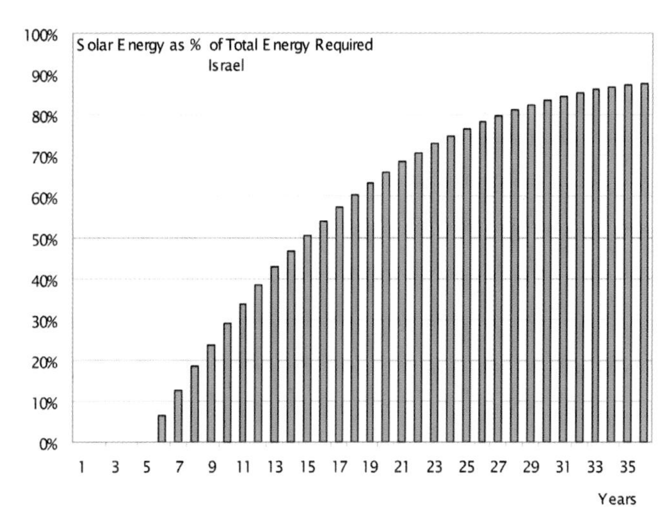

Figure 3.3 (b) Projected annual growth of percentage solar electricity generation in Israel for the first 36 years of a top-down VLS-PV programme, starting from 0 % in 2004–2008 (years 1 to 5) and reaching 87 % in year 36

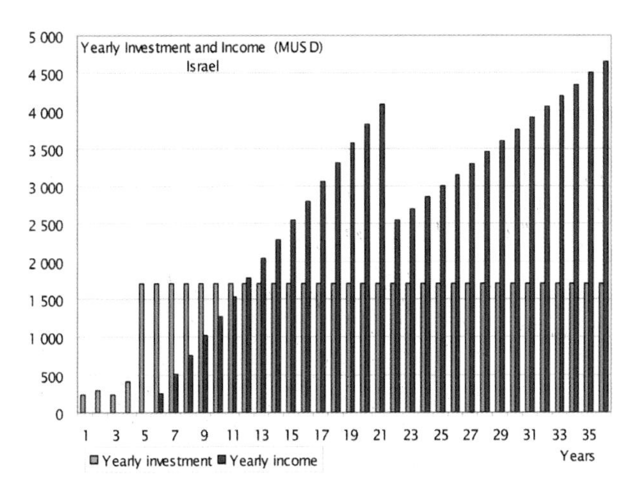

Figure 3.3 (c) Annual investments required for the top-down VLS-PV programme proposed for Israel and subsequent income from solar plants

Note: The first four years show manufacturing facility construction costs. Subsequent years show construction costs of an annual 1,5 GW of solar plants, each with 0,5 GW storage. Note that income, starting during year 6, enables all costs plus 3 % interest to have been paid off by year 21. The electricity price can then be dropped below the initial 9 US cents/kWh (naturally, with an accompanying temporary lowering of annual income).

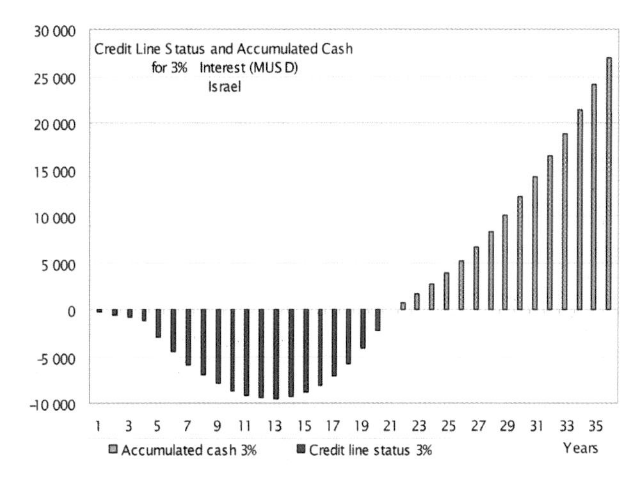

Figure 3.3 (d) Annual cash flow associated with investment, at 3 % interest, in a top-down VLS-PV programme in Israel

Note: Credit line reaches its maximum in year 13 and decreases thereafter due to steadily increasing revenue from solar plants. The debt is entirely paid off in year 21. Subsequent revenue covers all further VLS-PV plants without the need for further investment.

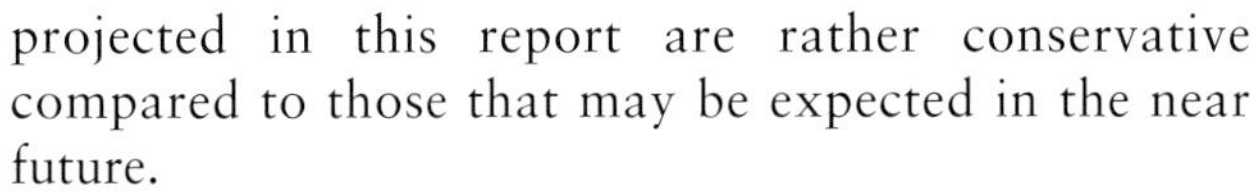

projected in this report are rather conservative compared to those that may be expected in the near future.

A third important discovery is that very large DC to AC inverters (in the MW range) can be mass produced for 40 USD/kW. This conclusion is based on a firm price quotation from a long-established manufacturer of power conditioning equipment. It is a full order of magnitude lower than the cost projection for inverters in our previous study.[3]

Yet another novel feature of the current study was the need to consider storage as an integral part of a VLS-PV system owing to reasons of supply stability. For this purpose we considered using vanadium redox flow batteries as the most promising new technology.

Finally, the most important result, which is extremely encouraging, is that even using our conservative figures, the cost of electricity production from such systems is directly competitive with the power generated from fossil fuels. The continued construction of fossil-burning plants could be discontinued in favour of VLS-PV construction. Furthermore, the solar solution is so cost effective that a top-down programme of the kind presented here is capable of largely replacing fossil fuel within a time scale of about 30 to 40 years in a self-sustaining manner. In other words, after approximately

20 years, the construction of all future VLS-PV plants (including the replacement of 30-year-old plants with new ones) can be funded entirely from the revenues of the existing plants.

With these considerations in mind, we now turn our attention to the other countries in the region.

3.8 OTHER MAJOR ELECTRICITY CONSUMERS IN THE MIDDLE EAST

As shown in Table 3.1, the nine major electricity consumers in the Middle East in decreasing order of production are Saudi Arabia, Iran, Turkey, Egypt, Israel, Iraq, the United Arab Emirates, Kuwait and Syria. Together, these states generate more than 90 % of the electricity in the Middle East. As previously emphasized, each state has its own economic considerations. The available solar radiation will also vary from state to state, and for geographically large countries may even vary from one region to another within the country. However, except where stated, and for the sake of simplicity, we shall adopt similar economic algorithms to those used in the Israel example and the same solar radiation levels that were used for the Negev calculations.

Specifically, we assume that the annual output of a VLS-PV plant in each of these countries will be a standard 2,0 TWh/GW. Our starting statistics for electricity requirements are the generating capacity and generation figures published for the year 2002,[2] and we assume that, on average, electricity requirements will rise linearly at a rate equal to the current population growth rate of each country. We assume that it takes four years to construct a VLS-PV plant production line and a fifth year to erect the first VLS-PV plant. We size the production line and VLS-PV plants in a manner that will enable one VLS-PV plant to be installed every year from year 5 onwards, and will result in an approximately 80 % solar share of total electricity by year 36. For all countries, we assume that production costs per GW will be similar to those assumed for the Israel study, and that the net value of the generated solar electricity is 8,5 US cents/kWh.

These simplifying assumptions are only useful for providing a first look at the economic and technical requirements for introducing VLS-PV in a top-down manner to each of these countries. However, they can easily be corrected and fine tuned if any country seriously considers implementing a VLS-PV programme. Figures 3.4 to 3.10 present comparative histograms of the economic considerations associated with introducing VLS-PV into the nine major electricity-producing countries in the region, using the top-down algorithm suggested for Israel, but rescaling it for the other countries in the manner described above.

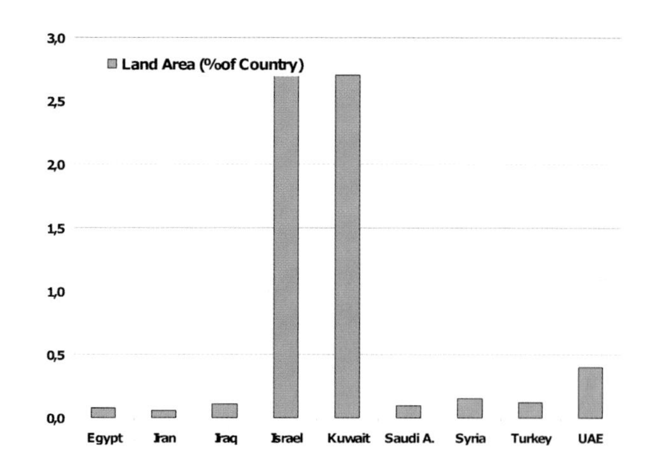

Figure 3.5 Percentage of each country's total land area

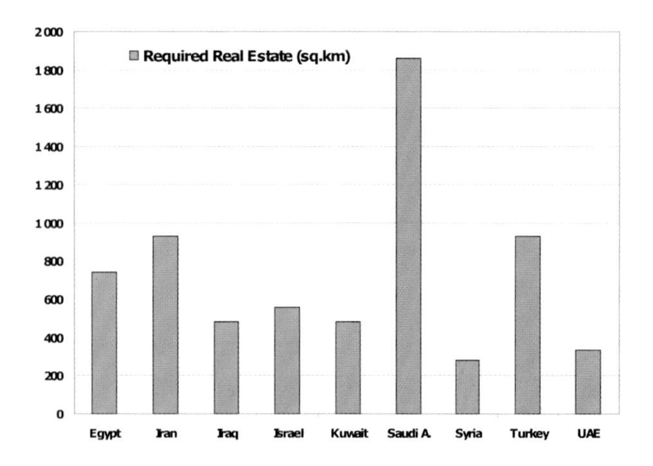

Figure 3.4 Real estate requirements (in square kilometres) for a top-down VLS-PV programme that would provide typically 80 % of each country's electrical requirements within 36 years for the nine largest electricity-producing countries in the Middle East

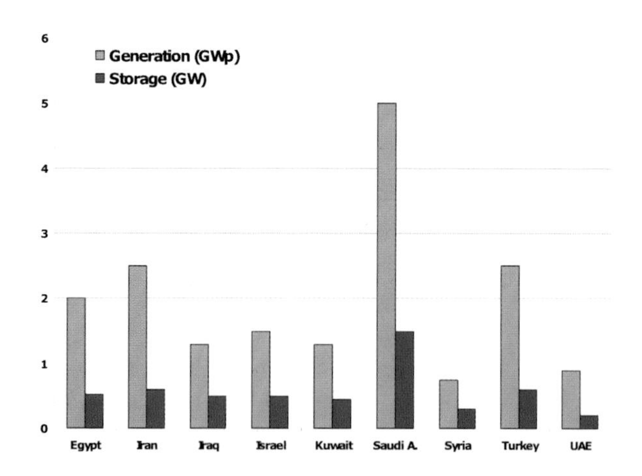

Figure 3.6 Size of proposed VLS-PV plants and storage units that would be erected annually in a top-down programme for the nine largest electricity-producing countries in the Middle East

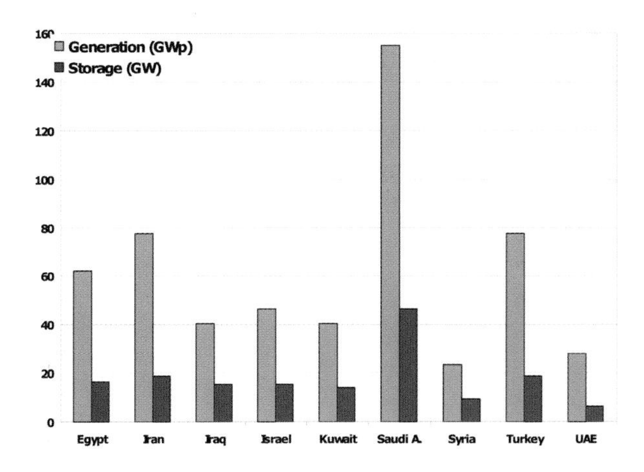

Figure 3.7 Total solar-generating and storage capacity at the end of 36 years of a top-down VLS-PV programme for the nine largest electricity-producing countries in the Middle East

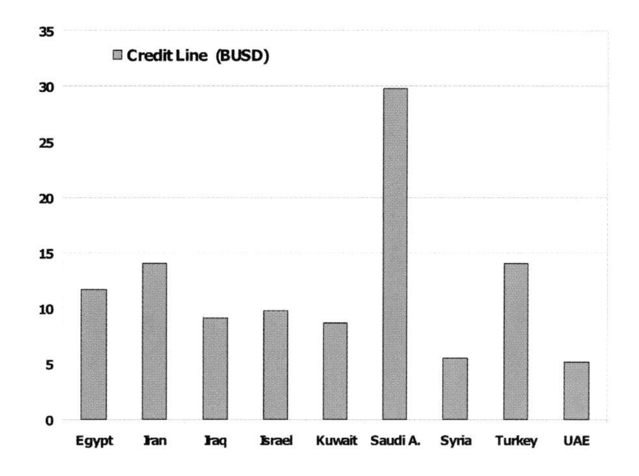

Figure 3.8 Maximum credit line required (at 3 % real interest) for carrying out a top-down VLS-PV programme for the nine largest electricity-producing countries in the Middle East

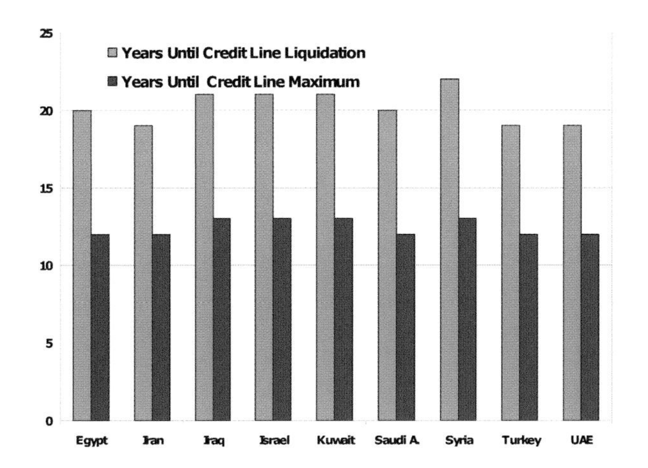

Figure 3.9 The numbers of years required for the credit line to reach its maximum and subsequently be completely liquidated by revenues from the VLS-PV plants for the nine largest electricity-producing countries in the Middle East

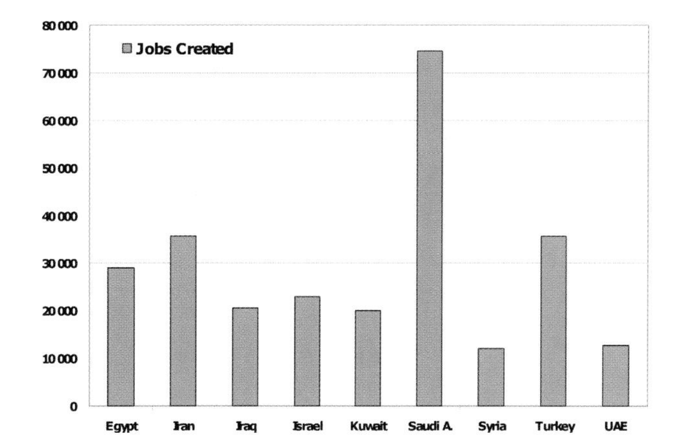

Figure 3.10 Total number of jobs created by a top-down VLS-PV programme for the nine largest electricity-producing countries in the Middle East

3.8.1 Saudi Arabia

The total generating capacity of Saudi Arabia in 2002 was 23,8 GW, of which 100 % was fossil generated. The total electricity production that year was 138,2 TWh. At an assumed annual growth rate of 3,3 %, which is equal to the 2003 population growth rate, the electricity requirements after 36 years will reach 444,8 TWh per year.

Our proposal is to annually install VLS-PV plants of 5,0 GW generating capacity and 1,5 GW storage capacity starting at year five of the programme. By year 36, the total installed solar generating capacity will accordingly be 155 GW and the total storage capacity will be 46,5 GW. The annual solar generation at that time will be 310 TWh. If the non-solar part remains unchanged, then the solar fraction of the country's generating capacity will be 87 %. Correspondingly, the percentage of annual electricity coming from solar will be 70 %.

Such a project would cover 1 860 km^2 of land. It would require a maximum credit line of 29 800 MUSD, which would be fully paid off, including total interest of 9 800 MUSD, after 20 years. The project would generate approximately 75 000 jobs.

Table 3.5 summarizes the details of all our economic calculations for Saudi Arabia. The graphs in Figures 3.11 (a) to (d) are the equivalent of Figures 3.3 (a) to (d) for Saudi Arabia.

Table 3.5 Expected economic benefits to Saudi Arabia of VLS-PV plant introduction during the first 36 years

Interest rate	3	%/year
Yearly added solar power	5,0	GW
Yearly added six-hour storage power	1,5	GW
Credit line capacity required for the entire project	29 788	MUSD
Interest paid	9 769	MUSD
Loan repaid after	20	Years
Total solar power installed	155	GW
Total storage power installed	46,5	GW
Electricity price after five years, when solar electricity sales start	9	US cents/kWh
Electricity price after 22 years, when all debts are paid off	5,5	US cents/kWh
Land area required for installation	1 860	km^2
Fraction of total national land area	0,095	%
Yearly manpower requirements for solar production	15 000	Jobs
Yearly manpower requirements for solar operation	38 750	Jobs
Yearly manpower requirements for storage production	4 500	Jobs
Yearly manpower requirements for storage operation	11 625	Jobs
Headquarters and engineering	4 650	Jobs
Total number of jobs after 36 years	74 525	Jobs

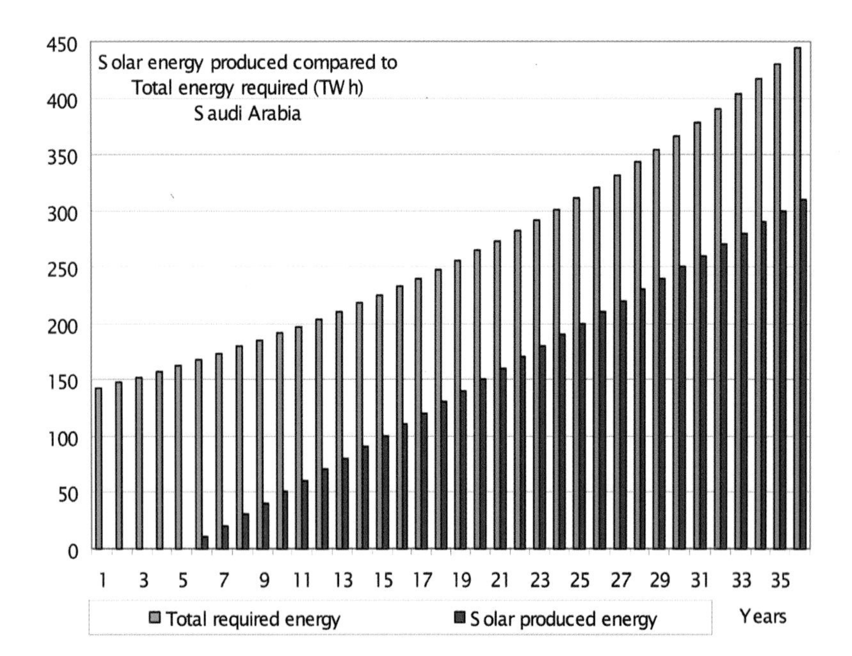

Figure 3.11 (a) Projected annual growth of electrical energy generation in Saudi Arabia for the first 36 years of a top-down VLS-PV programme, starting from an all-fossil 142,8 TWh per year in 2004 and reaching 444,8 TWh, of which 70 % is solar

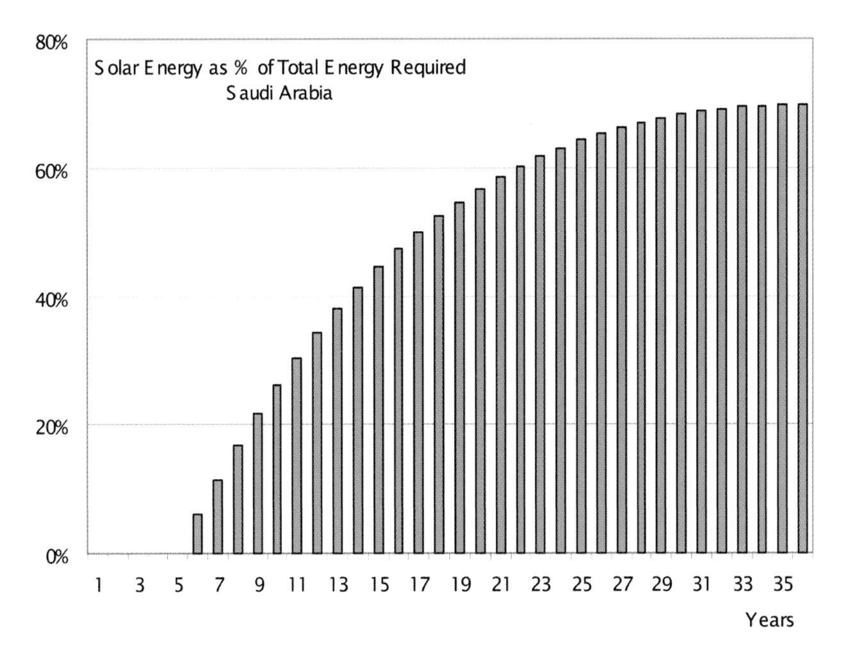

Figure 3.11 (b) Projected annual growth of the solar electricity-generation percentage in Saudi Arabia for the first 36 years of a top-down VLS-PV programme, starting from 0 % during 2004–2008 (years 1 to 5) and reaching 70 % in year 36

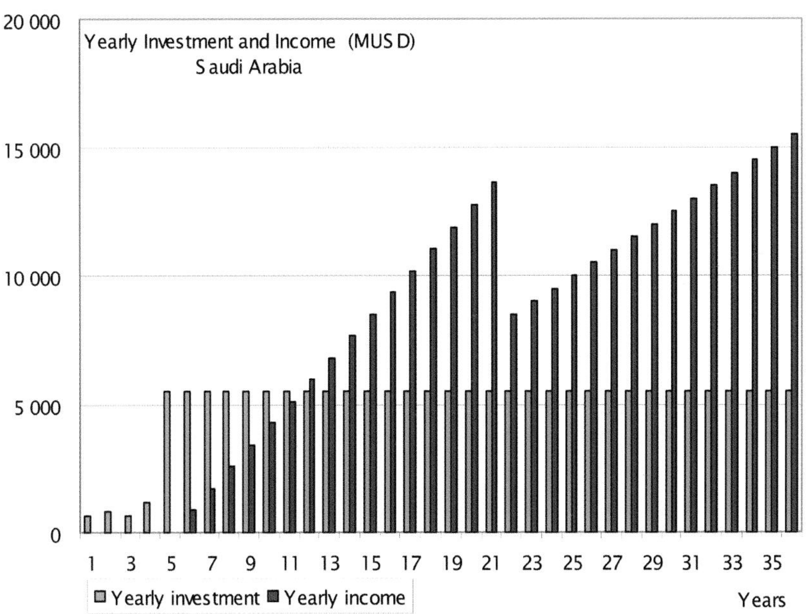

Figure 3.11 (c) Annual investments required for the top-down VLS-PV programme proposed for Saudi Arabia and subsequent income from solar plants

Note: The first four years show manufacturing facility construction costs. Subsequent years show construction costs of an annual 5 GW of solar plants, each with 1,5 GW storage. Note that income, starting at year 6, enables all costs plus 3 % interest to have been paid off by year 21. Electricity price can then be dropped below the initial 9 US cents/kWh (naturally, with an accompanying temporary lowering of annual income).

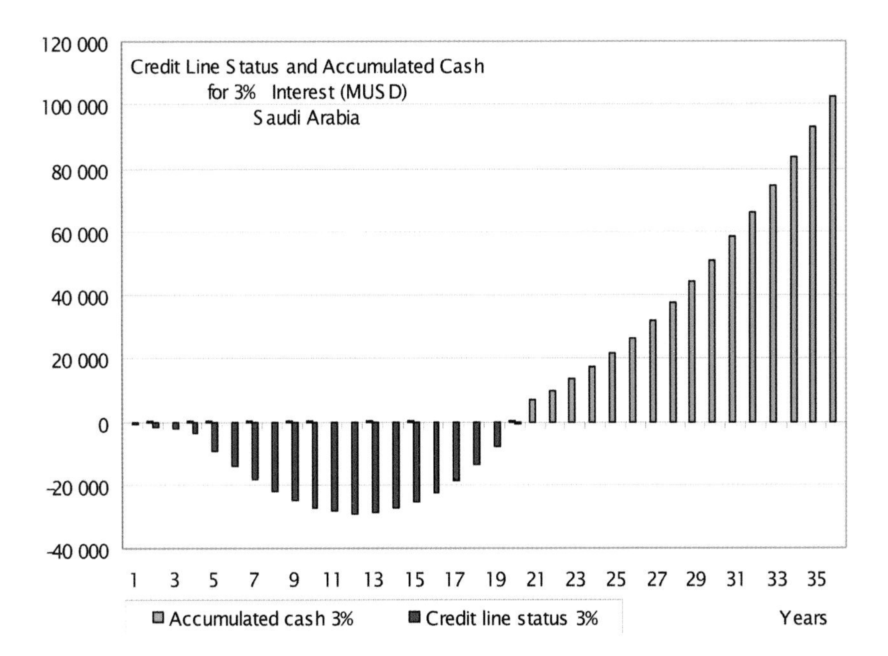

Figure 3.11 (d) Annual cash flow associated with investment, at 3 % interest, in a top-down VLS-PV programme in Saudi Arabia

Note: Credit line reaches its maximum in year 12 and decreases thereafter due to steadily increasing revenue from solar plants. The debt is entirely paid off in year 20. Subsequent revenue covers all further VLS-PV plants without need for further investment.

3.8.2 Iran

The total generating capacity of Iran in 2002 was 28 GW, of which 2,8 GW was hydro generated. The total electricity production that year was 129 TWh. At an assumed annual growth rate of 1,1 %, which is equal to the 2003 population growth rate, the electricity requirements after 36 years will reach 191,3 TWh per year.

Our proposal is to annually install VLS-PV plants of 2,5 GW generating capacity and 0,6 GW storage capacity starting at year five of the programme. By year 36, the total installed solar generating capacity will accordingly be 77,5 GW and the total storage capacity will be 18,6 GW. The annual solar generation at that time will be 155 TWh. If the non-solar part remains unchanged, then the solar fraction of the country's generating capacity will be 73 % (76 % will be from renewables if the present hydroelectric part is included). Correspondingly, the percentage of annual electricity coming from solar will be 81 %.

Such a project would cover 930 km^2 of land. It would require a maximum credit line of 14 000 MUSD, which would be fully paid off, including total interest of 4 500 MUSD after 19 years. The project would generate approximately 36 000 jobs.

Table 3.6 summarizes the details of all our economic calculations for Iran. The graphs in Figures 3.12 (a) to (d) are the equivalent of Figures 3.3 (a) to (d) for Iran.

Table 3.6 Expected economic benefits to Iran of VLS-PV plant introduction during the first 36 years

Interest rate	3	%/year
Yearly added solar power	2,5	GW
Yearly added six-hour storage power	0,6	GW
Credit line capacity required for the entire project	14 030	MUSD
Interest paid	4 494	MUSD
Loan repaid after	19	Years
Total solar power installed	77,5	GW
Total storage power installed	18,6	GW
Electricity price after five years, when solar electricity sales start	9	US cents/kWh
Electricity price after 22 years, when all debts are paid off	5,5	US cents/kWh
Land area required for installation	930	km^2
Fraction of total national land area	0,057	%
Yearly manpower requirements for solar production	7 500	Jobs
Yearly manpower requirements for solar operation	19 375	Jobs
Yearly manpower requirements for storage production	1 800	Jobs
Yearly manpower requirements for storage operation	4 650	Jobs
Headquarters and engineering	2 325	Jobs
Total number of jobs after 36 years	35 650	Jobs

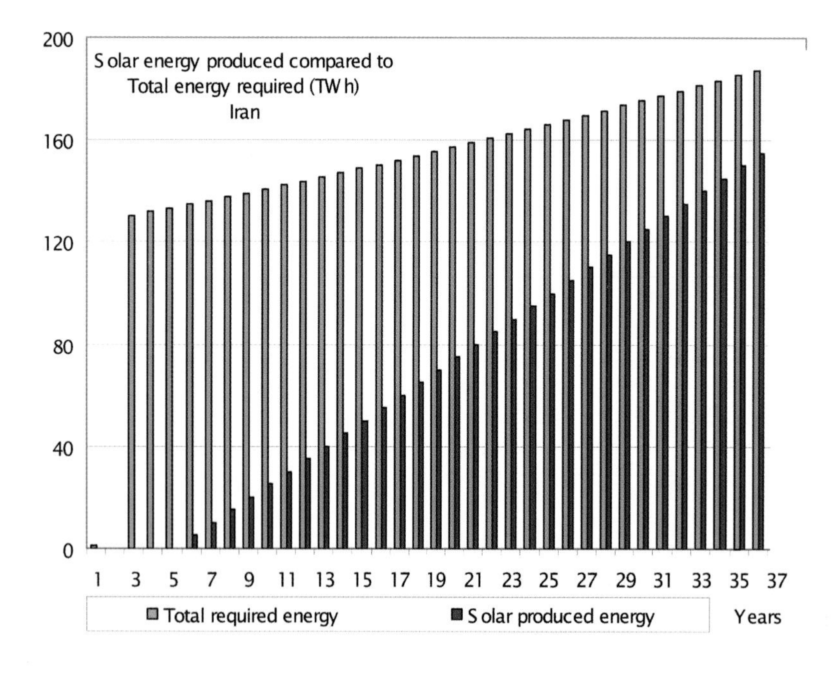

Figure 3.12 (a) Projected annual growth of electrical energy generation in Iran for the first 36 years of a top-down VLS-PV programme, starting from 130,4 TWh per year in 2004 and reaching 191,3 TWh, of which 81 % is solar

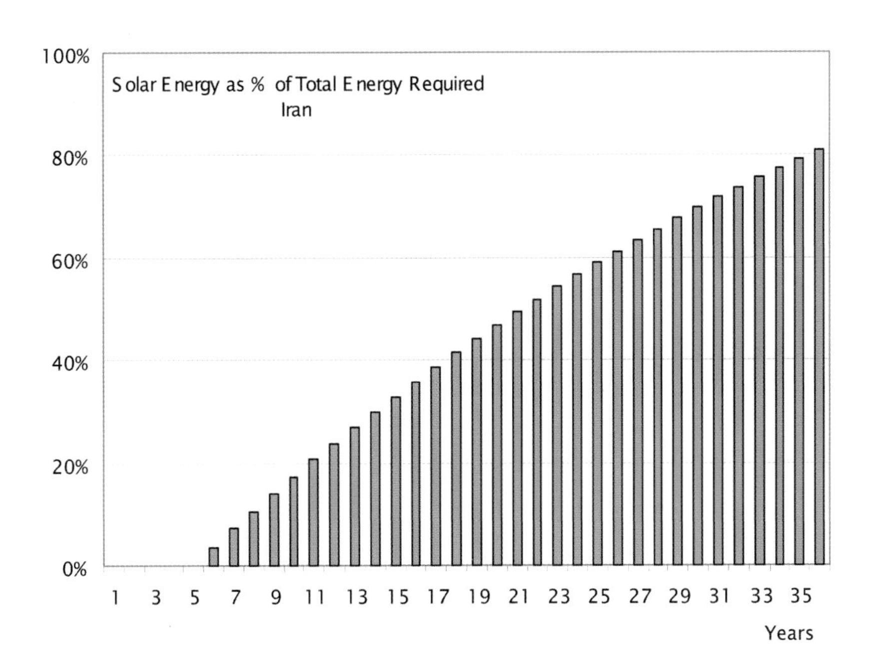

Figure 3.12 (b) Projected annual growth of solar electricity-generation percentage in Iran for the first 36 years of a top-down VLS-PV programme, starting from 0 % during 2004–2008 (years 1 to 5) and reaching 81 % in year 36

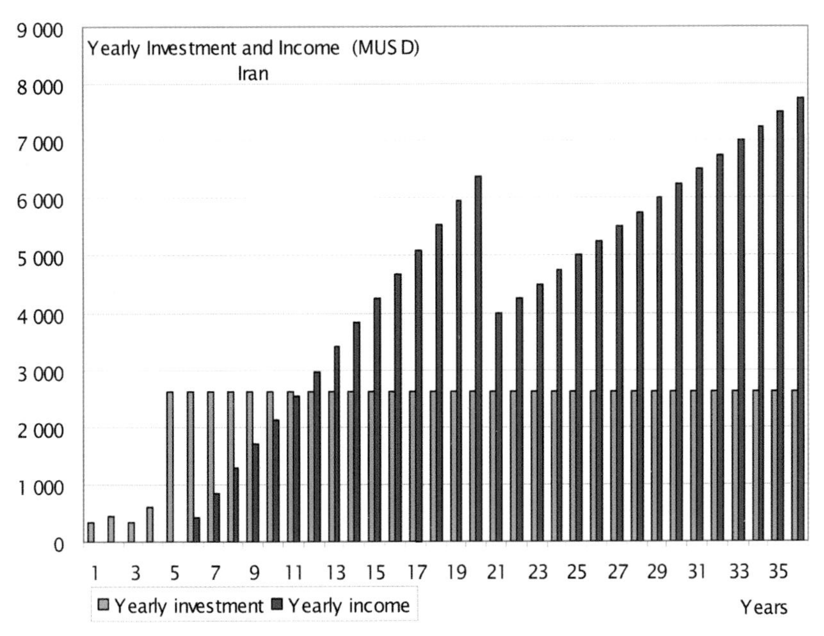

Figure 3.12 (c) Annual investments required for the top-down VLS-PV programme proposed for Iran and subsequent income from solar plants

Note: The first four years show manufacturing facility construction costs. Subsequent years show construction costs of 2,5 GW of solar plants annually with 0,6 GW storage. Note that income, starting in year 6, enables all costs plus 3 % interest to have been paid off by year 20. Electricity price can then be dropped below the initial 9 US cents/kWh (naturally, with an accompanying temporary lowering of annual income).

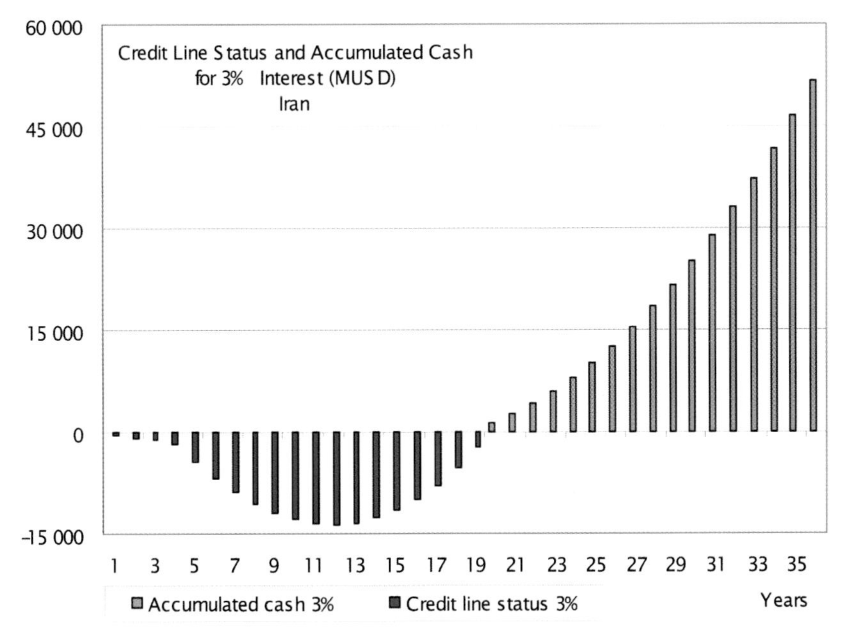

Figure 3.12 (d) Annual cash flow associated with investment, at 3 % interest, in a top-down VLS-PV programme in Iran

Note: Credit line reaches its maximum in year 13 and decreases thereafter due to steadily increasing revenue from solar plants. The debt is entirely paid off in year 20. Subsequent revenue covers all further VLS-PV plants without need for further investment.

3.8.3 Turkey

The total generating capacity of Turkey in 2002 was 28,3 GW, of which 11,7 GW was hydro generated. The total electricity production that year was 129 TWh. At an assumed annual growth rate of 1,2 %, which is equal to the 2003 population growth rate, the electricity requirements after 36 years will reach 189,4 TWh per year.

We propose annually installing VLS-PV plants of 2,5 GW generating capacity with 0,6 GW storage capacity starting at year five of the programme. By year 36, the total installed solar generating capacity will accordingly be 77,5 GW and the total storage capacity will be 18,6 GW. The annual solar generation at that time will be 155 TWh. If the non-solar part remains unchanged, then the solar fraction of the country's generating capacity will be 73 % (84 % will be from renewables if the present hydroelectric part is included). Correspondingly, the percentage of annual electricity coming from solar will be 82 %.

Such a project would cover 930 km^2 of land. It would require a maximum credit line of 14 000 MUSD, which would be fully paid off, including total interest of 4 500 MUSD, after 19 years. The project would generate approximately 36 000 jobs.

Table 3.7 summarizes the details of all our economic calculations for Turkey. The graphs in Figures 3.13 (a) to (d) are the equivalent of Figures 3.3 (a) to (d) for Turkey.

Table 3.7 Expected economic benefits to Turkey of VLS-PV plant introduction during the first 36 years

Interest rate	3	%/year
Yearly added solar power	2,5	GW
Yearly added six-hour storage power	0,6	GW
Credit line capacity required for the entire project	14 030	MUSD
Interest paid	4 494	MUSD
Loan repaid after	19	Years
Total solar power installed	77,5	GW
Total storage power installed	18,6	GW
Electricity price after five years, when solar electricity sales start	9	US cents/kWh
Electricity price after 22 years, when all debts are paid off	5,5	US cents/kWh
Land area required for installation	930	km^2
Fraction of total national land area	0,12	%
Yearly manpower requirements for solar production	7 500	Jobs
Yearly manpower requirements for solar operation	19 375	Jobs
Yearly manpower requirements for storage production	1 800	Jobs
Yearly manpower requirements for storage operation	4 650	Jobs
Headquarters and engineering	2 325	Jobs
Total number of jobs after 36 years	35 650	Jobs

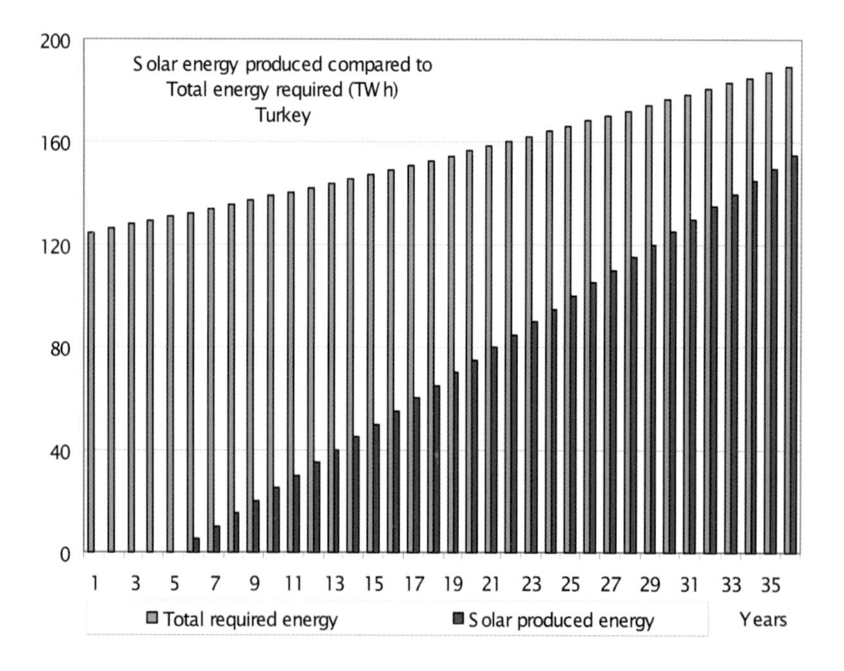

Figure 3.13 (a) Projected annual growth of electrical energy generation in Turkey for the first 36 years of a top-down VLS-PV programme, starting from 124,8 TWh per year in 2004 and reaching 189,4 TWh, of which 82 % is solar

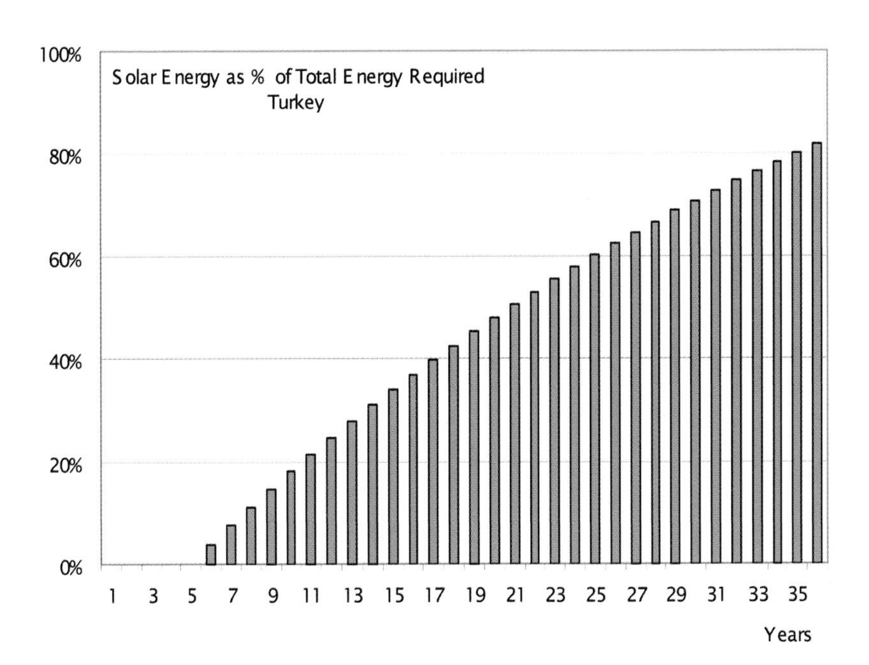

Figure 3.13 (b) Projected annual growth of solar electricity-generation percentage in Turkey for the first 36 years of a top-down VLS-PV programme, starting from 0 % during 2004–2008 (years 1 to 5) and reaching 82 % in year 36

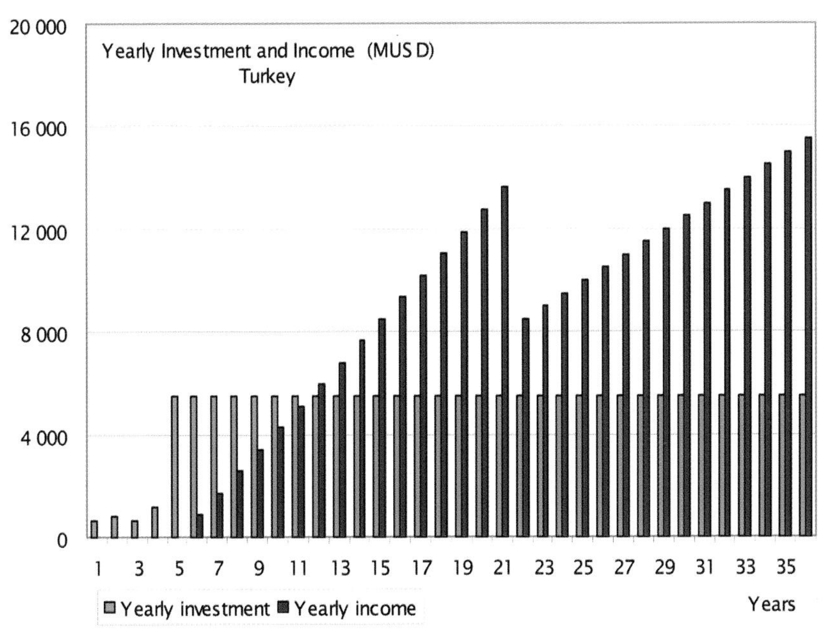

Figure 3.13 (c) Annual investments required for the top-down VLS-PV programme proposed for Turkey and subsequent income from solar plants

Note: The first four years show manufacturing facility construction costs. Subsequent years show construction costs of an annual 2,5 GW of solar plants with 0,6 GW storage. Note that income, starting in year 6, enables all costs plus 3 % interest to have been paid off by year 19. Electricity price can then be dropped below the initial 9 US cents/kWh (naturally, with an accompanying temporary lowering of annual income).

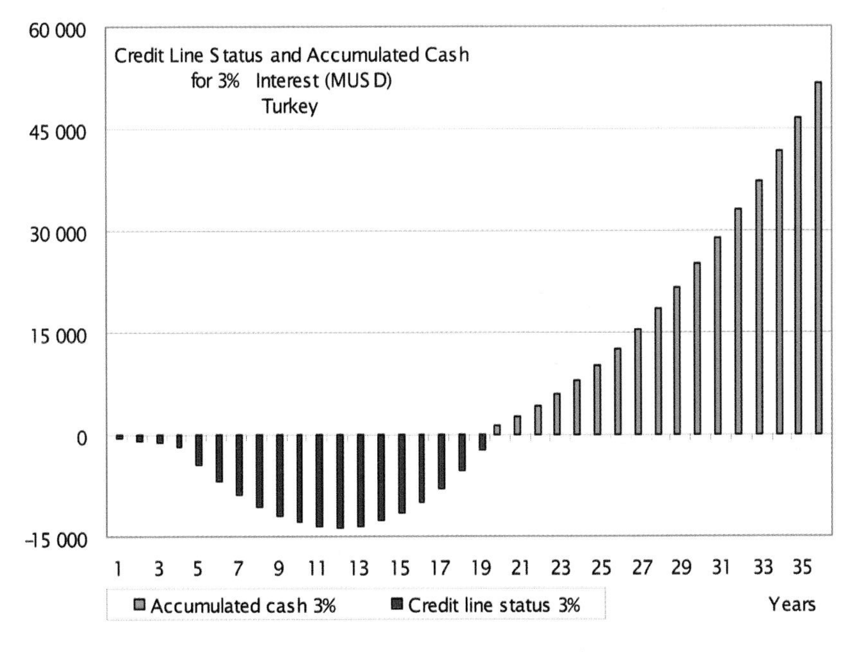

Figure 3.13 (d) Annual cash flow associated with investment, at 3 % interest, in a top-down VLS-PV programme in Turkey

Note: Credit line reaches its maximum in year 12 and decreases thereafter due to steadily increasing revenue from solar plants. The debt is entirely paid off in year 19. Subsequent revenue covers all further VLS-PV plants without need for further investment.

3.8.4 Egypt

The total generating capacity of Egypt in 2002 was 17,6 GW, of which 2,7 GW was hydro generated. The total electricity production that year was 81,3 TWh. At an assumed annual growth rate of 1,9 %, which is equal to the 2003 population growth rate, the electricity requirements after 36 years will reach 160,1 TWh per year.

Our proposal is to annually install VLS-PV plants of 2 GW generating capacity and 0,53 GW storage capacity starting at year five of the programme. By year 36, the total installed solar generating capacity will accordingly be 62 GW and the total storage capacity will be 16,3 GW. The annual solar generation at that time will be 124 TWh. If the non-solar part remains unchanged, then the solar fraction of the country's generating capacity will be 78 % (81 % will be from renewables if the current hydroelectric part is included). Correspondingly, the percentage of annual electricity coming from solar will be 77 %.

Such a project would cover 744 km² of land. It would require a maximum credit line of 11 700 MUSD, which would be fully paid off, including total interest of 3 800 MUSD, after 20 years. The project would generate approximately 29 000 jobs.

Table 3.8 summarizes the details of all our economic calculations for Egypt. The graphs in Figures 3.14 (a) to (d) are the equivalent of Figures 3.3 (a) to (d) for Egypt.

Table 3.8 Expected economic benefits to Egypt of VLS-PV plant introduction during the first 36 years

Interest rate	3	%/year
Yearly added solar power	2	GW
Yearly added six-hour storage power	0,525	GW
Credit line capacity required for the entire project	11 697	MUSD
Interest paid	3 825	MUSD
Loan repaid after	20	Years
Total solar power installed	62	GW
Total storage power installed	16,275	GW
Electricity price after five years, when solar electricity sales start	9	US cents/kWh
Electricity price after 22 years, when all debts are paid off	5,5	US cents/kWh
Land area required for installation	744	km²
Fraction of total national land area	0,075	%
Yearly manpower requirements for solar production	6 000	Jobs
Yearly manpower requirements for solar operation	15 500	Jobs
Yearly manpower requirements for storage production	1 575	Jobs
Yearly manpower requirements for storage operation	4 069	Jobs
Headquarters and engineering	1 860	Jobs
Total number of jobs after 36 years	29 004	Jobs

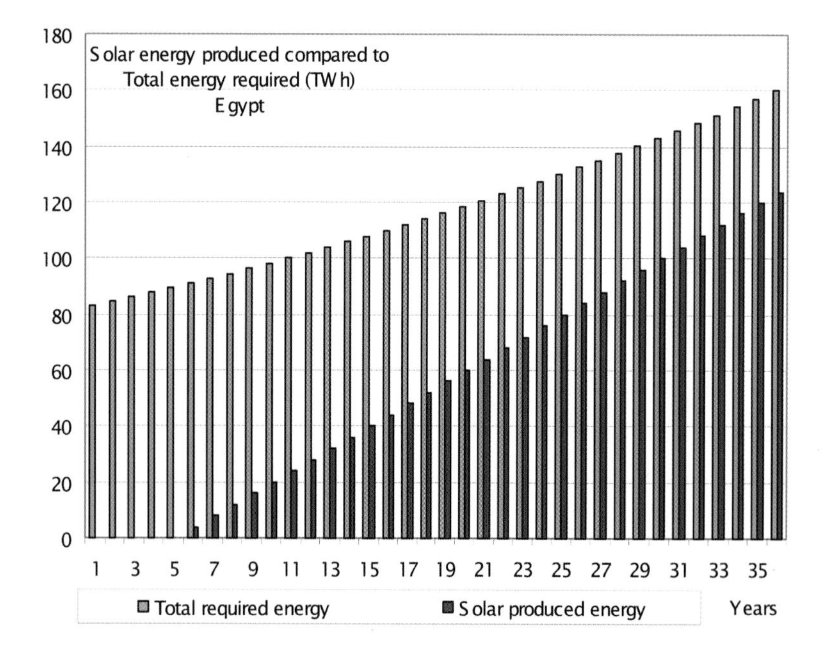

Figure 3.14 (a) Projected annual growth of electrical energy generation in Egypt for the first 36 years of a top-down VLS-PV programme, starting from 82,8 TWh per year in 2004 and reaching 160,1 TWh, of which 78 % is solar

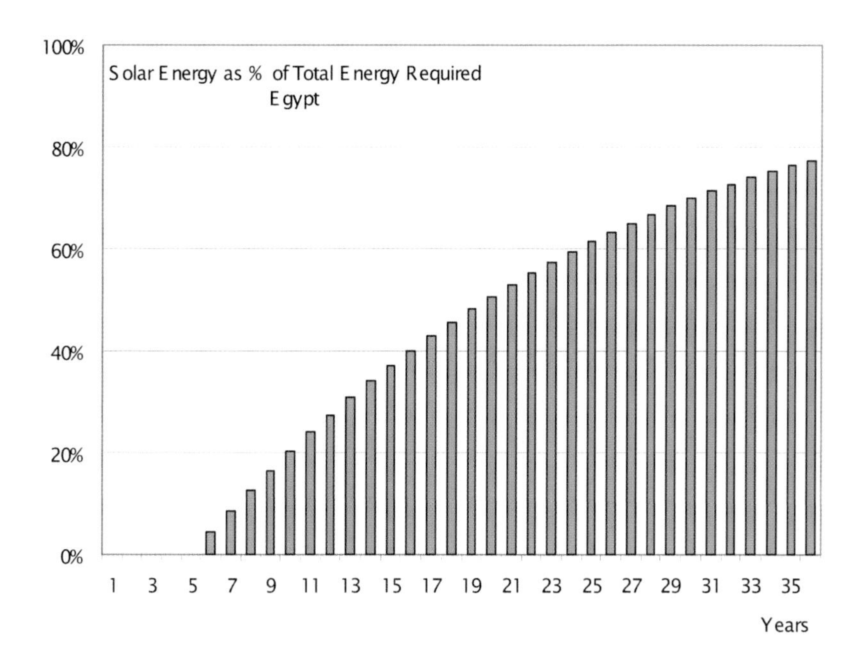

Figure 3.14 (b) Projected annual growth of solar electricity-generation percentage in Egypt for the first 36 years of a top-down VLS-PV programme, starting from 0 % during 2004–2008 (years 1 to 5), and reaching 78 % in year 36

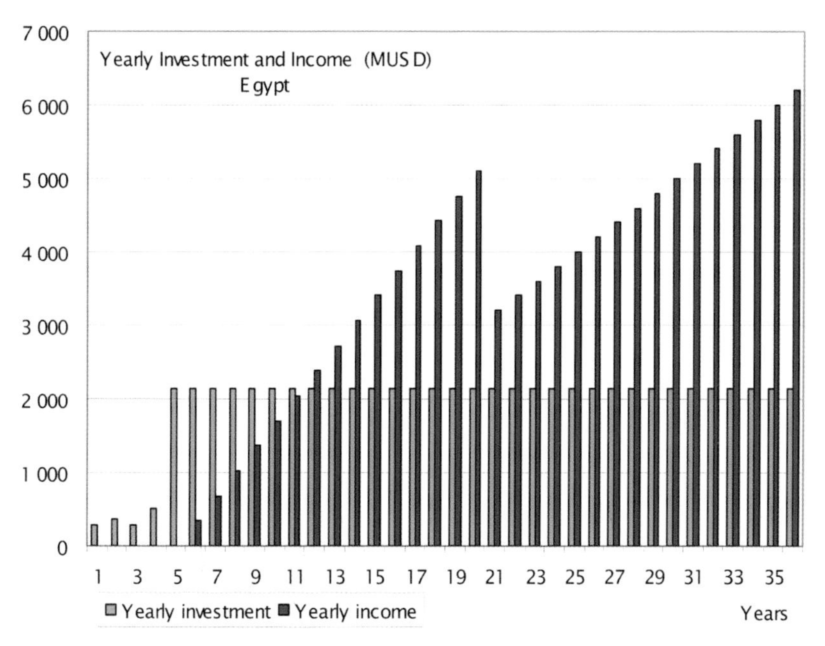

Figure 3.14 (c) Annual investments required for the top-down VLS-PV programme proposed for Egypt and subsequent income from solar plants

Note: The first four years show manufacturing facility construction costs. Subsequent years show construction costs of an annual 2 GW of solar plants with 0,525 GW storage. Note that income, starting in year 6, enables all costs plus 3 % interest to have been paid off by year 20. The electricity price can then be dropped below the initial 9 US cents/kWh (naturally, with an accompanying temporary lowering of annual income).

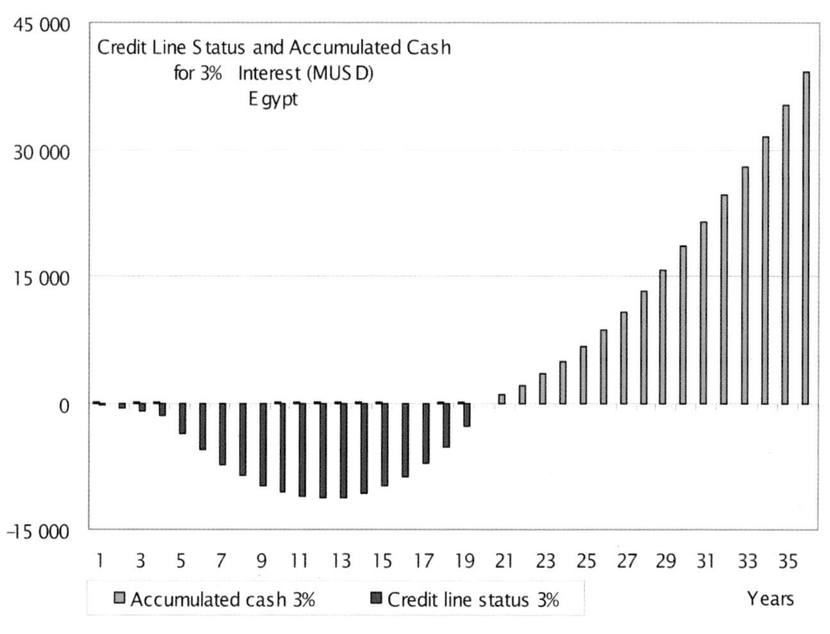

Figure 3.14 (d) Annual cash flow associated with investment, at 3 % interest, in a top-down VLS-PV programme in Egypt

Note: Credit line reaches its maximum in year 13 and decreases thereafter due to steadily increasing revenue from solar plants. The debt is entirely paid off in year 20. Subsequent revenue covers all further VLS-PV plants without need for further investment.

3.8.5 United Arab Emirates

The total generating capacity of the United Arab Emirates (UAE) in 2002 was 5,8 GW, of which 100 % was fossil generated. The total electricity production that year was 39,3 TWh. At an assumed annual growth rate of 1,6 %, which is equal to the 2003 population growth rate, the electricity requirements after 36 years will reach 69,6 TWh per year.

Our proposal is to annually install VLS-PV plants of 900 MW generating capacity and 200 MW storage capacity starting at year five of the programme. By year 36, the total installed solar generating capacity will accordingly be 27,9 GW and the total storage capacity will be 6,2 GW. The annual solar generation at that time will be 55,8 TWh. If the non-solar part remains unchanged, then the solar fraction of the country's generating capacity will be 83 %. Correspondingly, the percentage of annual electricity coming from solar will be 80 %.

Such a project would cover 335 km^2 of land. It would require a maximum credit line of 5 100 MUSD, which would be fully paid off, including total interest of 1 700 MUSD, after 19 years. The project would generate approximately 13 000 jobs.

Table 3.9 summarizes the details of all our economic calculations for the UAE. The graphs in Figures 3.15 (a) to (d) are the equivalent of Figures 3.3 (a) to (d) for the UAE.

Table 3.9 Expected economic benefits to the United Arab Emirates of VLS-PV plant introduction during the first 36 years

Interest rate	3	%/year
Yearly added solar power	0,9	GW
Yearly added six-hour storage power	0,2	GW
Credit line capacity required for the entire project	5 142	MUSD
Interest paid	1 696	MUSD
Loan repaid after	19	Years
Total solar power installed	27,9	GW
Total storage power installed	6,2	GW
Electricity price after five years, when solar electricity sales start	9	US cents/kWh
Electricity price after 22 years, when all debts are paid off	5,5	US cents/kWh
Land area required for installation	334,8	km^2
Fraction of total national land area	0,40	%
Yearly manpower requirements for solar production	2 700	Jobs
Yearly manpower requirements for solar operation	6 975	Jobs
Yearly manpower requirements for storage production	600	Jobs
Yearly manpower requirements for storage operation	1 550	Jobs
Headquarters and engineering	837	Jobs
Total number of jobs after 36 years	12 662	Jobs

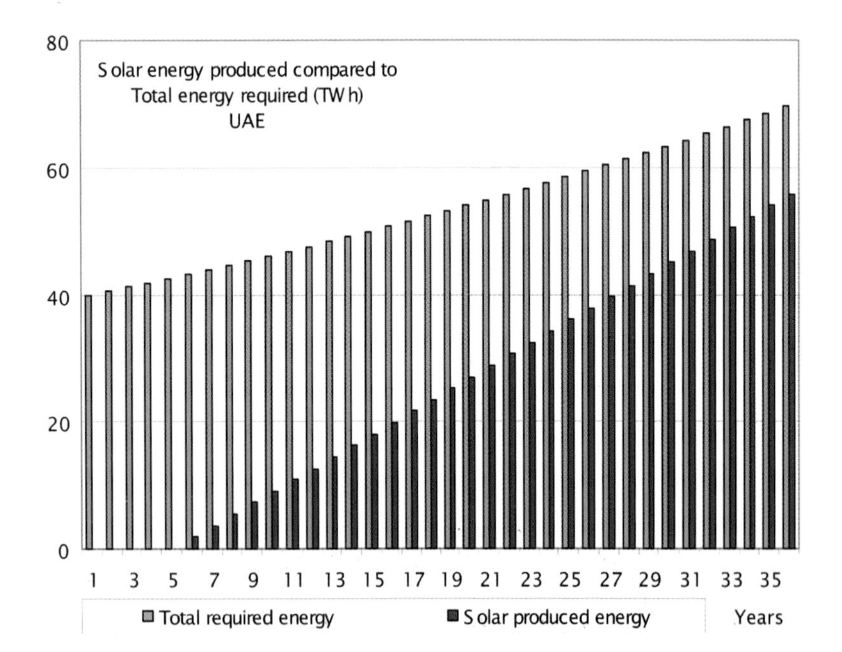

Figure 3.15 (a) Projected annual growth of electrical energy generation in the United Arab Emirates for the first 36 years of a top-down VLS-PV programme, starting from 39,9 TWh per year in 2004 and reaching 69,6 TWh, of which 80 % is solar

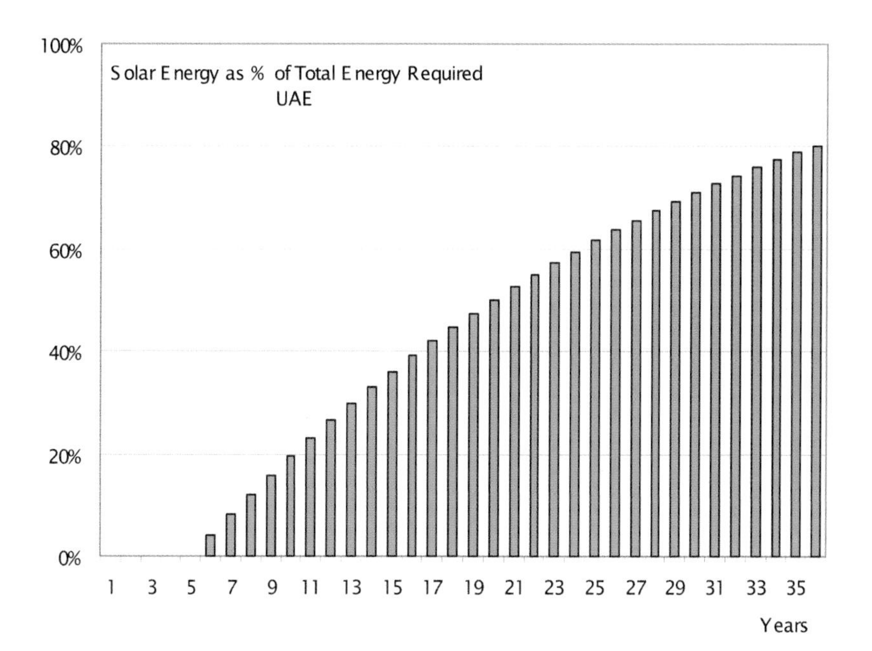

Figure 3.15 (b) Projected annual growth of solar electricity-generation percentage in the UAE for the first 36 years of a top-down VLS-PV programme, starting from 0 % during 2004–2008 (years 1 to 5) and reaching 80 % in year 36

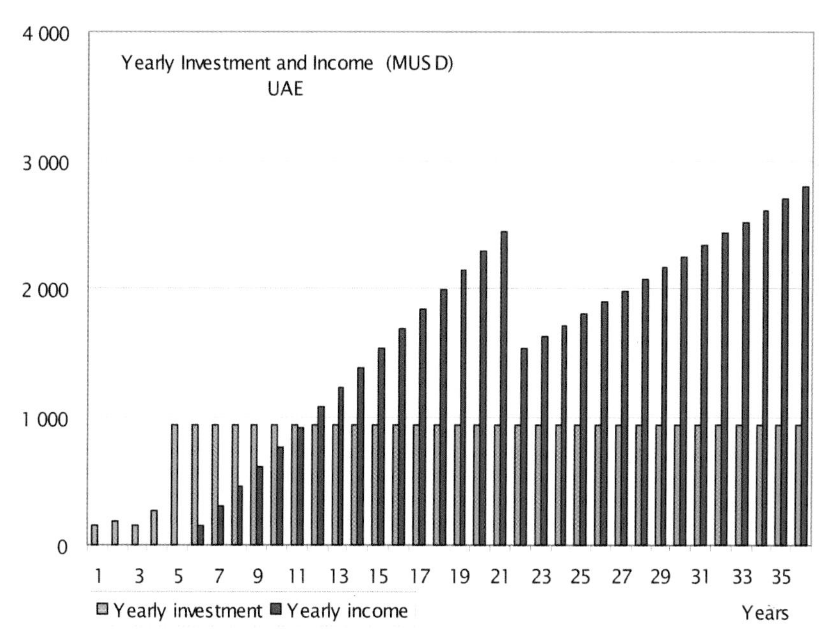

Figure 3.15 (c) Annual investments required for the top-down VLS-PV programme proposed for the UAE and subsequent income from solar plants

Note: The first four years show manufacturing facility construction costs. Subsequent years show construction costs of an annual 0,9 GW of solar plants with 0,2 GW storage. Note that income, starting in year six, enables all costs plus 3 % interest to have been paid off by year 20. The electricity price can then be dropped below the initial 9 US cents/kWh (naturally, with an accompanying temporary lowering of annual income).

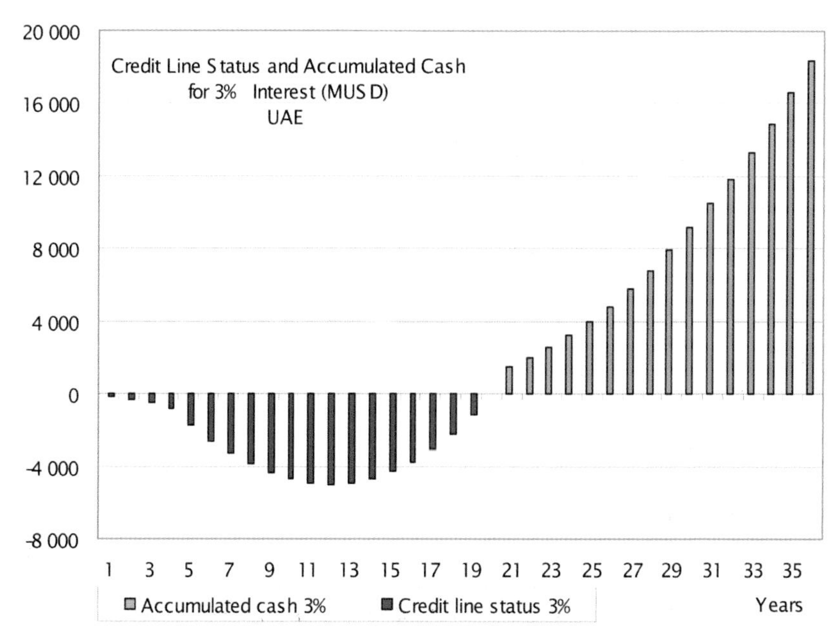

Figure 3.15 (d) Annual cash flow associated with investment, at 3 % interest, in a top-down VLS-PV programme in the UAE

Note: The credit line reaches its maximum in year 12 and decreases thereafter due to steadily increasing revenue from solar plants. The debt is entirely paid off in year 20. Subsequent revenue covers all further VLS-PV plants without need for further investment.

3.8.6 Iraq

The total generating capacity of Iraq in 2002 was 9,5 GW, of which 0,9 GW was hydro generated. The total electricity production that year was 34 TWh. At an assumed annual growth rate of 2,8 %, which is equal to the 2003 population growth rate, the electricity requirements after 36 years will reach 91,9 TWh per year.

Our proposal is to annually install VLS-PV plants of 1,3 GW generating capacity and 0,5 GW storage capacity starting at year five of the programme. By year 36, the total installed solar generating capacity will accordingly be 40,3 GW and the total storage capacity will be 15,5 GW. The annual solar generation at that time will be 80,6 TWh. If the non-solar part remains unchanged, then the solar fraction of the country's generating capacity will be 81 % (90 % will be from renewables if the present hydroelectric part is included). Correspondingly, the percentage of annual electricity coming from solar will be 88 %.

Such a project would cover 480 km^2 of land. It would require a maximum credit line of 9 100 MUSD, which would be fully paid off, including total interest of 3 300 MUSD, after 21 years. The project would generate approximately 21 000 jobs.

Table 3.10 summarizes the details of all our economic calculations for Iraq. The graphs in Figures 3.16 (a) to (d) are the equivalent of Figures 3.3 (a) to (d) for Iraq.

Table 3.10 Expected economic benefits to Iraq of VLS-PV plant introduction during the first 36 years

Interest rate	3	%/year
Yearly added solar power	1,3	GW
Yearly added six-hour storage power	0,5	GW
Credit line capacity required for the entire project	9 139	MUSD
Interest paid	3 302	MUSD
Loan repaid after	21	Years
Total solar power installed	40,3	GW
Total storage power installed	15,5	GW
Electricity price after five years, when solar electricity sales start	9	US cents/kWh
Electricity price after 22 years, when all debts are paid off	5,5	US cents/kWh
Land area required for installation	483,6	km^2
Fraction of total national land area	0,11	%
Yearly manpower requirements for solar production	3 900	Jobs
Yearly manpower requirements for solar operation	10 075	Jobs
Yearly manpower requirements for storage production	1 500	Jobs
Yearly manpower requirements for storage operation	3 875	Jobs
Headquarters and engineering	1 209	Jobs
Total number of jobs after 36 years	20 559	Jobs

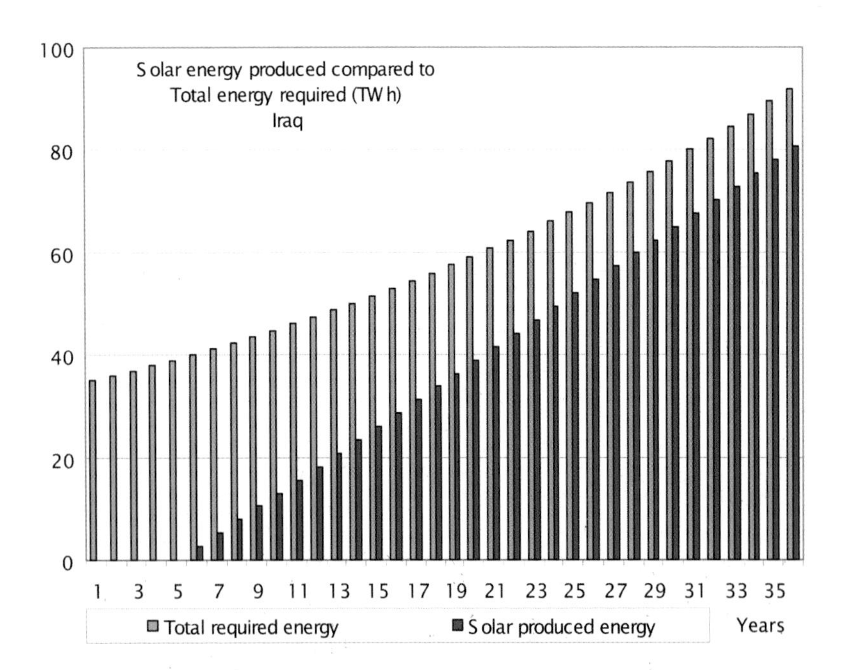

Figure 3.16 (a) Projected annual growth of electrical energy generation in Iraq for the first 36 years of a top-down VLS-PV programme, starting from 35 TWh per year in 2004 and reaching 91,9 TWh, of which 88 % is solar

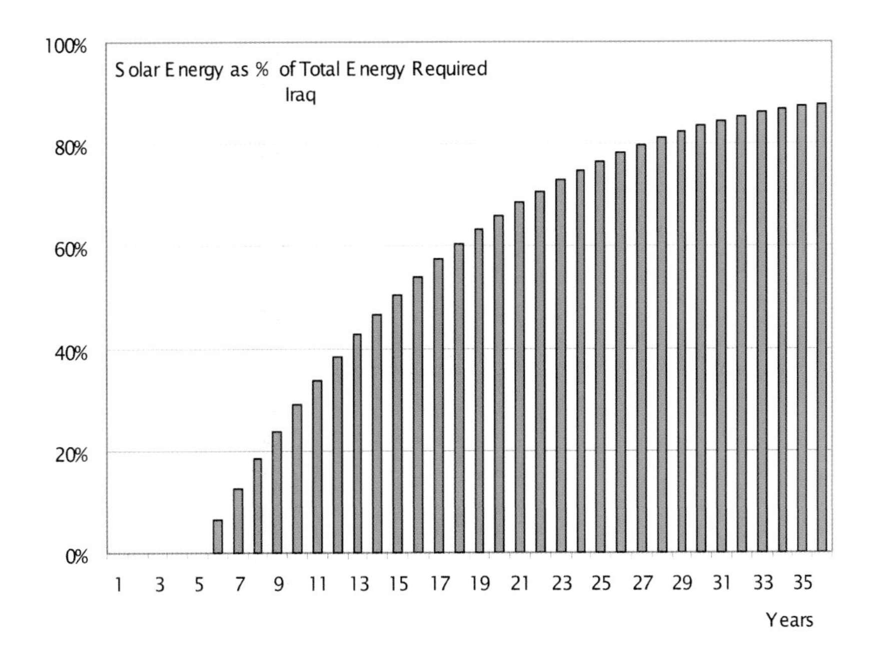

Figure 3.16 (b) Projected annual growth of solar electricity-generation percentage in Iraq for the first 36 years of a top-down VLS-PV programme, starting from 0 % during 2004–2008 (years 1 to 5) and reaching 88 % in year 36

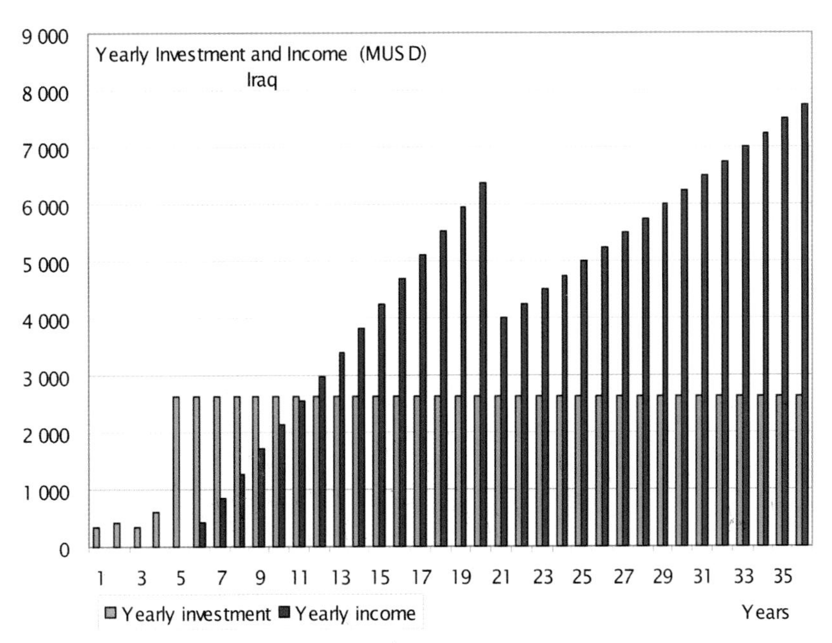

Figure 3.16 (c) Annual investments required for the top-down VLS-PV programme proposed for Iraq and subsequent income from solar plants

Note: The first four years show manufacturing facility construction costs. Subsequent years show construction costs of an annual 1,3 GW of solar plants with 0,5 GW storage. Note that income, starting in year 6, enables all costs plus 3 % interest to have been paid off by year 21. The electricity price can then be dropped below the initial 9 US cents/kWh (naturally, with an accompanying temporary lowering of annual income).

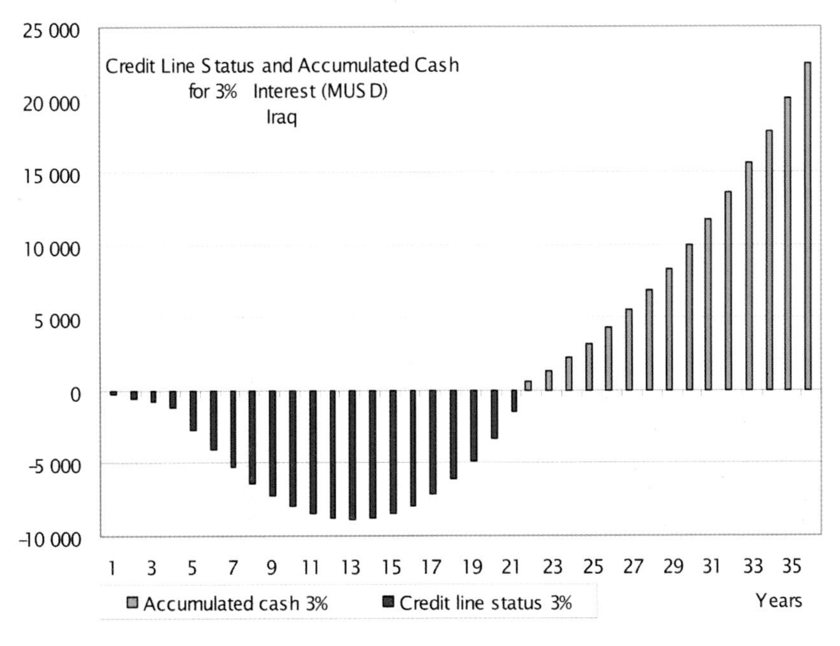

Figure 3.16 (d) Annual cash flow associated with investment, at 3 % interest, in a top-down VLS-PV programme in Iraq

Note: Credit line reaches its maximum in year 13 and decreases thereafter due to steadily increasing revenue from solar plants. The debt is entirely paid off in year 21. Subsequent revenue covers all further VLS-PV plants without need for further investment.

3.8.7 Kuwait

The total generating capacity of Kuwait in 2002 was 9,4 GW, of which 100 % was fossil generated. The total electricity production that year was 32,4 TWh. At an assumed annual growth rate of 3,3 %, which is equal to the 2003 population growth rate, the electricity requirements after 36 years will reach 104,3 TWh per year.

Our proposal is to annually install VLS-PV plants of 1,3 GW generating capacity and 450 MW storage capacity starting at year five of the programme. By year 36, the total installed solar generating capacity will accordingly be 40,3 GW and the total storage capacity will be 14,0 GW. The annual solar generation at that time will be 80,6 TWh. If the non-solar part remains unchanged, then the solar fraction of the country's generating capacity will be 81 %. Correspondingly, the percentage of annual electricity coming from solar will be 77 %.

Such a project would cover 480 km² of land. It would require a maximum credit line of 8 700 MUSD, which would be fully paid off, including total interest of 3 100 MUSD, after 21 years. The project would generate approximately 20 000 jobs.

Table 3.11 summarizes the details of all our economic calculations for Kuwait. The graphs in Figures 3.17 (a) to (d) are the equivalent of Figures 3.3 (a) to (d) for Kuwait.

Table 3.11 Expected economic benefits to Kuwait of VLS-PV plant introduction during the first 36 years

Interest rate	3	%/year
Yearly added solar power	1,3	GW
Yearly added six-hour storage power	0,45	GW
Credit line capacity required for the entire project	8 680	MUSD
Interest paid	3 065	MUSD
Loan repaid after	21	Years
Total solar power installed	40,3	GW
Total storage power installed	13,95	GW
Electricity price after five years, when solar electricity sales start	9	US cents/kWh
Electricity price after 22 years, when all debts are paid off	5,5	US cents/kWh
Land area required for installation	483,6	km²
Fraction of total national land area	2,7	%
Yearly manpower requirements for solar production	3 900	Jobs
Yearly manpower requirements for solar operation	10 075	Jobs
Yearly manpower requirements for storage production	1 350	Jobs
Yearly manpower requirements for storage operation	3 488	Jobs
Headquarters and engineering	1 209	Jobs
Total number of jobs after 36 years	20 022	Jobs

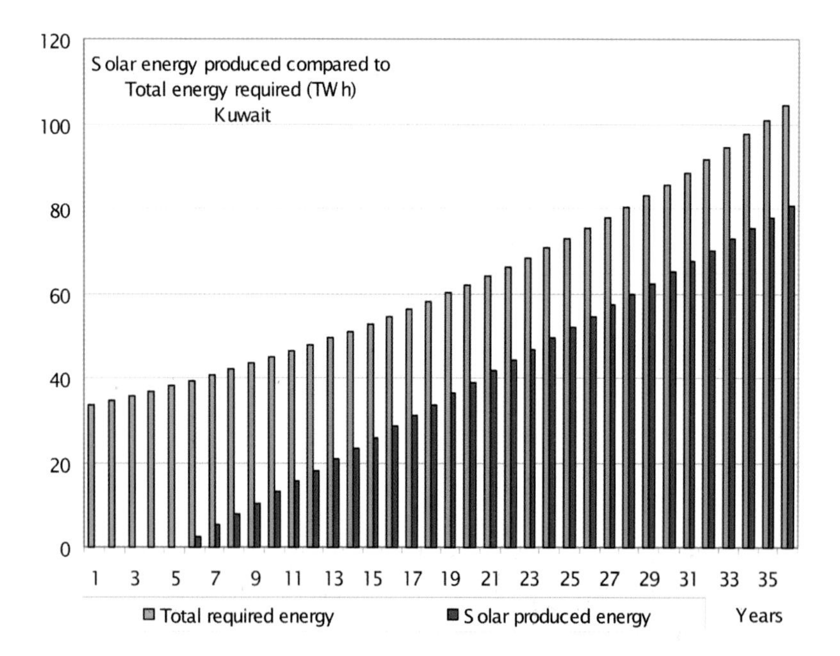

Figure 3.17 (a) Projected annual growth of electrical energy generation in Kuwait for the first 36 years of a top-down VLS-PV programme, starting from 33,5 TWh per year in 2004 and reaching 104,3 TWh, of which 77 % is solar

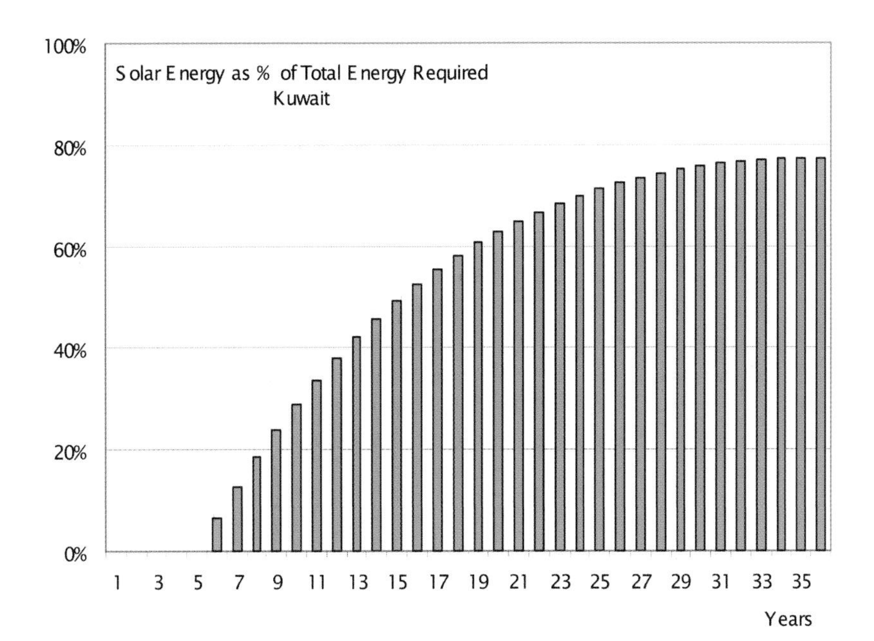

Figure 3.17 (b) Projected annual growth of solar electricity-generation percentage in Kuwait for the first 36 years of a top-down VLS-PV programme, starting from 0 % during 2004–2008 (years 1 to 5) and reaching 77 % in year 36

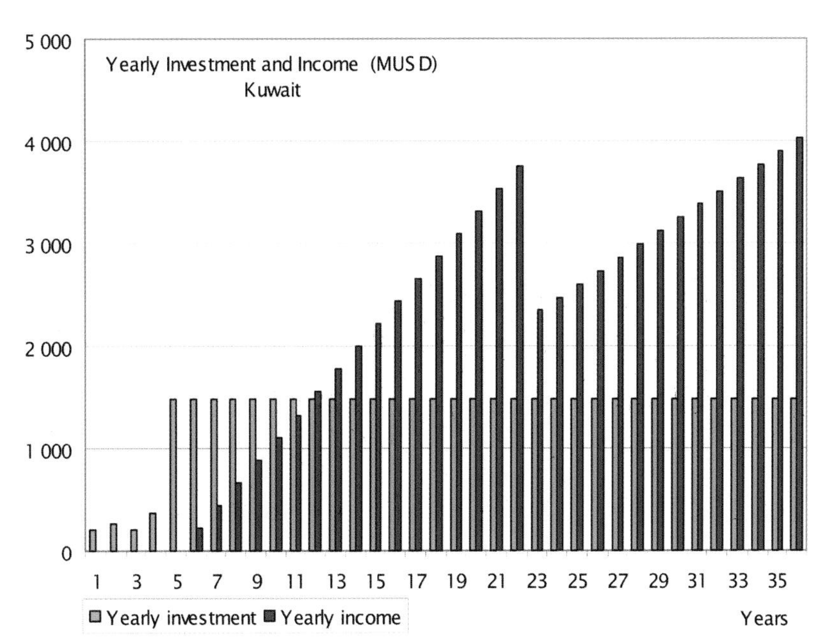

Figure 3.17 (c) Annual investments required for the top-down VLS-PV programme proposed for Kuwait and subsequent income from solar plants

Note: The first four years show manufacturing facility construction costs. Subsequent years show construction costs of an annual 1,3 GW of solar plants with 0,45 GW storage. Note that income, starting in year 6, enables all costs plus 3 % interest to have been paid off by year 21. The electricity price can then be dropped below the initial 9 US cents/kWh (naturally, with an accompanying temporary lowering of annual income).

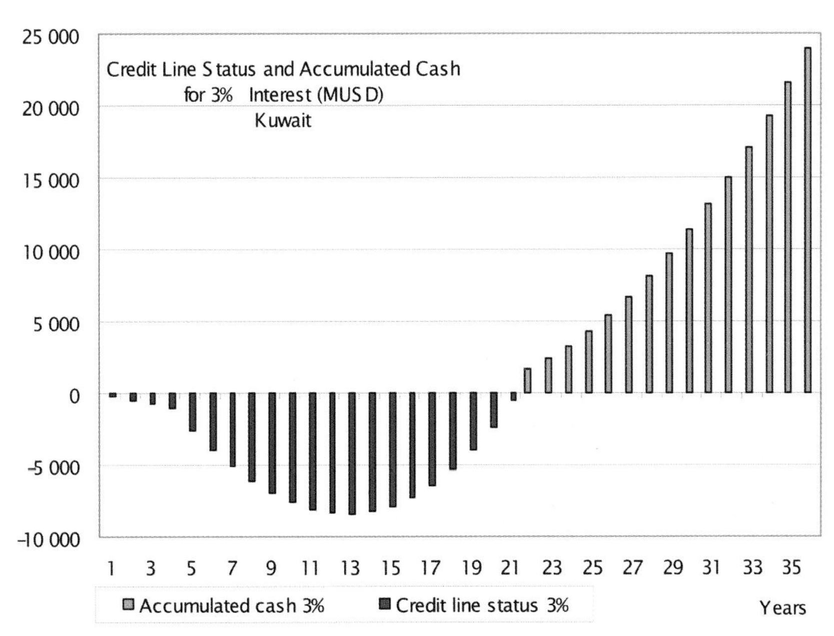

Figure 3.17 (d) Annual cash flow associated with investment, at 3 % interest, in a top-down VLS-PV programme in Kuwait

Note: Credit line reaches its maximum in year 13 and decreases thereafter due to steadily increasing revenue from solar plants. The debt is entirely paid off in year 21. Subsequent revenue covers all further VLS-PV plants without need for further investment.

3.8.8 Syria

The total generating capacity of Syria in 2002 was 7,6 GW, of which 1,5 GW was hydro generated. The total electricity production that year was 26,1 TWh. At an assumed annual growth rate of 2,5 %, which is equal to the 2003 population growth rate, the electricity requirements after 36 years will reach 63,5 TWh per year.

Our proposal is to annually install VLS-PV plants of 750 MW generating capacity and 300 MW storage capacity starting at year five of the programme. By year 36, the total installed solar generating capacity will accordingly be 23,3 GW and the total storage capacity will be 9,3 GW. The annual solar generation at that time will be 46,5 TWh. If the non-solar part remains unchanged, then the solar fraction of the country's generating capacity will be 75 % (80 % will be from renewables if the present hydroelectric part is included). Correspondingly, the percentage of annual electricity coming from solar will be 73 %.

Such a project would cover 280 km^2 of land. It would require a maximum credit line of 5 500 MUSD, which would be fully paid off, including total interest of 2 100 MUSD, after 22 years. The project would generate approximately 12 000 jobs.

Table 3.12 summarizes the details of all our economic calculations for Syria. The graphs in Figures 3.18 (a) to (d) are the equivalent of Figures 3.3 (a) to (d) for Syria.

Table 3.12 Expected economic benefits to Syria of VLS-PV plant introduction during the first 36 years

Interest rate	3	%/year
Yearly added solar power	0,75	GW
Yearly added six-hour storage power	0,3	GW
Credit line capacity required for the entire project	5 514	MUSD
Interest paid	2 061	MUSD
Loan repaid after	22	Years
Total solar power installed	23,25	GW
Total storage power installed	9,3	GW
Electricity price after five years, when solar electricity sales start	9	US cents/kWh
Electricity price after 22 years, when all debts are paid off	5,5	US cents/kWh
Land area required for installation	279	km^2
Fraction of total national land area	0,15 %	%
Yearly manpower requirements for solar production	2 250	Jobs
Yearly manpower requirements for solar operation	5 813	Jobs
Yearly manpower requirements for storage production	900	Jobs
Yearly manpower requirements for storage operation	2 325	Jobs
Headquarters and engineering	698	Jobs
Total number of jobs after 36 years	11 985	Jobs

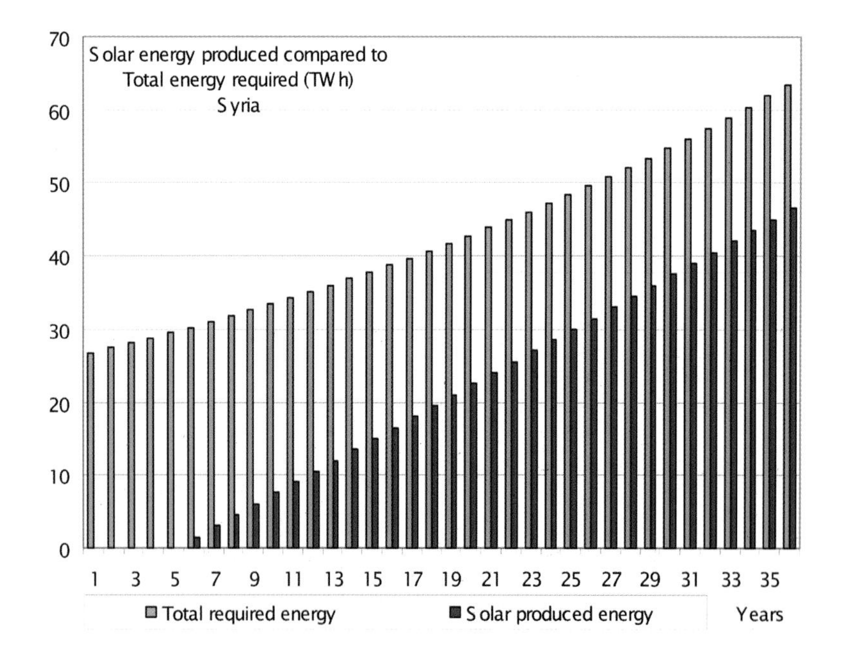

Figure 3.18(a) Projected annual growth of electrical energy generation in Syria for the first 36 years of a top-down VLS-PV programme, starting from 26,8 TWh per year in 2004 and reaching 63,5 TWh, of which 73 % is solar

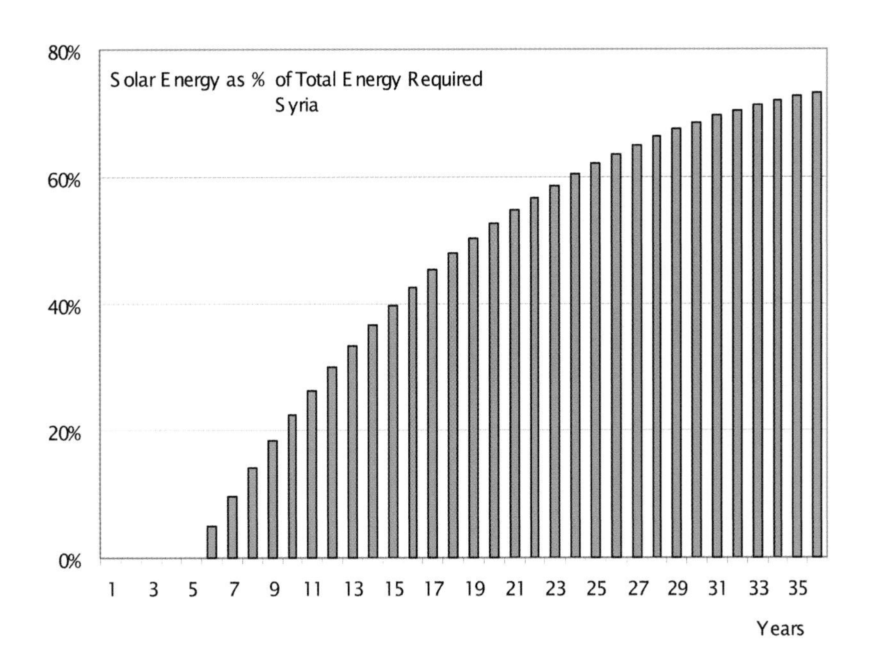

Figure 3.18 (b) Projected annual growth of solar electricity-generation percentage in Syria for the first 36 years of a top-down VLS-PV programme, starting from 0 % during 2004–2008 (years 1 to 5) and reaching 73 % in year 36

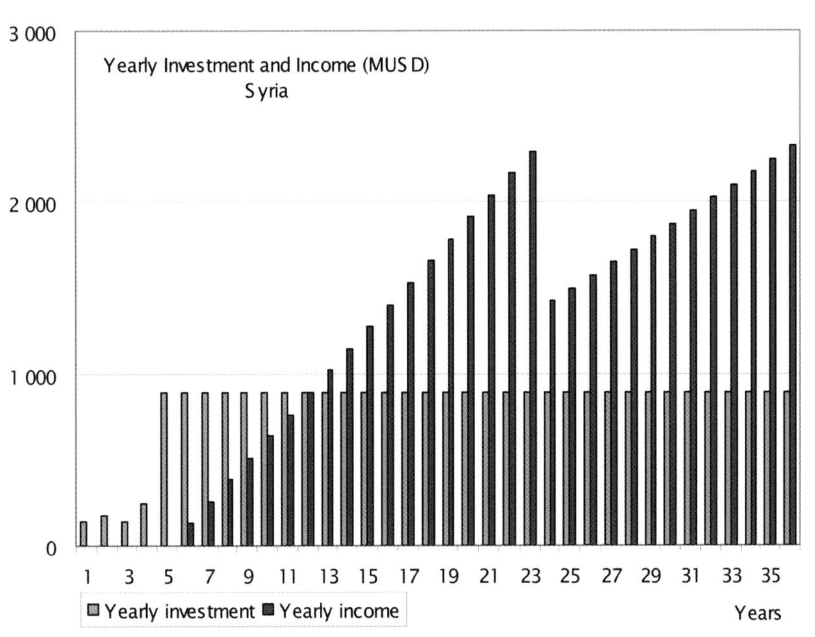

Figure 3.18 (c) Annual investments required for the top-down VLS-PV programme proposed for Syria and subsequent income from solar plants

Note: The first four years show manufacturing facility construction costs. Subsequent years show construction costs of an annual 0,75 GW of solar plants with 0,3 GW storage. Note that income, starting in year 6, enables all costs plus 3 % interest to have been paid off by year 22. The electricity price can then be dropped below the initial 9 US cents/kWh (naturally, with an accompanying temporary lowering of annual income).

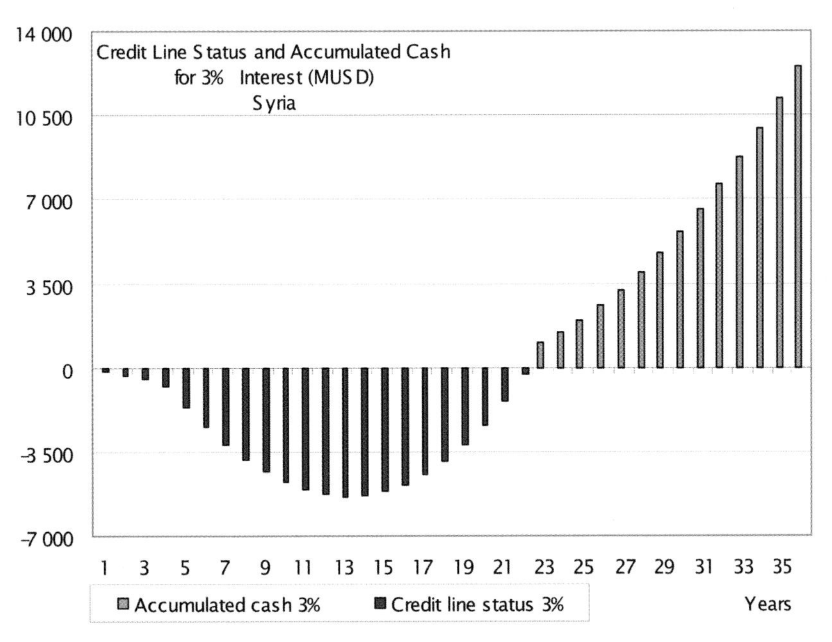

Figure 3.18 (d) Annual cash flow associated with investment, at 3 % interest, in a top-down VLS-PV programme in Syria

Note: Credit line reaches its maximum in year 13 and decreases thereafter due to steadily increasing revenue from solar plants. The debt is entirely paid off in year 22. Subsequent revenue covers all further VLS-PV plants without need for further investment.

3.9 SUMMARY AND CONCLUSIONS

The top-down approach to the introduction of VLS-PV that we have adopted for our Middle East study has consisted of four parts.

First, we studied the current electricity requirements and land availability of all countries in the region, with the specific aim of being able to provide some 80 % of their total electricity needs, from solar energy, within a time period of 36 years. For all of the major electricity producing countries it was concluded that land area considerations should present no obstacles.

Second, we studied *existing* CPV technology at the system-component level, specifically, CPV cells, collector-tracker assemblies, storage batteries and inverters, considering the expected costs involved in mass production. These costs included the VLS-PV plants and necessary mass production facilities for their manufacture. It was concluded that in Israel, VLS-PV plants would cost no more than 850 USD/kW, and that production facilities, capable of an annual throughput of 1,5 GW collectors and 0,5 GW storage, would cost approximately 1 170 MUSD.

Third, we studied the kind of investment that would be necessary to create a production facility in four years, the first VLS-PV during the fifth year, and one successive new VLS-PV plant every year thereafter. Assuming an open credit line being made available by the government or investors, at a 3 % real rate of interest, it was concluded that, in the Israeli case:

- The credit line would reach its maximum value in the 13th year.
- The maximum required credit would be equal to the cost of approximately ten fossil-fuelled plants.
- The credit-line *plus interest* would be fully paid off by electricity revenues after 21 years.
- By that time, revenues would be sufficiently high to enable both the continued annual production of VLS-PV plants with no further investment, *and* the decommissioning and replacement of old plants after 30 years of service.

It is important to point out that after the initial investment has been paid off, the price of electricity no longer depends upon any factors relating to its generation. It becomes, in effect, a purely arbitrary figure that can be fixed at any desired level. For our examples we arbitrarily fixed it at US5,5 cents/kWh. If it is deemed desirable to continue installing VLS-PV plants at the rate of one per year, though, then the price of electricity can be lowered to a figure enabling the annual net revenue from sales to precisely cover the cost of one new VLS-PV plant. Similarly, if it becomes necessary to replace old plants after 30 years of service, it is sufficient to determine the electricity price during the 29th year at a level that will cover the cost of constructing two new VLS-PV plants the following year, etc. Simple arithmetic will, in both of these examples, show that the required electricity price will be less than the 9 US cents/kWh that we adopted for our calculations.

In the fourth part of this study, we repeated the Israeli calculations for the other major electricity producers in the region, making simplifying assumptions that were specified in each case. Given uncertainties surrounding local electricity prices, labour costs and production/consumption growth rates, our results for these countries should be regarded as indicative rather than definitive. The precise assumptions we made were clearly stated so that any Middle Eastern country seriously interested in embarking upon a VLS-PV programme can redo the calculations after inserting the appropriate values of the relevant parameters.

A number of far-reaching conclusions can be drawn from the results of this study. First, VLS-PV plants can yield electricity at costs fully competitive with fossil fuel. Second, one may think in terms of typically 80 % of a country's entire electricity requirements coming from solar energy within a time period of 30 to 40 years. Third, VLS-PV plants turn out to be *triply renewable*. In addition to the normal sense in which solar is deemed to be a renewable energy, the revenues from this top-down approach are sufficient to completely finance the continued annual construction of VLS-PV plants *and* the replacement of 30-year-old VLS-PV plants with new ones *without* any further investment.

There are also firm grounds for even more optimism. First, we have assumed the use of CPV cells of 32 % nominal efficiency, leading to 25 % system efficiency. Cells with demonstrated higher efficiencies already exist, and efficiencies approaching 50 % may eventually be achieved. Successive generations of VLS-PV plants will cost less USD/kW as time proceeds, resulting in correspondingly lower electricity costs.

Second, since a top-down VLS-PV programme is self-sustaining, and no fuel costs are involved, the asymptotic cost of electricity is essentially the operation and maintenance cost of the plants, estimated at 0,5 US cents/kWh. At such low electricity costs, electrolysis would become the technology of choice for hydrogen production. This would enable the large-scale replacement of hydrocarbon transportation fuels, on a global scale, with environmentally benign hydrogen.

Third, major advances may be expected in the area of high temperature superconducting (HTSC) technology. Cost estimates made for the Negev,[5] using existing HTSC cables, suggest that the new infrastructure of a few hundred kilometres necessary to convey solar-generated power from VLS-PV plants in the Negev to the population centres in the north of the country would add approximately 0,07 US cents/kWh to the cost of electricity when averaged over 30 years. Clearly, with further advances, it will be possible for substantial parts of the less sunny regions to use electricity generated in the deserts of the world.

In conclusion, the present top-down study has provided a strong indication that VLS-PV could directly compete with fossil fuel as the principal source of electricity for any country in the Middle East, and an investment scheme has been suggested for its implementation.

REFERENCES

1 CIA (2004) *The World Factbook 2004*, ISSN 1553-8133, www.cia.gov/cia/publications/factbook

2 'Official energy statistics from the US government', 1 June 2004, www.eia.doe.gov/emeu/international/electric.html

3 Kurokawa, K. (ed) (2003) *Energy from the Desert: Feasibility of Very Large Scale Photovoltaic Power Generation (VLS-PV) Systems*, James and James, London, p149

4 Faiman, D. (2003) 'Large area concentrators', *Proceedings of the Second Workshop: The Path to Ultra-High Efficient Photovoltaics* (ed H. Ossenbrink and A. Jaeger-Waldau), 3–4 October 2002, JRC-Ispra, Italy, pp110–117

5 Raviv, D. and Rosenstreich, R. (2004) 'A paradigm change in replacing fossil fuel with solar energy', *Proceedings of the 12th Sdeh Boker Symposium on Solar Electricity Production* (ed D. Faiman), 23–24 February, Ben-Gurion University Publication No DSEEP 04/25), pp45–61

6 Faiman, D, Feuermann, D., Ibbetson, P., Zemel, A., Ianetz, A., Liubansky, V. and Setter, I. (2001) *Data Processing for the Negev Radiation Survey: Seventh Year: Part 3 – Typical Meteorological Year v 3.0*, Israel Ministry of Energy and Infrastructure Publication RD-13-01, Jerusalem, December 2001

7 NREL (2004) *Solar Energy Technologies Program: Multi-year Technical Plan 2003–2007 and Beyond*, NREL Report No DOE/GO-102004-1775, January 2004

8 Green, M. A., Emery, K., King, D. L., Igari, S. and Warta, W. (2004) 'Solar cell efficiency tables (version 23)', *Progress in Photovoltaics*, vol 12, pp55–62

9 Israel Electric Corporation (2004) *Annual Report for 2004*, Israel Electric Corporation

10 Hayden, H., Johnston, P., Garboushian, V. and Roubideaux, D. (2002) 'APS installation and operation of 300 kW of Amonix high concentration PV systems', *Proceedings of the 29th IEEE-PVSC*, New Orleans, LA, 19–24 May, pp1362–1365

CHAPTER FOUR

The Asian region: Project proposals of VLS-PV on the Gobi Desert

4.1 DEMONSTRATIVE RESEARCH PROJECT FOR VLS-PV IN THE GOBI DESERT OF MONGOLIA

4.1.1 Introduction

Research and development (R&D) on photovoltaics (PV) and governmental support for deploying PV systems are active on a global scale. So far, PV systems are only complementary power sources for existing large scale centralized power plants due to the unstable output and expensive generation cost of PV systems. PV should become a main energy source over the long range, considering the future cost reduction in PV modules from further R&D and market growth, as well as global energy and environmental issues. At the same time, future increase in consumption due to economic growth is expected in emerging countries, especially in Asia.

Rich solar energy can obviously be found in deserts and arid land. Introducing PV systems in these areas may enable supplying environmentally friendly energy to people living in severe conditions, and changing desert into productive land by using PV power. Success in large scale PV application in desert areas, combined with creating an appropriate business model and well-organized capacity-building, will lead to its widespread use. Consequently, in the future, PV will evolve into very large scale photovoltaic power generation (VLS-PV) networks that can contribute to both global energy and environmental issues.

Mongolia has a vast Gobi Desert area in the south and south-east. There are two types of electricity users in Mongolia: nomadic families and users through transmission networks. While electrification using PV for nomadic families has been implemented, an existing electricity network now supports the Mongolian economy.

In the vast Mongolian plateau, coal, copper, lead and rare metals such as gold and silver have been discovered and some have been developed for mining. Coal, in particular, has been mined through the ages for domestic electricity and heat, and copper for export. Erdenet Copper Mines, the largest open-pit copper mine in Asia, has over 500 million tonnes of copper. There are abundant mineral resources in the Gobi Desert area. Major deposits of copper (Cu) and gold (Au) were recently discovered in Oyu Tolgoi and Tsagaansuvraga in south and south-east Mongolia. Their development is possible with the development of a reliable energy supply to these sites. The mining sites are located in very remote areas and a significant amount of new infrastructure, especially electricity and water supply, as well as rail transportation systems, will have to be established for the copper and gold to reach current marks. It is expected that such development will be valuable under the national policy of Mongolia. Large copper and gold mines will be established at Oyu Tolgoi and Tsagaansuvraga, resulting in the consumption of electricity for the Gobi Desert area growing quickly.

Electricity networks (transmission lines) have only been constructed in specific regions such as those centring on Ulaanbaatar, the capital of Mongolia. The transmission lines have been essentially constructed along a railway connecting Atlanbulug, on the northern border, with Zamiin Uud, on the south-eastern border, through Ulaanbaatar. The railway plays a very important role in the Mongolian economy. Therefore, these areas along the railway and transmission lines are expected to be developed further in the future. However, electricity for these areas is generated by coal at Ulaanbaatar, polluting the surrounding air. Installing large scale carbon-free renewable electricity such as VLS-PV systems could contribute to protecting against air pollution and supporting regional development.

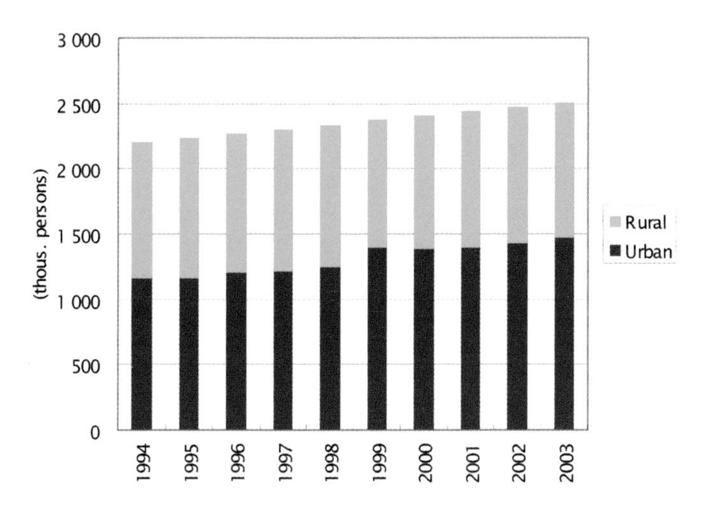

Figure 4.1 Trends in population by area[1]

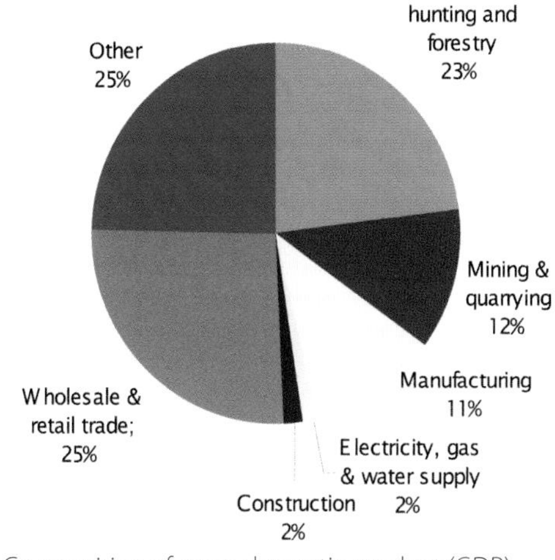

Figure 4.2 Composition of gross domestic product (GDP) by industry[1]

4.1.2 General information on Mongolia and specified areas

Mongolia is landlocked in the highlands of the Central Asian continent, with a total land area of 1 564 million km^2, and the greater part of the country is grass and arid land. The terrain consists of vast semi-desert and desert plains, mountains in the west and south-west, and the Gobi Desert in the south and south east. The climate is desert, continental (large daily and seasonal temperature ranges), and cool and dry. The precipitation is concentrated between June and August; the major part of the country has annual precipitation of less than 400 mm. The north has more precipitation, while the southern Gobi Desert only has below 50 mm – the precipitation differs considerably from region to region.

The population was approximately 2,5 million in 2003, with approximately 1,5 million in urban areas (see Figure 4.1), especially 0,9 million in Ulaanbaatar. The main industries based on gross domestic product (GDP) are wholesale and retail trade; agriculture, hunting and forestry; transport, storage and communication; mining and quarrying; and manufacturing (see Figure 4.2).

Figure 4.3 and Table 4.1 are showing existing power plants in Mongolia. Figure 4.4 is showing the electrification map of Mongolia. The power sector of Mongolia is currently operated by state-owned enterprises under supervision of the Ministry of Fuel and Energy. There are three main power grids: Central Energy System (CES), linking Ulaanbaatar with the iron-making city of Darkhan, the copper-mining city of Erdenet and the coal-mining city of Baganuur; East Energy System (EES), centred in Choibalsan; and Western Energy System (WES), with a constant supply from Russia. In all of these systems, state-owned enterprises engage in the power and heat supply business. Rural areas not connected with these systems have publicly operated diesel power stations in *aimags* (provinces) and *soums* (villages). The above three power grids are not interconnected.

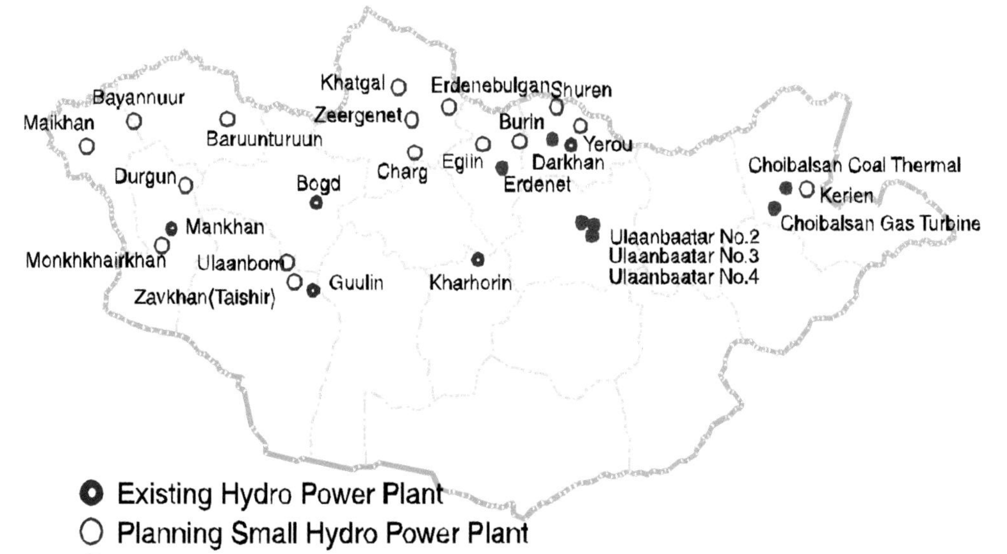

Figure 4.3 Location of the existing power plants[3]

Table 4.1 Existing power plants in Mongolia (as of 2000)[4]

Item	Power station number	Rated power (MW)	Interconnection with Russia
CES thermal (coal fired)	5	796,0	Connected when peak supply
Ulaanbaatar	No 2	24,0	
Ulaanbaatar	No 3	148,0	
Ulaanbaatar	No 4	540,0	
Darkhan		48,0	
Erdenet		36,0	
WES thermal (diesel for emergency)	3	27,6	Connected for all power supply
EES thermal (coal fired)	1	36,0	Independent
Province diesel	14	91,4	Independent
Mini-hydro	5	3,1	Independent
Total	28	954,1	

CES, the largest power grid, has thermal power plants No 2 (24 MW), No 3 (148 MW) and No 4 (540 MW) in Ulaanbaatar, supplying power and heat. Darkhan has Darkhan Power Plant (48 MW), and Erdenet has Erdenet Power Plant (36 MW). The total installed capacity of CES is 796 MW. EES has Choibalsan Power Plant (36 MW), while WES is supplied with electricity from Russia, keeping a diesel power plant on standby for emergency. The total installed capacity of Mongolia is 960,1 MW including Dalandzadgad Power Plant (6 MW), commissioned in 2000 in the southern Gobi region.

All of the power plants on the three main power grids are coal fired using domestic coal. Heavy oil used for starting up power plants and diesel oil used for diesel power plants come entirely from Russia. As coal-fired power plants have a limited ability to match the loads, CES imports electricity from Russia to meet peak demand.

At present, the most important energy resource in Mongolia is coal, and coal consumption is increasing year by year. While Ulaanbaatar consumes large amounts of coal, the mines are far from Ulaanbaatar and most coal is transported by railway. Most electricity is generated in Ulaanbaatar by coal. The electricity is delivered to some areas far from Ulaanbaatar and the delivery distance is likely to be over 1 000 km.

In 2003, the power supply amounted to 2 512 GWh compared with 2 478 GWh in 1990, when market-oriented economic reform began, taking over ten years to recover to the same level. This illustrates the economic collapse due to losing Soviet support and the withdrawal of Soviet troops, who were large consumers of energy during the communist era, which greatly affected the energy sector. Station loss and transmission/distribution loss continued to be large, which is probably due to ageing facilities and power theft, and, in rural areas, to an imbalance between the phases of power distribution. Incidentally, the transmission/distribution loss of the central grid amounted to 21,9 % in 2003 and the loss in rural areas was 40 to 50 %, which is extremely high.

According to the master plan developed with the assistance of the Asian Development Bank,[6] demand is scheduled to increase at an annual average growth rate of 2,9 % between 2001 and 2020 (see Table 4.3).

As mentioned earlier, the railway is part of the most important infrastructure in Mongolia and most transportation is via railway (see Figures 4.6 to 4.8). The railway has been constructed from Atlanbulug to Zamiin Uud, going through Ulaanbaatar. There are some cities in this area that are expected to develop as core sites of economic activity of the Mongolian south-

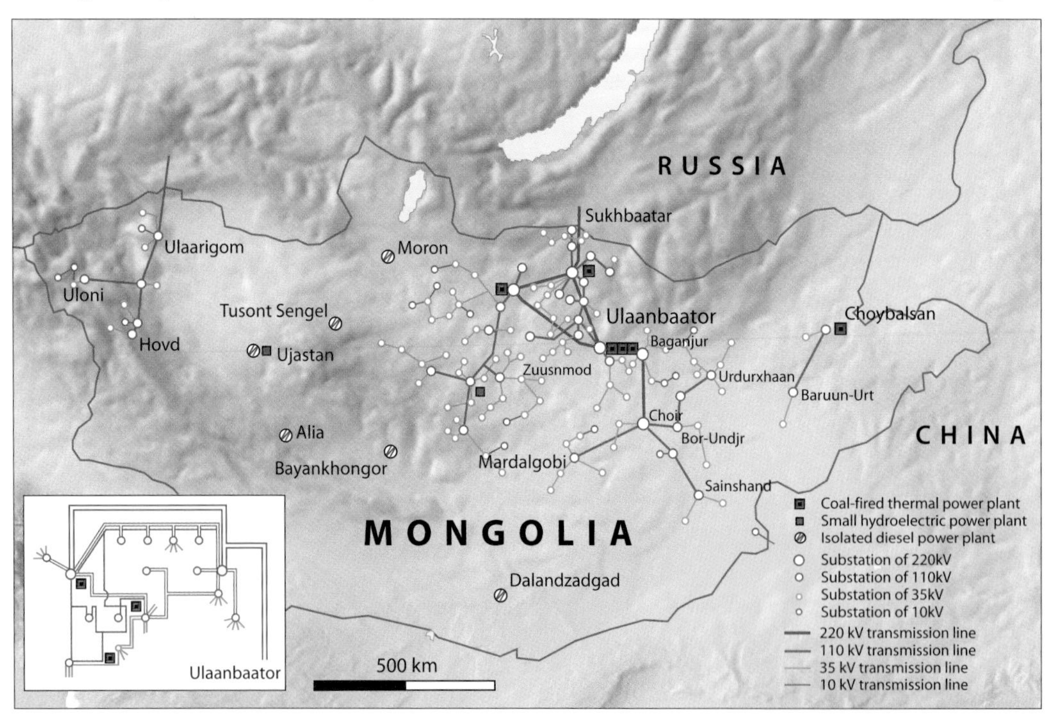

Figure 4.4 Electrification map of Mongolia[5]

Table 4.2 Trends in electricity balance in Mongolia (GWh)[1,2]

	Total	Gross generation	Import	Total consumption	Losses in transmission and distribution	Internal station use	Export
1994	2 930	2 715	215	1 861	472	597	
1995	3 009	2 628	381	1 909	502	598	
1996	2 997	2 614	383	1 936	482	579	
1997	3 096	2 720	376	1 939	507	608	42
1998	3 042	2 675	367	1 929	465	588	60
1999	3 045	2 842	203	1 867	509	610	59
2000	3 127	2 946	181	1 910	576	616	25
2001	3 213	3 017	196	1 948	603	644	18
2002	3 279	3 112	167	2 032	583	649	16
2003	3 309	3 138	171	2 195	489	618	7

Table 4.3 Electricity load demand 2005–2020 (GWh)[6]

	2000	2005	2010	2015	2020
Central Energy System (CES)					
Ulaanbaatar	915,2	979,3	1 124,2	1 338,6	1 626,4
Outside Ulaanbaatar	703,4	761,5	852,0	941,8	1 015,5
Erdenet	772,1	823,6	878,6	937,2	999,7
Station use	618,0	641,1	713,7	804,4	910,4
Total CES	2 849,7	3 205,5	3 568,4	4 021,9	4 552,0
Peak load (MW)	526,0	571,2	638,9	724,3	824,5
Western Energy System (WES)	22,7	25,3	28,2	31,8	36,0
Peak load (MW)	8,1	9,0	10,0	11,3	12,8
Eastern Energy System (EES)	58,7	65,5	73,0	82,3	93,1
Peak load (MW)	11,0	12,3	13,7	15,4	17,4
Total	2 931,1	3 296,4	3 669,6	4 136,0	4 681,1

ern and south-east region (see Table 4.4). Atlanbulug is located on the northern border between Russia and Mongolia in a free trade area. Sainshand is the biggest city in this area, and the main industry is livestock and trade. Zamiin Uud is located on the south-eastern border, facing China. Although the population is small compared with other cities, Zamiin Uud will develop as a second free-trade area in the near future. In some cities, coal is a main industry. Coal mined here has been carried to Ulaanbaatar and used for power plants.

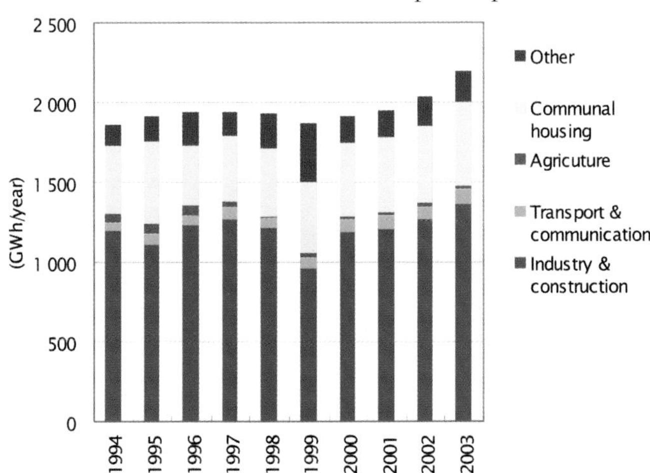

Figure 4.5 Trends in electricity consumption by sector[1,2]

4.1.3 Overview of the project proposed

This project has been developed as a conceptual design and used for a trial calculation of the power generation and construction costs for VLS-PV systems in world deserts. The case study option results for VLS-PV in desert areas showed that the Gobi Desert area of Mongolia is one of the most promising candidate sites for introducing the 100 MW class VLS-PV specified by the International Energy Agency (IEA) PVPS Programme Task 8 study.[9] However, the harsh meteorological environment of the Gobi Desert might seriously affect the PV system's performance and design specifications.

The VLS-PV scheme is a project that has not been carried out before. In order to achieve VLS-PV, a sustainable development scheme will be required. We

Table 4.4 Population and industry of cities along the railway[7]

Cities along a railway	Population	Industry
1 Sainshand	18 290	Trade, livestock
2 Choir	13 026	Coal, livestock
3 Zamiin Uud	5 486	Trade, livestock
4 Baganuur	17 120	Coal
5 Mandalgobi	14 517	Livestock
6 Bor-Undur	6 406	Coal

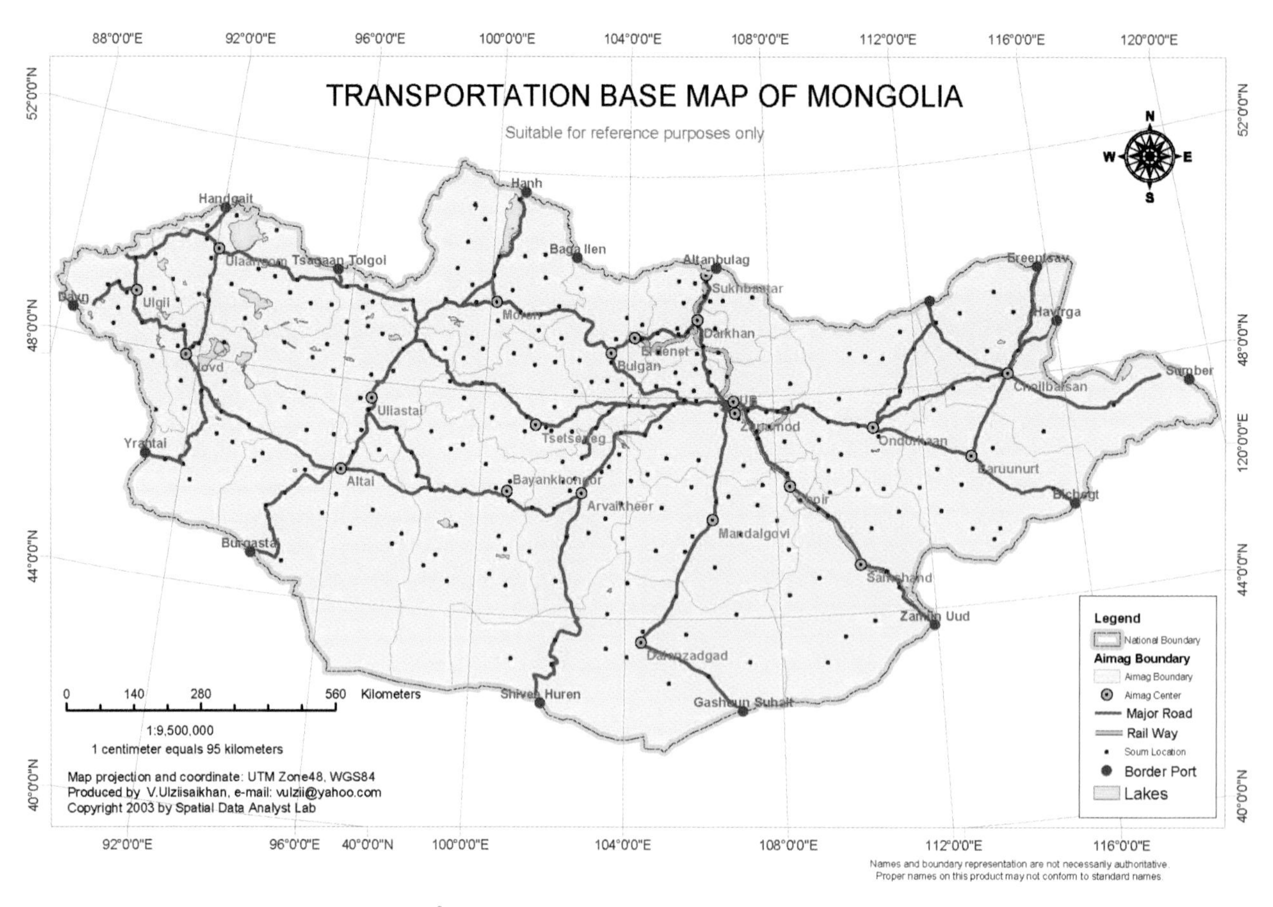

Figure 4.6 Transportation base map of Mongolia[8]

have proposed a step-by-step approach to achieving VLS-PV.[9] In analysing possible approaches and basic concepts for VLS-PV development, the following four stages were suggested by the IEA PVPS Task 8 study.[9] There are many technical and non-technical aspects that should be considered in each stage:

1 S-0 stage: R&D stage;
2 S-1 stage: pilot stage;
3 S-2 stage: demonstration stage;
4 S-3 stage: deployment stage.

We propose a demonstrative research project in the areas along the railway and discuss a future possibility for VLS-PV in the Gobi Desert, in Mongolia. The proposed project will include three phases as follows (see Table 4.5 and Figure 4.9). The potential sites in the Gobi Desert area along the railway were identified using long-term meteorological observation data conducted over the last 30 years. Grid access, as well as favourable market, economic, climatic and weather conditions, prevail in southern Mongolia – hence the choice of the candidate sites for the development of the VLS-PV system in the Gobi desert.

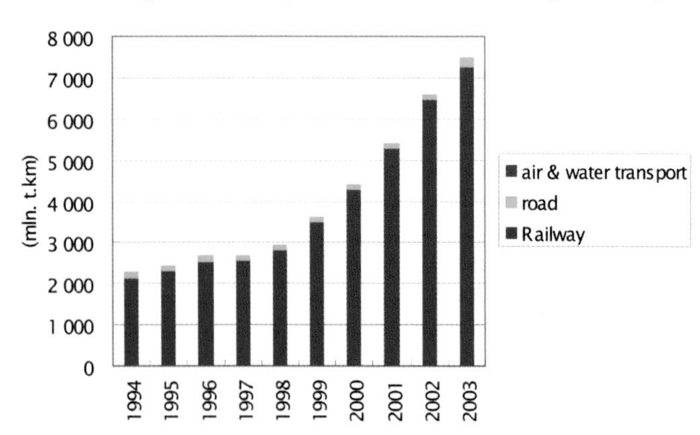

Figure 4.7 Trends in amounts of freight transportation (t.km)[1,2]

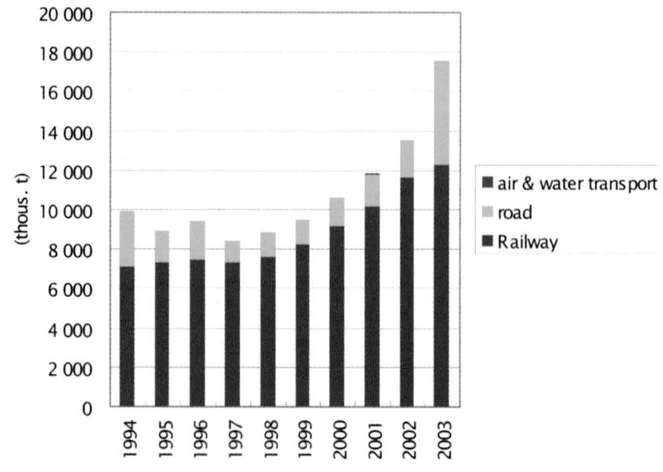

Figure 4.8 Trends in amounts of freight transportation (t)[1,2]

Table 4.5 Proposed projects of VLS-PV development in Mongolia

	Location	Capacity	Period	Demands
First phase: R&D/pilot stage	Sainshand	1 MW	2006–	Households and public welfare (significant level compared to the peak demand and electricity usage in Sainshand city)
Second phase: demonstration stage	Four sites along the railway: 1 Sainshand 2 Zumiin Uud 3 Choir 4 Bor-Undur	10 MW/site (total: 40 MW)	2010–	Industry (surpasses the peak demand and almost equivalent to electricity usage around these locations)
Third phase: deployment stage	Five sites along the railway: 1 Sainshand 2 Zumiin Uud 3 Choir 4 Bor-Undur 5 Mandalgobi One site between Oyu Tolgoi and Tsagaansuvrage	100 MW/site (sub-total: 500 MW) and 500 MW (total: 1 GW)	2020–	Power supply (almost double the peak demand and significant level compared to electricity usage in Mongolia)

Outline of the first phase

It is expected that the first phase will start in 2006 and take four to five years. The project site will be Sainshand and the capacity of the PV system will be 1 MW. The assumed demands are households and public welfare needs in the region. The project has benefits beyond electricity. Apart from creating jobs and employment, the tourism industry will also benefit.

This phase is a pilot phase to research and find out what a suitable PV system is in the installation

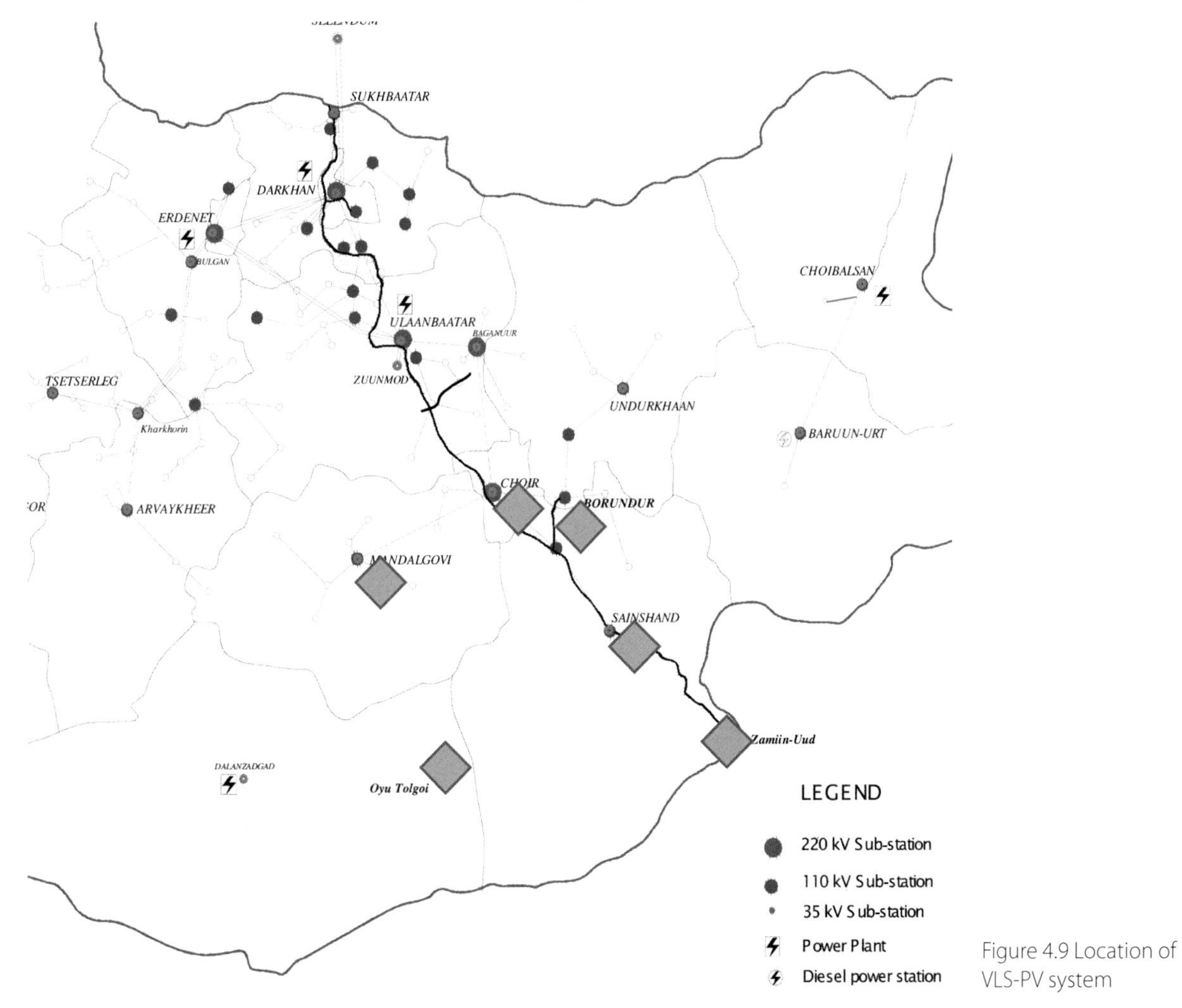

Figure 4.9 Location of VLS-PV system

Figure 4.10 Image of PV system layout of the first phase

environment (including operation condition), to develop operation and maintenance (O&M) techniques of large scale PV systems, and to specify items for the demonstration phase and capacity-building for VLS-PV development in the region (see Figure 4.10). The capacity of the pilot plant is approaching a significant level compared to the peak demand and electricity usage in Sainshand. As specific technological items, other factors such as micro-grid/distribution system, energy storage and special concerns about cold desert climate are considered. Detailed R&D items for the first stage are described in section 4.1.4 'Preliminary demonstrative research project for VLS-PV in Mongolia'.

Collaborative research has been carried out for a few years between the Tokyo University of Agriculture and Technology (TUAT) and the National Institute of Advanced Industrial Science and Technology (AIST) in Japan and the National University of Mongolia. The results of the collaborative research will be very useful for the first phase of the project and will serve as groundwork for VLS-PV system design specifications.

The project cost of the first phase is estimated at 8 MUSD. There is still work to be done to optimize construction materials in order to minimize capital costs. Since this phase is an R&D stage to research and find out a suitable PV system for the harsh desert environment, the cost may be very high because some additional measurement equipment will be involved to monitor technical data. However, an interest-free loan to demonstrate the PV system and support developing countries will be expected to apply to this. It is thought that calculating the generation cost is not that important at this phase. However, if assuming no interest, 30 years' depreciation, 0,8 performance ratio, and 0,5 %/year degradation rate, the generation cost is calculated to be 17–21 US cents/kWh.

The Law for the Promotion of Renewable Energy Utilization and the National Renewable Energy Programme have recently been drafted and submitted to the government for approval by the parliament. Final approval of these two documents will provide a positive effect associated with taxes and other funding that will assist in developing the VLS-PV systems.

Outline of the second phase

Based on the results from the first phase, the project will shift to the second phase: the demonstration stage of the VLS-PV in the Gobi Desert.

In this stage, 10 MW PV systems will be installed in Sainshand, Zumiin Uud, Choir and Bor-Undur along the railway line. These sites are important cities and the scale is classified as medium to large scale in Mongolia. The total capacity of PV systems installed will reach 40 MW, and the demands assumed are to supply industry sectors, such as mining, located near the sites.

The second phase is a demonstration stage to develop PV system design specifications and O&M techniques of VLS-PV systems and to control grid connection with existing transmission lines. Averaged annual electricity production is expected to be 54 to 66 GWh/year, which corresponds to 2,4 to 3,0 % of the total electricity consumption provided through the utility grid network and to 1,6 to 2,0 % of the total amount of electricity generated by power plants. The plant capacity will surpass the peak demand and is almost equivalent to electricity usage assumed around the locations. It is thought that the rate may be suitable for demonstrative research on achieving VLS-PV, which has a capacity of 10 MW up to several GW. However, the electricity transmission system and energy storage should also be considered. Figure 4.11 provides an example of a demonstration site for a large scale PV system with batteries, which has been carried out at AIST, Japan.

For the second phase, it is intended that PV modules

Characteristic	NaS	Redox
Capacity (kW)	2 000	170
Storage Time (h)	7,2	8
Energy Stored (kWh)	14 400	1 360
Weight (ton)	173	260

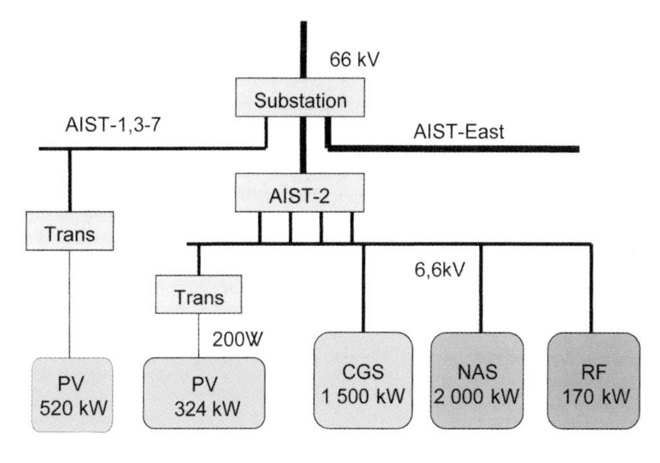

Figure 4.11 2 MW sodium-sulphur battery and 170 kW vanadium redox flow battery with 1 MW PV facilities at the National Institute of Advanced Industrial Science and Technology (AIST), Japan[10]

can be manufactured locally in Mongolia and be used as part of 10 MW (40 MW total) PV systems. At present, there is one PV manufacturing facility owned by the government of Mongolia and the production capacity of the facility is 1,5 MW/year; but this project may become the firebrand for the capacity expansion. In the next phase, PV module manufacturing facilities will be constructed and operated near the site where VLS-PV will be installed. VLS-PV development schemes, including social development, will be designed in detail in this phase.

The second phase will start in 2010. It is expected that by 2010 PV system cost will decrease to 3 USD to 4 USD/W. The total investment for the second phase, including construction and O&M, is forecast to be 192 USD to 256 MUSD. When assuming a 2 % interest rate, 30 years' depreciation, 1,4 % property tax, 0,85 performance ratio, and 0,5 %/year degradation rate, the generation cost is calculated to be 9,7 to 15,9 US cents/kWh. The cost will be quite high compared with the electricity tariff in Mongolia, which was approximately 4 US cents/kWh in 2003.

However, there is a possibility of introducing some incentives and added values to VLS-PV. Considering the large losses and internal use at power plants, the reduction in coal consumption will be greater than that corresponding to power consumption. This means that electricity generated by VLS-PV provides an environmental role in reducing carbon dioxide (CO_2) emissions and in saving coal. Renewable energy development is also a promising way of ensuring social development and is one of the most important policies in Mongolia. As a result, there may be a tax allowance for renewable energy. If there were, for example, a CO_2 reduction value (green certification value) and an exemption from property tax, the actual generation cost would be decreased by a considerable 5 to 10 US cents/kWh. Considering the recent trend of electricity price hikes, VLS-PV has the possibility of being economically feasible by providing additional support.

Outline of the third phase

The project will then shift to the third phase (deployment stage). In this stage, 10 MW PV systems will be augmented to a 100 MW VLS-PV system, and one new 100 MW system will be constructed in Mandalgobi. Besides these 100 MW VLS-PV systems, another 500 MW VLS-PV system will be constructed between Oyu Tolgoi and Tsagaansuvraga, which are located in Umnugobi and Dornogobi provinces.

The capacity of the plant is approaching double the peak demand, and is at a significant level compared to electricity usage in Mongolia, while estimated generation cost is similar to the generation cost of the existing plant. Required R&D items will be to make new stable loads, develop transmission systems and operate PV plants and the existing plant economically.

Mandalgobi is a middle-sized city located in the Gobi Desert, and a 220 kV transmission line has already been built from Choir. Transmission lines along the railway line should be uprated because the electricity voltage will be over the existing capacity. A PV module manufacturing facility will also be built in Sainshand.

At Oyu Tolgoi and Tsagaansuvraga, it has recently been decided to establish large copper (Cu) and gold (Au) mines. The consumption of electricity for the Cu–Au mines will grow rapidly. Oyu Tolgoi is located approximately 230 km from Dalanzadgad and 340 km from Sainshand. Tsagaansuvraga is located approximately 160 km from Sainshand and 350 km from Dalanzadgad. It is assumed that PV modules manufactured at Sainshand will be used for a 500 MW VLS-PV system. There are abundant mineral resources in the Gobi Desert and such development will be valuable under a national policy in Mongolia.

In addition to the development, there is a possibility that the railway may be extended from Sainshand to Oyu Tolgoi and Tsagaansuvraga in order to carry the mining products. Then, the railway is also available for carrying PV modules from Sainshand.

A concrete sustainable development scheme for this stage has not yet been designed; but it is intended that the capacity of the total VLS-PV system will reach 1 GW.

4.1.4 Preliminary demonstrative research project for VLS-PV in Mongolia

It is expected that the first phase will start in the near term and will take four to five years. The project site will be at Sainshand and the PV system capacity will be 1 MW. Collaborative research activity between the AIST and the TUAT in Japan and the National University of Mongolia (NUM) has also been implemented at Sainshand. It is possible to locate these projects as preliminary demonstrative research projects for VLS-PV. Based on the results obtained by these projects and related investigations, a concrete demonstration project for VLS-PV will be launched in Mongolia.

Project proposal for research and development (R&D)/pilot stage: Detailed R&D issues of the first phase

The first phase is a pilot stage to research and find a suitable PV system for the installation environment (including operation conditions), to develop O&M techniques for large scale PV systems, and to specify items for the demonstration stage and capacity-building for VLS-PV development in the region. The project site will be at Sainshand and the PV system capacity will be 1 MW.

Objectives

The objectives of the first phase will be as follows:

- Design and install 1 MW medium-sized demonstrative PV systems and their possible application, such as water pumping and purification, into existing weak networks in selected sites around the Sainshand area in the Gobi Desert.
- Find out the optimal total design of the PV system and application suitable for the Gobi Desert by analysing its operational performance and climate data.
- Investigate possible local environmental, economic and social impacts of such kinds of PV installation.
- Investigate general technical guidelines for the design and operation of the VLS-PV system and a business model for sustainable operation and future growth that can help local capacity-building.

Steps and schedule

The project will consist of the following three steps and is intended to start in 2006 and end in 2009/2010:

- Step 1: feasibility step (one year):
  - activity no 11: selection of a specific site for the project;
  - activity no 12: PV system design and its application.
- Step 2: demonstration step (three years):
  - activity no 21: demonstrative operation and evaluation of medium scale PV system (including construction);
  - activity no 22: research for environmental, economic and social impacts.
- Step 3: transition step (two years):
  - activity no 31: capacity-building in the region;
  - activity no 32: design the details of the VLS-PV development scheme.

Location

Sainshand city, located on the eastern side of the Gobi Desert in Mongolia, has been chosen as a project site for the first phase. Sainshand city is easily accessible from Ulaanbaatar, as well as from China and Russia by train (Siberian Railway).

Technical and economical description

Step 1: feasibility step (one year)

Activity no 11: selection of a specific site for the project.

Minimum requirements for the project site may be as follows:

- ground suitable for installing PV systems;
- sufficient potential for solar irradiation;

- enough habitation and electricity demand;
- available water resources (amount, level and quality of groundwater);
- accessible transportation; and
- human resources.

Based on a preparatory survey of the following items, a project site satisfying the requirements will be selected. In order to ensure sufficient data on climate, topography and soils, short-term (0,5 to 1 year) preliminary data monitoring might be recommended:

- Ambient temperature and humidity, which affects the performance and durability of system components:
  - Generally, the components have individual working temperature and humidity specifications where manufacturers guarantee their normal operations by presuming usual climate conditions. However, the climate of the Gobi Desert is expected to be harsh and demanding beyond such conditions. These data are necessary for selecting the system components, system design and output estimation.
- Irradiation and continual days of non-sunshine:
  - Irradiation data, such as an annual statistical data set of hourly global requirements, are critical. Or it might be better to obtain the individual components of the irradiation, which are beam irradiation and scattered irradiation. These data are used in order to discuss different types of PV system applications, such as fixed flat-plate, tracking and concentrator applications. The continual days without sunshine dominate the selection of the required battery capacity.
- Wind speed and direction, which are used for designing array support and foundation:
  - Long-term statistical data on maximum, average and minimum wind speeds are desirable, but those for a limited period are acceptable. Further information on sand storms should be investigated.
- Rainfall and snowfall:
  - Little rainfall is expected in desert areas, but we should pay attention to the rainfall. Snowfall is rather important when considering the latitude of the Gobi Desert.
- Soil and groundwater condition:
  - The nature of the soil also influences the design of array support and foundations. Considering possible changes in the behaviour of groundwater by land coverage with PV arrays, including salinization, information on soil conditions and groundwater levels, may be helpful. Information on the quality and level of groundwater is useful in designing groundwater pumps and water purification.

Activity no 12: PV system design and its application.
Expected PV system components are as follows. Based on information obtained from activity no 11, the total system design and cost estimation will be discussed in detail:

- PV system:
  - The PV system capacity will be assumed to be 1 MW. However, the capacity depends upon other factors, such as habitation scale, required power for irrigation and available financial resources. Nevertheless, certain land coverings by PV arrays may be needed to evaluate the shading effect.
  - Array type will be fixed flat plate and the orientation may be south. The inclination is dependent upon the site.
  - Low-cost materials may be applied for array support, such as wooden array supports and sun-dried blocks.
- Balance of system (BOS):
  - This comprises inverters (power conditioners), batteries and other electrical apparatus.
- Application:
  - This involves, for example, groundwater pumps, water purification systems, sprinklers and the tourism industry.

Step 2: demonstration step (three years)

Activity no 21: demonstrative operation and evaluation of medium scale PV systems (including construction):

- Construction of the PV system and its applications:
  - As described, locally available materials and products are to be used.
  - Irrigation water is supplied to the land area around the PV system. The effects and influences of watering the bare land surfaces, land surfaces covered by PV arrays and other elements will then be measured.
- Equipping data-monitoring systems:
  - This comprises climate data such as irradiation, temperature, wind and rainfall.
  - This involves the technological performance of PV systems, batteries and applications.
  - Here, the environmental effect is considered, such as the water balance and any change in vegetation and groundwater levels.
- Evaluation of the PV system and its application:
  - The operational performance of the PV system and its applications are evaluated by using data from observations.
  - The issue of long-term degradation of PV modules is critical for the sustainable growth of PV systems from a long-range viewpoint. For this purpose, an additional exposure facility for various types of PV modules and arrays – crystalline silicon (c-Si), polycrystalline silicon (p-Si), amorphous silicon (a-Si), copper indium

gallium diselenide (CIGS), bifacial c-Si, concentrator, etc. – is installed close to the main PV system. Current–voltage (I–V) characteristics of each PV module and array will be periodically measured and evaluated.

Table 4.6 Installed instrument for research in Sainshand

Solar-powered battery system (85 Ah)	
Meteorological equipment	Pyranometer
	Albedo meter
	Vane anemometer
	Thermometer
	Hygrometer
Tested module	BP single crystalline PV module (SP75), 75 W
	Sharp polycrystalline PV module (NE-80A1H), 80 W
Measurement device	I–V curve tracer
	Programmable data logger
	Removable data memory
	Thermocouple thermometer

Activity no 22: Research for environmental, economic and social impacts:

- Irrigation:
  - Quality of purified water: is it suitable or not?
  - Amount of water supply: is it enough, too much or too little?
  - Employed irrigation technique: is it good or bad?
  - Selection of vegetation: is it appropriate or not?
- Shading effect by PV array:
  - Comparative evaluation: compare the results of with and without water supply, and bare land surface and land surface under PV array.

Step 3: transition step (two years)

Activity no 31: capacity-building in the region:

- General guideline for the design and operation of the PV system, combined with applications such as irrigation systems:
  - required data set for system design;
  - procedure of system design;
  - common database used for system design;
  - technical operation and maintenance.
- Technology transfer: VLS-PV research lab:
  - technical research and training centre:
    - device research;
    - cell/module characterization;
    - nationwide weather observation;
    - system engineering: system performance evaluation, education course, national technical guidelines and regulation;
  - special demonstrative PV facility: aggregation of various small-scale PV installation:
    - device research;
    - different types of PV modules;
    - different types of PV systems: fixed flat plate, vertical installation with bifacial module, tracking and concentrator;
    - useful for education and tracking.

Activity no 32: incorporate detail on the VLS-PV development scheme within the design:

- Business model for sustainable operation and future growth towards VLS-PV:
  - possible management system and financial support;
  - outline and contents of the second phase.

Figure 4.12 Installed instrument

Table 4.7 Measurement parameters

Meteorological data	Characteristics of PV module
Global horizontal irradiation	Short circuit curent (I_{sc})
Global in-plane irradiation (south face 45°)	Open circuit voltage (V_{oc})
Albedo	Maximum output current (I_{pm})
Temperature	Maximum output voltage (V_{pm})
Wind condition	Temperature of back surface
Humidity	

Collaborative research between Japan and Mongolia in Sainshand

As mentioned earlier, the collaborative research between TUAT and AIST in Japan and the National University of Mongolia has been carried out for several years. This research will be very useful for the first phase of the project and will serve as groundwork for VLS-PV system design specifications. Some results will be introduced in the following section.

Objectives

The objectives of the research are as follows:

- Set up solar batteries, measurement equipment (I–V curve tracer, etc.) and meteorological observation equipment.
- Investigate the characteristics of PV systems for the desert environment.
- Verify power-generation simulation technology for achieving VLS-PV.
- Clarify the specification requirements for the system design.
- Measure the I–V characteristics of the solar panel with weather conditions such as the quantity of solar radiation and the temperature.
- Clarify real environmental characteristics in the Gobi Desert.

Installed instrument and parameters measured

For the research, the instruments shown in Table 4.6 and Figure 4.12 have been installed in the meteorological observation bureau (latitude: N44°53', longitude: E110°10') in Sainshand.

The parameters shown in Table 4.7 are measured at ten-minute intervals, and a data logger and storage memory module that can record the data for two months or more are available.

Examples of results measured

Various data and parameters have been measured and observed. However, most results are tentative because the research is still ongoing. Here, results on meteorological data and PV module performance ratio are introduced.

Meteorological data

Annual irradiation in Sainshand is approximately 4,66 kWh/m^2/day (1 700 kWh/m^2/year) on a horizontal surface and 5,82 kWh/m^2/day (2 100 kWh/m^2/year), with a tilted angle of 45° (see Table 4.8).

PV module performance ratio

Figures 4.13 and 4.14 show PV module performance ratios originally measured and those to which a correction for temperature is given. The performance ratios of PV module 1 have been exceeding 1,0 in winter and

Table 4.8 Measured meteorological data in Sainshand

	Global horizontal irradiation (kW/m^2/day)	Global in-plane irradiation (kW/m^2/day)	Albedo (%)	Average wind speed (m/s)	Averaged temperature (°C)	Humidity (%)
October 2002	3,68	5,60	0,30	3,7	2,2	47,4
November 2002	2,68	5,22	0,32	3,2	–8,4	53,8
December 2002	2,14	4,16	0,38	2,1	–18,0	64,8
January 2003	2,67	4,89	0,39	2,0	–19,4	78,6
February 2003	3,60	5,79	0,40	2,5	–13,4	75,6
March 2003	4,63	6,16	0,28	2,6	–3,4	49,7
April 2003	5,72	6,21	0,28	3,7	8,1	33,3
May 2003	6,45	6,18	0,27	3,6	15,4	34,5
June 2003	6,63	6,02	0,28	3,3	21,0	38,2
July 2003	6,40	5,85	0,27	3,0	24,3	42,2
August 2003	6,00	6,18	0,27	2,9	20,9	43,1
September 2003	5,04	6,30	0,28	2,6	15,8	47,9
October 2003	4,05	6,45	0,29	2,9	6,1	38,5
November 2003	2,56	4,88	0,32	2,9	–7,8	55,9
December 2003	2,16	4,89	0,32	3,2	–15,4	63,4
January 2004	2,45	5,11	0,30	2,7	–15,8	59,4
February 2004	3,31	5,65	0,30	4,1	–8,5	46,0
Average in 2003	4,66	5,82	0,31	3,01	0,22	51,32

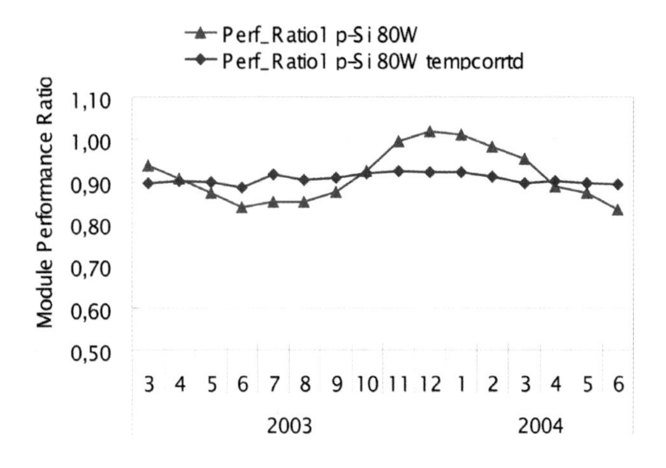

Figure 4.13 Performance ratio of PV module 1 (poly-Si)[11]

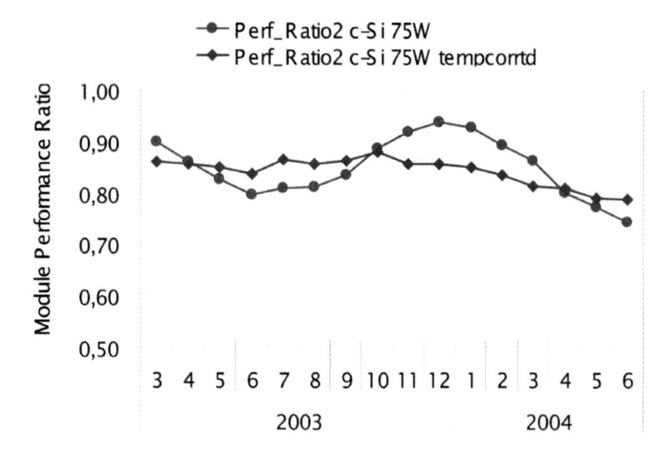

Figure 4.14 Performance ratio of PV module 2 (mono-Si)[11]

have fallen to 0,83 in summer (see Figure 4.13). The performance ratios that have been corrected for temperature are an almost constant 0,9, and the temperature is clearly the key factor for influencing the performance ratio. The performance ratios of PV module 2 show the same tendency as PV module 1, but those corrected for temperature show a tendency to fall from 0,85 (see Figure 4.14).

4.1.5 Recommendations

The VLS-PV system has a huge generating capacity, and step-by-step enlargement will be an effective way of achieving GW scale while preventing the financial, technological and environmental risks caused by rapid development. Since VLS-PV is a very large scale and long-range project that has not been implemented before, it is essential to follow the R&D perspective. In addition, the local benefits from a long-term sustainable operation of a VLS-PV system should be discussed.

The VLS-PV system may contribute both to protecting against air pollution and to supporting regional development. Apart from creating jobs and employment, the tourism industry will also benefit. Thus, the project has benefits beyond electricity, and there is a possibility of introducing some incentives and added values to VLS-PV. If a CO_2 reduction value (green certification value) and an exemption from property tax were utilized, for example, the actual generation cost would be decreased to almost the same level as existing power plants. Considering the recent trend of electricity price hikes, VLS-PV has the possibility of being economically feasible by providing additional support.

It is expected that the first phase will start in the near term, taking four to five years, and that a demonstration project for VLS-PV will be launched in Mongolia. A concrete sustainable development scheme has not yet been designed for the third phase; but it is intended that the capacity of the total VLS-PV system will reach 1 GW.

Renewable energy development is a promising way of ensuring social development and one of the most important policies in Mongolia. The Law for the Promotion of Renewable Energy Utilization and the National Renewable Energy Programme have recently been drafted and submitted to the government for the approval of parliament. Final approval of these two documents will provide a positive impact associated with taxes and other funding that will assist in developing VLS-PV systems.

REFERENCES

1. National Statistical Office of Mongolia (2004) *Mongolian Statistical Yearbook 2003–2004*, National Statistical Office of Mongolia, Ulaanbaatar, Mongolia
2. National Statistical Office of Mongolia (2004) 'Mongolia in a market system', *Statistical Yearbook 1998–2002*, National Statistical Office of Mongolia, Ulaanbaatar, Mongolia
3. Nippon Koei Co (2000) *Master Plan Study for Rural Power Supply by Renewable Energy in Mongolia*, July, JICA (Japan International Cooperation Agency), Tokyo, Japan
4. Energy Authority of Mongolia (MOID) (2000) *Energy Sector of Mongolia*, MOID, Ulaanbaatar, Mongolia
5. KfW (2001) 'Sector Programme Energy I project report', January, KfW, Frankfurt, Germany
6. ADB (2002) *Final Report for Capacity Building in Energy Planning*, ADB, March
7. National Statistical Office of Mongolia (2001) *Population and Housing Census 2000 – Statistical Booklet: Dornogobi*, National Statistical Office of Mongolia, Ulaanbaatar, Mongolia
8. Spatial Data Analyst Lab of Mongolia (2001) *Road Map of Mongolia*, Geodesy and Cartography Office of Mongolia, Ulaanbaatar, Mongolia
9. Kurokawa, K. (ed) (2003) *Energy from the Desert: Feasibility of Very Large Scale Photovoltaic Power Generation (VLS-PV) Systems*, James and James, London

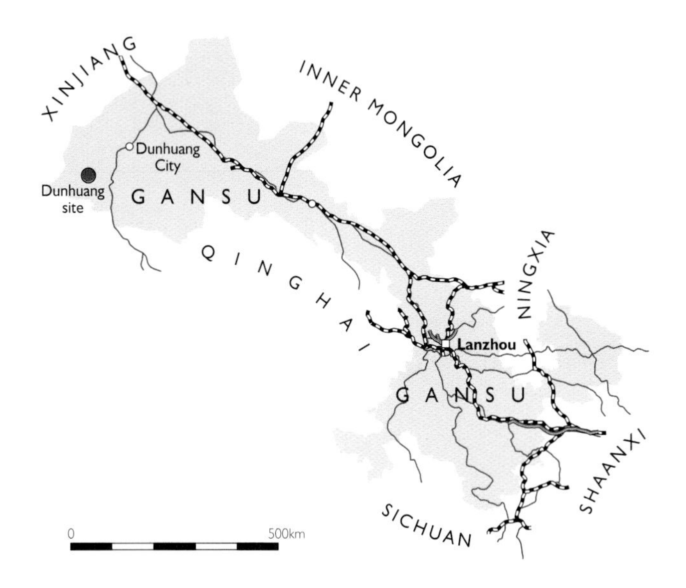

Figure 4.15 Location of Dunhuang, Gansu

Figure 4.16 Gobi area in Dunhuang

10. National Institute of Advanced Industrial Science and Technology (AIST)
11. Result of joint research projects between TUAT, AIST and NUM, 2004–2005

4.2 FEASIBILITY STUDY ON AN 8 MW LARGE SCALE PV SYSTEM IN DUNHUANG, CHINA

4.2.1 Introduction

Energy shortages and environmental pollution have become the bottleneck of social and economic development in China. Improving the current structure of the energy supply and promoting the utilization of renewable energy are effective solutions. Photovoltaic power generation is clean energy without emitting greenhouse gases. China has huge desert lands, such as the Gobi Desert, providing the possibility of PV systems with large or very large scale applications. Only large scale applications can reduce the PV cost to the level of traditional electric power.

In early 2004, supported by the World Wide Fund for Nature (WWF) and the government of Gansu Province, an expert team headed by the Electric Engineering Research Institute of Academy of Science started work on the feasibility study of large scale photovoltaic power generation (LS-PV) in the Gobi Desert. The expert team visited many potential remote places, and finally four sites were chosen as the most suitable in which to install LS-PV. Qiliying, Dunhuang, is one of the four sites and is planned to be the location for an 8 MW PV system power plant. The expert team has since started to work on the feasibility study. Experts involved in the job include PV experts from the Electric Engineering Research Institute, experts from financial agencies, experts from the Electric Power Bureau of Dunhuang, and government officers from the local Development and Reform Commission of Gansu.

An evaluation meeting for the 8 MW LS-PV feasibility study was held at Dunhuang, Gansu Province, PRC (China) from 21 to 24 September 2004. The feasibility study was well appraised by the meeting. The formal proposal of the 8 MW LS-PV project was submitted to the central government for approval by the Gansu provincial government in November 2004 after the meeting.

4.2.2 General information on the site

The Gobi area in Gansu is about 18 000 km^2. This area can be used to build a 500 GW VLS-PV, which is more than the entire power capacity in China today. The targeted place for 8 MW LS-PV in the Gobi Desert is at Qiliying, 13 km away from Dunhuang city. The latitude of the place is N40°39', with a longitude of E94°31' and an elevation of 1200 m. It is only 5 km from

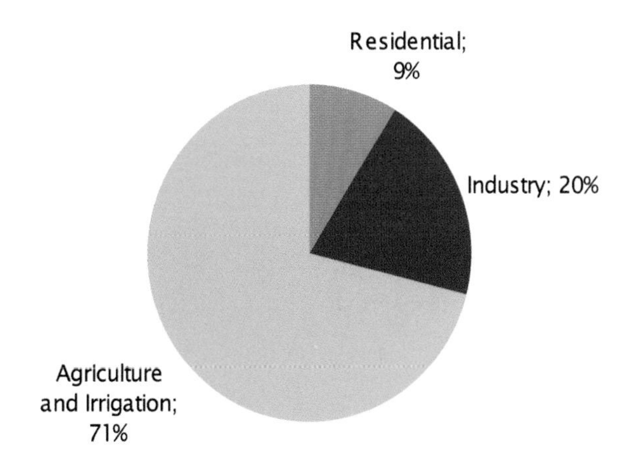

Figure 4.17 The share of electricity consumption in Dunhuang, 2003

Table 4.9 Electricity consumption in Dunhuang in 2003 (GWh/year)

Sector	Electricity consumption
Residential	116
Industry	252
Agriculture and irrigation	918
Total	1 286

Table 4.10 Electricity price by sector (yuan/kWh)

Sector	Electricity price
Residential	0,470
Industry	0,316
Agriculture	0,394
Business	0,712

Note: 1 yuan = approximately 0,12 USD

Table 4.11 Irradiation data in Dunhuang

	Horizontal (MJ/m²/month)	In-plane (40°) (MJ/m²/month)	In-plane (40°) (kWh/m²/day)
January	334,08	558,30	5,00
February	390,51	562,49	5,58
March	582,79	715,84	6,41
April	595,01	628,84	5,82
May	857,34	811,81	7,27
June	856,88	771,03	7,14
July	805,45	742,68	6,65
August	744,97	760,67	6,82
September	587,63	700,94	6,49
October	481,11	683,00	6,12
November	341,20	583,44	5,40
December	295,60	543,78	4,87
Total (annual)	6 872,57	8 042,82	2 234,12

Qiliying to the 6 000 kVA/35 kV transformer station, so it will not be that costly to build a high voltage transmission line.

There are only five micro-hydropower stations with 10 MW in total capacity in Dunhuang city and the annual output is 35 GWh/year, which is 2,7 % of electricity consumption in the city. Most electricity is provided from other cities or other provinces through the utility grid.

In Dunhuang, electricity is mainly consumed by agriculture and farm irrigation during the daytime, matching well with solar irradiation. The share of electricity consumption is shown in Figure 4.17 and Table 4.9.

The feed-in tariff for small hydropower is between 0,2 to 0,3 yuan/kWh and the electricity price to the end users varies by sectors, as shown in Table 4.10.

Table 4.11 and Figure 4.18 show irradiation data in Dunhuang. Annual horizontal irradiation is 6 873 MJ/m²/year and in-plain irradiation facing south with a 40° tilted angle is 8 083 MJ/m²/year (2 234 kWh/m²/year).

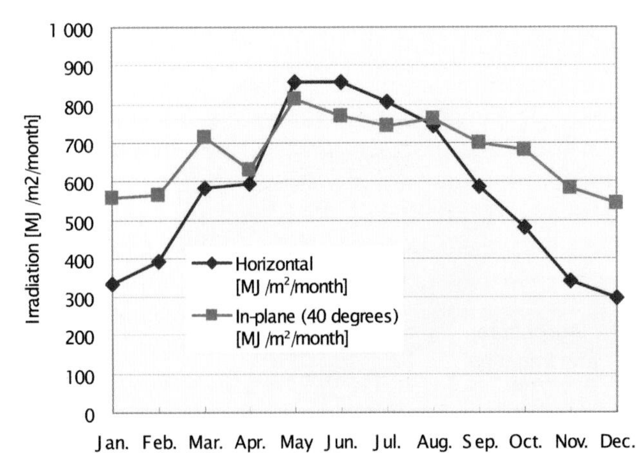

Figure 4.18 Irradiation data in Dunhuang

4.2.3 Technical profile

System design

The 8 MW PV system will be divided into eight sub-stations of 1 MW each. Each 1 MW sub-station will feed the generated electricity to the high-voltage grid (35 000 V) through a 1 000 kVA transformer. Each 1MW sub-station will be divided into five channels with 200 kW for each. Each 200 kW PV channel will be equipped with a grid-connected inverter to convert the DC power from the PV into three-phase AC power for the primary of the 1 000 kVA transformer.

Each 1MW sub-station and each 200 kW channel will be independent of other channels. Such design offers the following advantages:

- easier troubleshooting and maintenance;
- flexibility for potential investors;
- allows different type of PV systems to be installed and compared.

Figures 4.19 and 4.20 outline the principle of the system.

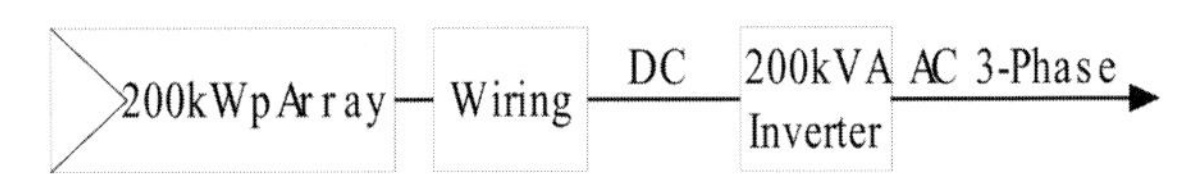

Figure 4.19 200 kW PV channel

Requirement of PV system components

Based on the above design plan, the requirements of PV system components are discussed. The specifications of PV modules and inverters proposed are shown in Tables 4.12 and 4.13. A PV sub-array consists of 18 PV modules in a series and 2 series in parallel, connected with one inverter. 35 sub-arrays make one 200 kW PV channel, and 5 PV channels make one 1MW PV sub-station.

Therefore, an 8 MW PV system requires 50 400 pieces of PV modules and 40 sets of 200 kVA inverters.

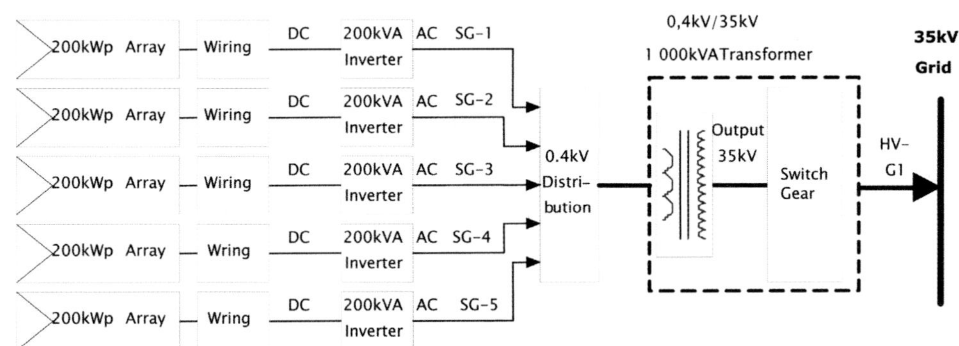

Figure 4.20 1 MW PV sub-station

Table 4.12 Specifications of proposed PV module

Pm (Wp)	160
Isc (A)	5,08
Voc (V)	43,6
Vm (V)	35,6
Im (A)	4,5
Size (mm)	1 587 × 790
Weight (kg)	15

Table 4.13 Specifications of proposed DC/AC inverter

Rated power (kVA)	200
Maximum power (kVA)	245
Rated input DC voltage (V)	640
Maximum input DC voltage (V)	800
Maximum power point tracking (MPPT) voltage (V)	450–800
Rated input DC current (A)	315
Maximum input DC current (A)	400
AC output	Three-phase 415V ± 10 %
Output frequency	50 Hz
Efficiency	10 % load: 91 %
	50 % load: 96 %
	100 % load: 96 %
Size (mm)	2 000 × 2 100 × 800
Weight	1400 kg

Eight sets of 0,4 kV/35 kV (1 000 kVA) transformers are also required (see Table 4.14).

Table 4.14 Requirements of PV system components

PV modules (160 W)	50 400 pieces
Inverter (200 kVA)	40 sets
Transformer (0,4 kV/35 kV (1 000 kVA))	8 sets

System efficiency and annual output

In order to estimate the system efficiency of the 8 MW PV system, the following elements should be considered:

- PV array efficiency = 0,85, considering:
  - PV module mismatching loss;
  - dust shading loss;
  - temperature loss;
  - DC wiring loss.
- Inverter efficiency = 0,95.
- AC efficiency = 0,95.

Therefore, the system efficiency is assumed to be 0,77. Using the efficiency and annual in-plain irradiation facing south with a 40° tilted angle, the annual output is calculated to be 13 761 MWh/year.

4.2.3.4 Field requirement

Figure 4.21 shows the configuration of a PV sub-array. The distance between PV sub-arrays is 6,5 m, which is calculated by the formula:

$$D = \frac{0{,}707\,H}{\tan\left(\arcsin\left(0{,}648\cos ø - 0{,}399\sin ø\right)\right)}.$$

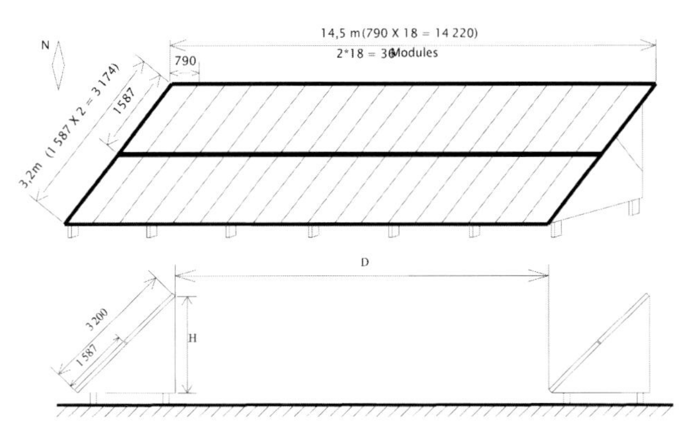

Figure 4.21 Configuration of PV sub-array

Figure 4.22 shows a field layout for a 1 MW PV sub-station consisting of five 200 kW PV channels. The required area for a 1 MW PV sub-station is approximately 23 200 m^2.

Based on such array layouts, and by considering additional area for some buildings in the field, the area required for an 8 MW PV system is estimated at approximately 308 000 m^2. Figure 4.23 shows a field design for an 8 MW PV system.

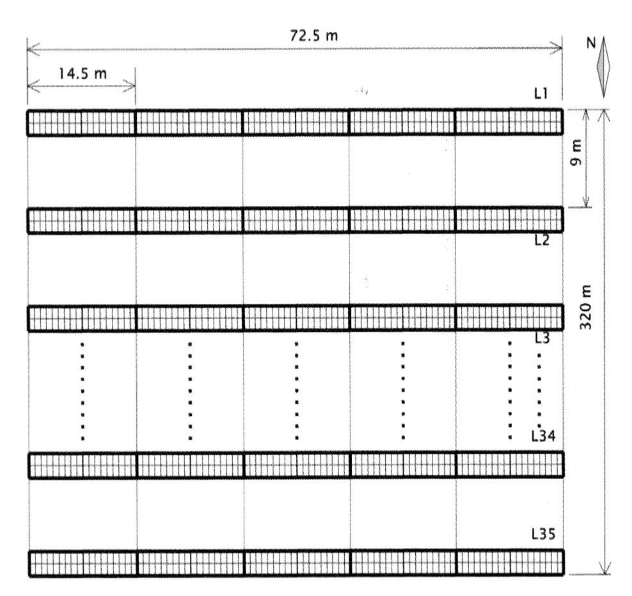

Figure 4.22 Field design for 1 MW PV sub-station

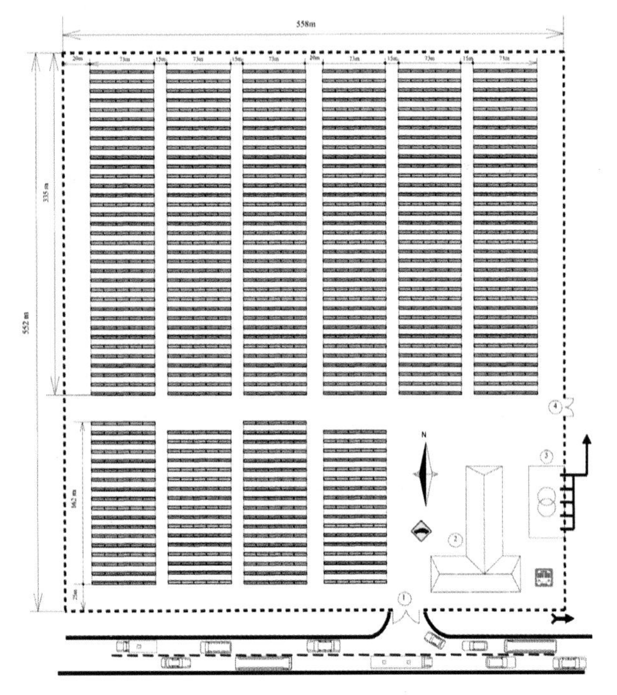

Figure 4.23 Field design for 8 MW LS-PV system

4.2.4 Financial feasibility

Table 4.15 shows a capital investment estimate for the 8 MW LS-PV system.

The total capital investment is 322,47 million yuan (approximately 38,70 MUSD), and 86 % of total investment is for PV system equipment, such as the PV module, inverter and transformer. However, it is expected that 96,74 million yuan (approximately 11,61 MUSD) (30 % of the total capital) will be grant provided by the central government of China, and the real required capital will be 225,73 million yuan (approximately 27,09 MUSD).

Table 4.15 Capital investment for 8 MW LS-PV system

	Investment (million yuan)	Share (%)
Equipment	277,02	85,91
PV module	*236,8*	*73,43*
Inverter	*34,2*	*10,61*
Transformer	*3,52*	*1,09*
Test and monitoring	*2,5*	*0,78*
Civil construction	15,56	4,83
Transportation and installation	7,35	2,34
Feasibility study and preliminary investment	7,0	2,17
Miscellaneous	15,36	4,65
Total	322,47	100

Note: 1 yuan = approximately 0,12 USD

The Gansu grid company will guarantee 1,683 yuan/kWh (0,202 USD/kWh) as the feed-in tariff, and annual income for the PV system will be 23,16 million yuan (approximately 2,78 MUSD), which is 13 761 MWh/year × 1,683 yuan/kWh. The tariff purchased by the grid company will be added on to all the electricity consumed in Gansu Province, and the electricity consumption in Gansu Province was 340 × 10^8 kWh in 2002. Therefore, the additional tariff will be 0,000 68 yuan/kWh (0,000 082 USD/kWh), which is 23,16 million yuan/340 × 10^8 kWh. For a family who consumes 2 kWh/day, the annual consumption will be 730 kWh, and they will only need to pay 0,5 yuan/year (0,06 USD/year) in addition.

4.2.5 Recommendations

Solar energy is a non-polluting, CO_2-free energy source, which provides good prospects for development in China. The vast deserts in the western parts of China have provided for a new area of development for solar energy. The main obstacle for developing large scale solar power plants is the high cost involved. Large scale market development and industrialization of solar power are important in removing this obstacle. Government policy support and encouragement of solar PV technology are key to the viability of large scale solar power plants.

The proposed 8 MW LS-PV plant in Dunhuang city is considered the first pilot project in China of the Great Desert Solar PV Programme, proposed by the WWF and the expert group. Development of further large scale PV systems in other regions is also being discussed. It has been proposed that 30 GW of solar PV power-generation capacity could be developed by 2020 if government incentives are developed and put in place. This could enable China to become a world leader in solar power development.

CHAPTER FIVE

The Oceania region: Realizing a VLS-PV power generation system at Perenjori

5.1 INTRODUCTION

Perenjori is a small township approximately 350 km north-east of Perth in the wheat belt of Western Australia and situated at S29° latitude and E116,2° longitude. The required land to set up a VLS-PV power-generation project can be obtained at a reasonably low price or leased for 30 to 50 years from local farmers. The land is flat and suitable for mounting the structure or installing solar PV power-generation projects.

Several issues arise in terms of achieving a very large scale solar photovoltaic power-generation project at Perenjori. Although the Great Sandy Desert receives more solar radiation than Perenjori, the overall economy of setting up the project and generation costs will be less at Perenjori due to its proximity to enough loads and the availability of the local grid.

Figure 5.1 Location map of Perenjori

5.2 POTENTIAL OF VLS-PV PROJECT AT PERENJORI

5.2.1 Power requirement and creating extra load in the region

The current load of Perenjori is approximately 3 MW, and the local grid of Western Power meets it. However, as Perenjori is at the fringe of the grid, the power quality is very poor, with a low voltage supply, regular voltage fluctuations, and brownouts and blackouts.

Setting up any grid-connected power-generation systems at the end point of the grid is always beneficial for improving local grid quality.

Given that the load is now almost negligible for the proposed very large scale photovoltaic power generation (VLS-PV) at Perenjori, it is necessary to analyse the situation and to determine whether new loads can be created to justify setting up large power-generation projects at Perenjori.

The discussions held with the officials of the Shire of Perenjori have revealed strong interest in promoting the mining industry in the region, provided that sufficient and good quality power is available for mining operations. The mining projects as depicted in Table 5.1 are anticipated to operate in the region, within a radius of 100 km or so, during the next three to five years.

Furthermore, a number of other mining companies have also shown an interest in setting up mining operations in the Perenjori region. Therefore, over the next 10 to 15 years there will be a load of the order of 1 GW.

5.2.2 Basic meteorological data for designing and sizing the project

There are no solar radiation, temperature or wind speed data available for Perenjori. The beam radiation (direct), number of sunshine hours and temperature

Table 5.1 Mining projects anticipated in the region

Mining company	Expected load requirement	Distance from Perenjori
Aurora Gold Mines	30 MW	40–70 km
Mid West Corporation	35 MW	50–80 km
Mount Gibson Iron	60–80 MW	50–70 km
Galabar Mines	5–10 MW	96 km
Morawa Mines	100–150 MW	40–50 km

data are the main critical data for designing a solar power system based on the crystalline solar cell modules.

Although there are no solar radiation data available (global or beam), the images shown in Figure 5.2 estimate the number of sunshine hours in Australia. This figure indicates that Perenjori has good sunshine for seven to nine months, although the level of radiation is not known.

The Sustainable Energy Development Office (SEDO) of the government of Western Australia was asked to fund a meteorological weather monitoring station at Perenjori, but refused due to lack of funds. Funding could be sought again in the future or other sources of funding could be explored.

5.2.3 Possible scenario for realizing the VLS-PV project

The size of a VLS-PV system may range from 10 MW (pilot) to 1 or several GW (commercial, consisting of one plant or an aggregation of a number of units, distributed in the same region and operating together).

Figure 5.3 provides a rough idea of how a VLS-PV project can be realized within 100 km of Perenjori over the next 15 years, aggregating to a capacity of over 1 GW.

5.2.4 Grid connections versus stand-alone PV power-generation systems

A new grid at Perenjori has to be created to sustain a load of the order of 1 GW. The capacity of the existing grid is only 33 kVA and is not suitable to handle a large load from VLS-PV systems set up in the region. The funding of a new grid will be a crucial issue.

It will be possible to install several stand-alone solar PV systems as per the load requirement of each individual mining operation in the region. Then, when there are three or four projects set up in the region, they can be interconnected by creating a small local grid.

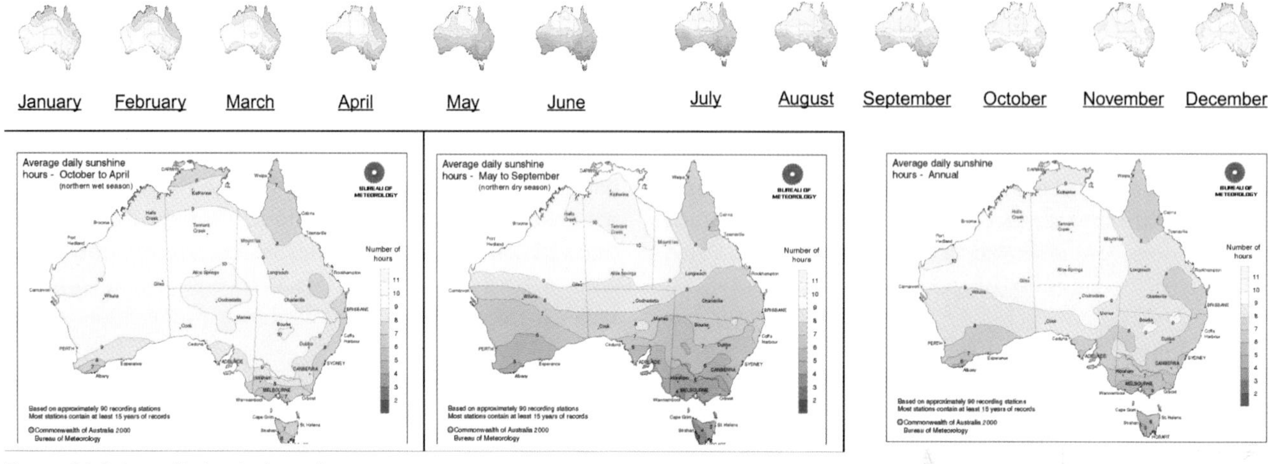

Figure 5.2 Solar radiation in Australia

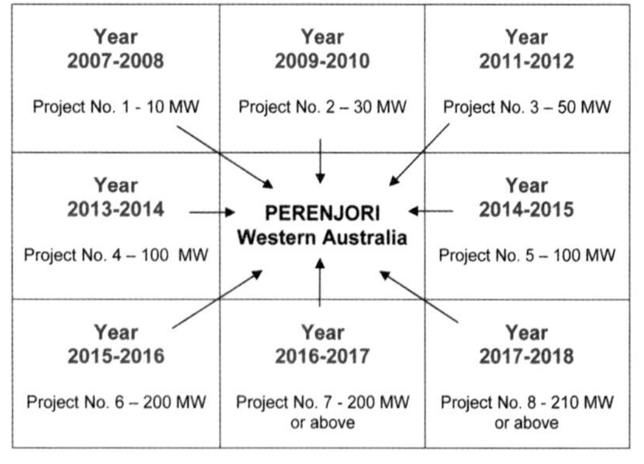

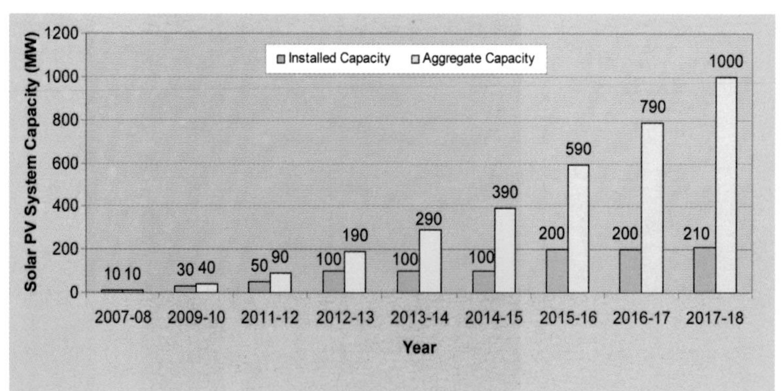

Figure 5.3 Possible scenario to realize VLS-PV at Perenjori

5.2.5 Funding of VLS-PV power-generation projects

Currently, solar PV power-generation projects are not economically viable without subsidies. However, the increase in the efficiency of solar cells and mass production of solar modules will make the cost of solar power generation attractive enough to install solar PV power-generation projects without subsidies until the final commercial stage of the VLS-PV project.

Private investors, as well as a number of ethical funds, would be interested in financing VLS-PV power-generation projects in the Perenjori region. International banking organizations can also come forward to lend funds for such large projects.

5.2.6 Pathway to setting up a local PV manufacturing facility

The following companies in Western Australia are now engaged in solar photovoltaics (PV):

- Solco Ltd (solar pumps manufacturing);
- Solar Cells Pty Ltd (marketing of solar PV devices/systems);
- WA Solar Supplies (marketing of solar PV modules and devices); and
- W D Moore and Co (solar pumps manufacturing).

Figure 5.4 depicts a possible scenario for developing a local solar PV manufacturing industry.

There is not yet any solar modules manufacturer in Western Australia, but there is an interstate manufacturer, BP Solar in Sydney, and a number of distributors/dealers of modules representing international solar modules manufacturers. However, Prime Solar Pty Ltd is a Perth-based company, which envisages setting up of large polysilicon and silicon wafers manufacturing facilities in the Saxony-Anhalt State of Germany.

The VLS-PV project can provide a pathway to setting up of a silicon polysilicon, silicon wafers, solar cells and modules manufacturing facilities either by Prime Solar Pty Ltd or by another company in Western Australia to supply the solar modules for solar power generation projects in the Perenjori regions.

5.2.7 Rainwater collection from PV modules

Rainwater is a free source of nearly pure water. For centuries, the world has relied upon harvesting rainwater to supply water for household, landscape and agricultural uses. Before urban water systems were developed, rainwater was collected (mostly from roofs) and stored in cisterns or storage tanks.

Today, many parts of the world, including the entire continent of Australia, promote rainwater as the principal means of supplying household water. Rainwater is one of the purest sources of water available. Its quality almost always exceeds that of ground or surface water. It does not come into contact with soil or rocks, where it can dissolve minerals and salts, nor does it come into contact with many of the pollutants

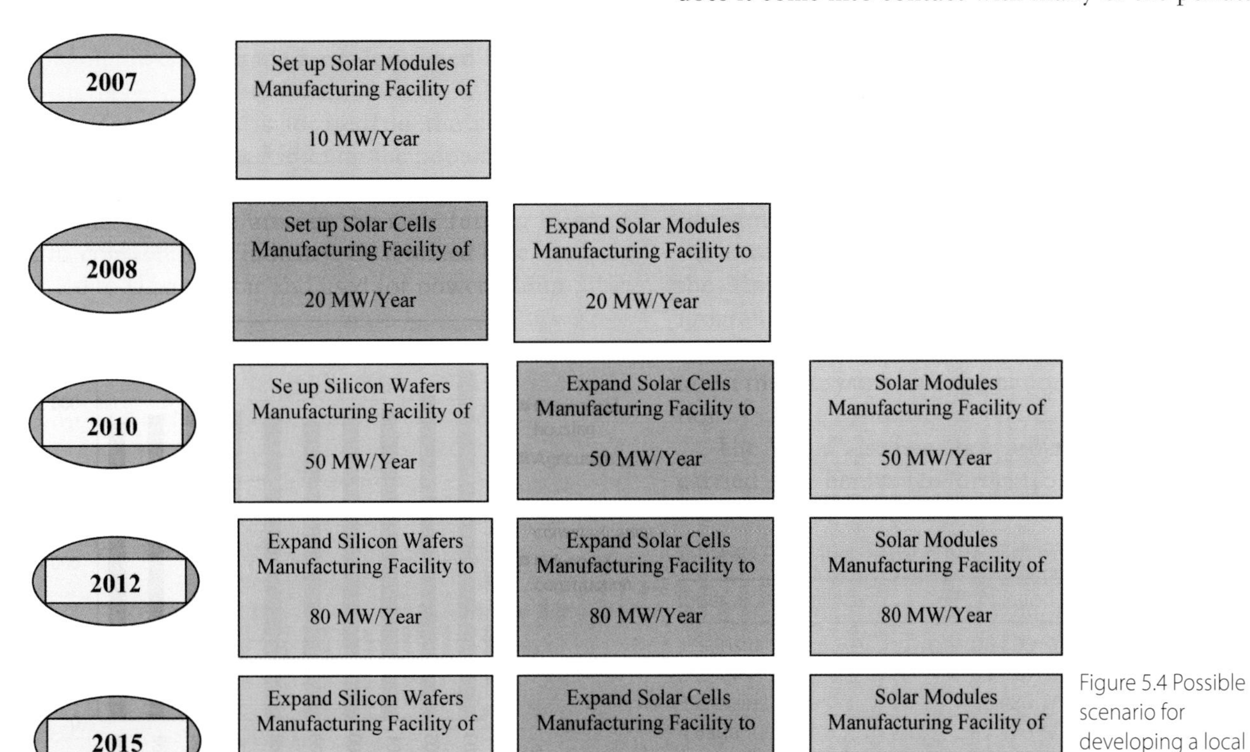

Figure 5.4 Possible scenario for developing a local PV manufacturing industry

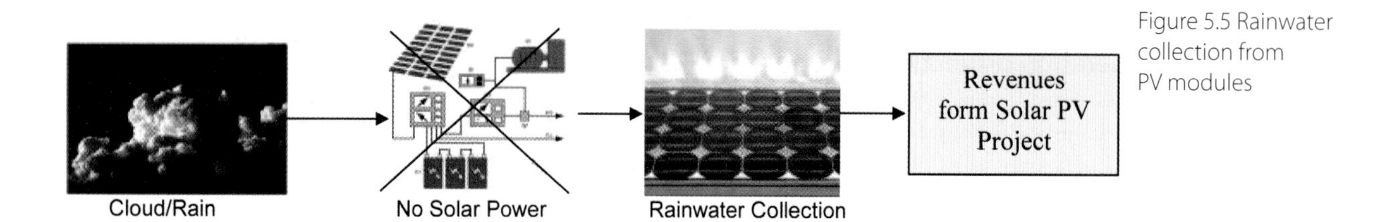

Figure 5.5 Rainwater collection from PV modules

that are often discharged into local surface waters or contaminate groundwater supplies. However, rainwater quality is influenced by where it falls. Rainfall in areas where heavy industry or crop dusting are prevalent may not have the same purity as rain falling in other areas.

The VLS-PV projects will deploy large PV module areas of approximately 10 million m^2 for 1 GW systems. These areas can easily be used to collect the rainwater from the solar modules.

Rainwater collection from a solar PV system can be an additional source of revenue generation, compensating, to some extent, for the high cost of power generation from solar PV systems.

This idea was floated by the president of the Shire of Perenjori since there is an average 300 to 330 mm/annum of rainfall during the monsoon months (June to August). When there is no sun during the monsoon season, the PV modules can become the source for harvesting rainwater.

5.2.8 Greenhouse gas emission abatement

Large scale power generation through photovoltaic systems is expected to become more economical in the next decade, particularly when the environmental costs of producing electricity from fossil fuels are considered in power system planning.

Many environmental issues should be resolved in the future, most of which have been caused by human activities. Global warming is one of the most important, and one major cause is carbon dioxide (CO_2) emissions from fossil fuel combustion. Desertification is another key issue, especially when thinking about VLS-PV, and is primarily a problem of sustainable development. It is a matter of addressing poverty and human well-being, as well as preserving the environment. Social and economic issues, including food security, migration and political stability, are closely linked to land degradation, as are such environmental issues as climate change, biological diversity and freshwater supply.

VLS-PV may result in a variety of environmental impacts that would contribute to environmental issues in the future, both directly and indirectly. For example, since VLS-PV supplies electric power without the combustion of fossil fuels, it will be possible to significantly reduce CO_2 emissions, providing that VLS-PV replaces conventional fossil-fuelled power plants. Emissions of sulphur dioxide (SO_2) and nitrogen oxide (NO_x) will also be reduced. Furthermore, while the unsustainable use of fuelwood and charcoal is a major cause of land degradation, the energy derived from installing VLS-PV in a desert area can be a replacement energy source. VLS-PV will also reduce direct sunshine onto the land surface, although it is now unclear whether this will be useful for preventing desertification and land degradation. Since such impacts will differ according to the site where VLS-PV is introduced, a precise evaluation of microclimatic changes around the installation area will be needed.

Furthermore, VLS-PV will be an international cooperative project, and the early stage of its development may be adopted as a small scale clean development mechanism (CDM) project within the framework of the Kyoto Protocol.

5.3 PILOT 10 MW PV POWER-GENERATION PROJECT AT PERENJORI

As the first step in establishing a VLS–PV system at Perenjori, a project for installing a 10 MW pilot power-generation system is proposed.

5.3.1 Estimated project cost

The estimated project cost of a 10 MW PV power-generation project at Perenjori will be of the order of 60 million AUD (approximately 45,6 MUSD), with the cost breakdowns depicted in Table 5.2.

Figure 5.6 shows the percentage cost distribution of components for a 10 MW PV power project to be installed at Perenjori. Almost 70 % of the cost of the project is for PV modules alone.

5.3.2 Levelized cost of power generation

In order to calculate the power generation cost from a 10 MW pilot PV power plant at Perenjori, the assumptions shown in Table 5.3 were used.

Table 5.2 Estimated project cost for 10 MW PV power-generation system (AUD)

Components	Unit cost	Total cost
PV modules	4,4 AUD/W	44 000 000
Mounting structure with single-axis tracking	10 % of modules	4 400 000
Inverter(s)	0,50 AUD/W	5 000 000
Transformers and cabling	4 % of modules	1 760 000
Installation and commissioning	7 % of modules	3 080 000
Land	Lump sum	500 000
Miscellaneous, including transportation to site	Lump sum	1 260 000
Total		60 000 000

Note: 1 AUD = approximately 0,76 USD

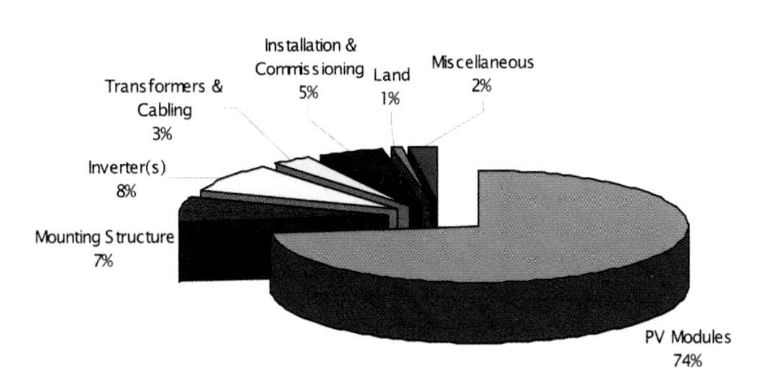

Figure 5.6 Cost composition of 10 MW PV power-generation system project

Table 5.3 Assumptions for 10 MW pilot PV power plant

Cost of solar modules	4,40 AUD/W
Type of modules	Crystalline
Type of system	Single-axis tracking array
Life of solar modules	30 years
Salvage value of modules	Nil
Cost of supportive frames and trackers	10 % of the module cost
Cost of inverter	0,50 AUD/W
Efficiency of inverter	91 %
Other power losses	8 %
Installation and commissioning cost	7 % of the module cost
Transformers and other electrical equipment	4 % of the module cost
Miscellaneous cost	2,1 % of the module cost
Financial cost	6 % interest rate
Loan term	Throughout the life of the system
Land area	25 ha
Land cost	500 000 AUD
Average yearly number of sunshine hours	5,9 hours/day
Subsidy on the system	50 % of the project cost minus land cost
Insurance premium	0,5 %/year of the modules, inverter and frames
Operation and maintenance cost	0,5 % of the modules, inverter and frames
Additional gain in output due to tracking	21 %
Renewable energy certificate trading price	0,04 AUD/kWh

Note: 1 AUD = approximately 0,76 USD

5.3.3 Estimated annual power generation

Although there are no solar radiation and temperature data available for the proposed location of Perenjori, Table 5.4 estimates the power to be generated from the 10 MW solar photovoltaic single sun-tracking power-generation system at Perenjori.

The total power generation will depend upon the number of sunshine hours when solar radiation intensity is more than 1 000 W/m^2. An average annual 5,90 sunshine hours can be conservatively estimated.

Installing single-axis sun trackers has been proposed to achieve a reasonable amount of power per kilowatt of solar PV modules. Therefore, a 21 % greater output in comparison to a fixed-array solar PV system is forecasted.

In addition, 8 % power losses are estimated due to interconnections, degradation of solar module efficiency and the toleration of the electrical equipment efficiency, such as transformers.

The annual cost for a 10 MW solar PV project at a 6 % interest rate and a project life of 30 years has been calculated in Table 5.5.

The interest rate on the borrowing is a crucial factor and affects the cost of power generation. In the present case, a 6 % interest rate is used; but it can be as low as 3 % if this project is funded by an overseas banking organization. In Table 5.6, a comparison with different interest rates illustrates the power generation cost per kWh. Examples of the life of the plant for 20, 25 and 30 years have also been provided to calculate the power generation costs.

Table 5.4 Estimated power generation from 10 MW pilot PV power plant

Item	Parameters	Unit
Size of the plant	10 000	kW
Average annual number of sunshine hours	5,90	Hours/day
Inverter efficiency	91	%
Other power losses, including degradation of solar cells	8	%
Average power generation (fixed array)	18 029 102	kWh/year
Average power generation (single-axis tracking array)	21 815 213	kWh/year

Figure 5.7 shows the levelized life-cycle power generation cost against different interest rates for borrowing funds and different project lives.

Table 5.5 Annual cost for 10 MW pilot PV power plant (AUD)

Item	Annual cost at 6 % interest rate
Depreciation/year on the capital investment	1 008 333
Financial cost per year	1 815 000
Insurance premium per year	297 500
O&M expenses per year	297 500
Total	3 418 333

Note: 1 AUD = approximately 0,76 USD

Table 5.6 Generation cost of 10 MW pilot PV power plant

Subsidized project cost	Interest rate	Generation cost for project life (AUD/kWh)		
		20 Years	25 Years	30 Years
30 250 000	7 %	0,154	0,140	0,131
30 250 000	6 %	0,140	0,126	0,117
30 250 000	5 %	0,126	0,112	0,103
30 250 000	4 %	0,112	0,098	0,089
30 250 000	3 %	0,098	0,084	0,075

Note: 1 AUD = approximately 0,76 USD

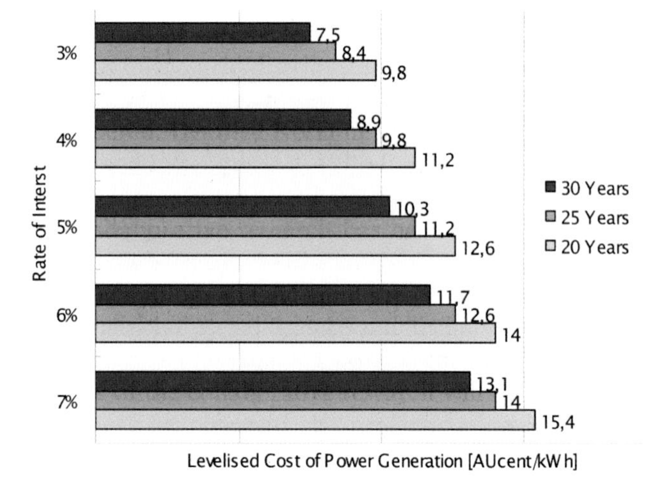

Figure 5.7 Levelized life-cycle generation cost of 10 MW PV power plant

Note: 1 AUD = approximately 0,76 USD

5.3.4 Economics of rainwater collection from PV modules

A rainwater collection mechanism will be developed to harvest the rainwater falling on the solar PV modules since there is an average of 300 to 330 mm/annum of rainfall at Perenjori. This water will be bottled and sold to generate additional revenues from the solar PV system.

We can use this formula to calculate a worse case scenario for collecting rain in Perenjori:

- 100 000 m^2 of catchment area;
- 300 mm of rainfall in the year;
- 75 % collection efficiency:
  - 22 500 000 litres of rainwater per year = 100 000 m^2 solar collectors area × 300 mm of rainfall/year × 75 % collection efficiency.

Table 5.7 indicates the extent of additional revenues that can be generated from the PV system. The cost of bottling has been estimated at 0,25 AUD/litre (0,19 USD/litre), which includes the depreciation on a water collection and storage facility, depreciation on plant and machinery, electricity, labour, plastic bottles, marketing costs and other factors.

Table 5.7 Estimated additional revenues by rainwater collection from PV modules

Selling price (AUD/litre)	Quantity (Litres/year)	Cost of bottling (AUD/litre)	Revenues (AUD)
0,30	22 500 000	0,25	1 125 000
0,35	22 500 000	0,25	2 250 000
0,40	22 500 000	0,25	3 375 000
0,45	22 500 000	0,25	4 500 000
0,50	22 500 000	0,25	5 625 000
0,55	22 500 000	0,25	6 750 000

Note: 1 AUD = approximately 0,76 USD

Although the market selling price for bottled rainwater is approximately 1,00 AUD/litre (0,76 USD/litre), considering a real example, we have tabulated the rate of bottled rainwater from this plant to 0,35 AUD/litre (0,27 USD/litre) (a gross profit margin of 0,1 AUD/litre, or 0,076 USD/litre). Therefore, collecting rainwater can generate additional gross revenues of 2 250 000 AUD (approximately 1 710 000 USD) or net revenues after tax of 1 575 000 AUD (approximately 1 197 000 USD).

Rainwater collection should definitely be experimented with at Perenjori, which can generate additional revenues to compensate for the cost of power generation by at least 0,072 AUD/kWh (0,055 USD/kWh), making solar PV systems cost effective and attractive to investors.

Table 5.8 Effective generation cost considering revenues from rainwater collection

Rate of interest	Effective power generation cost (AUD/kWh)
7 %	0,058
6 %	0,044
5 %	0,031
4 %	0,017
3 %	0,003

Note: 1 AUD = approximately 0,76 USD

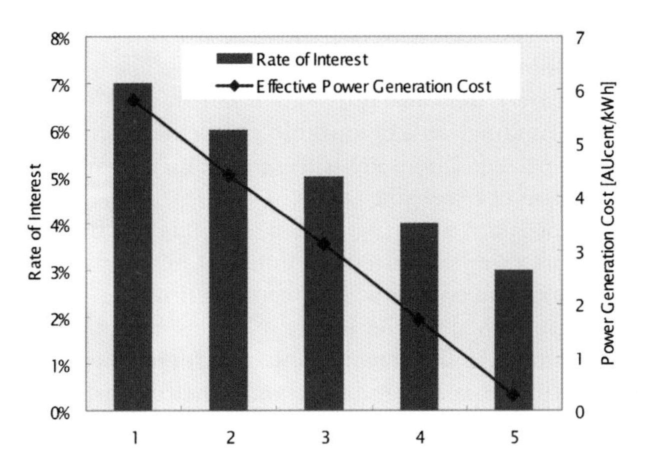

Figure 5.8 Effective generation cost considering revenues from rainwater collection

Note: 1 AUD = approximately 0,76 USD

Therefore, the cost of power from this project can be as low as 0,003 AUD/kWh (0,002 3 USD/kWh) at a 3 % rate of interest, which is mainly due to the rainwater harvesting. However, we strongly recommend carrying out a detailed study on the cost of rainwater harvesting before drawing conclusions about the net benefits.

5.4 ANALYSIS OF GOVERNMENT SUBSIDY PROGRAMMES

The Australian Greenhouse Office (AGO) was set up as a federal government agency at Canberra to promote using renewable energy in Australia. AGO has several programmes to support the installation of renewable-based power-generation projects.

The Sustainable Energy Development Office (SEDO) of the government of Western Australia is responsible to the AGO in implementing programmes and handles the grants/subsidies on behalf of the AGO in the state of Western Australia.

5.4.1 Renewable Remote Power-Generation Programme (RRRGP)

In areas of Australia not serviced by a main electricity grid, electricity generated from renewable sources is often effective in reducing reliance on diesel for electricity generation. The Renewable Remote Power-Generation Programme (RRPGP) provides financial support to increase the use of renewable energy generation in remote parts of Australia that currently rely on diesel for electricity generation.

The objective of the RRPGP is to increase the uptake of renewable energy technologies in remote areas of Australia, which will:

- help in providing an effective electricity supply to remote users;
- assist the development of the Australian renewable energy industry;
- help meet the energy infrastructure needs of indigenous communities; and
- lead to long-term greenhouse gas reductions.

Over 200 million AUD (approximately 152 MUSD) were available over the four-year life of the RRPGP. Programme funds are still available to participating states and territories on the basis of the relevant diesel fuel excise paid in each state or territory by public generators in the financial years of 2000–2001 to 2003–2004.

The funds are available to the participating jurisdictions to fund approved sub-programmes or major projects. Potentially eligible installations are those for which renewable energy generation replaces all or some of the diesel used for off-grid electricity generation. RRPGP funding may also be available for new off-grid installations where it can be demonstrated that the energy source would otherwise have been diesel.

The RRPGP may provide support for up to 50 % of the capital costs of renewable generation equipment. Capital costs encompass expenditure on:

- renewable energy-generating equipment, such as photovoltaic arrays, wind turbines and hydro units;
- enabling equipment (equipment necessary to make usable electricity available from the renewable energy-generating equipment, such as inverters, control and monitoring equipment, and batteries); and
- essential non-equipment requirements, such as installation, design and project management costs.

Under SEDO's Rural Renewable Energy Program for Medium Projects (30 kW to 2 MW) designed for grid-connected renewable energy power systems in specific rural or fringe-of-grid areas of Western Power's electricity grid (the South West Interconnected System) in Western Australia the subsidies can be as high as 70 % of the project cost.

5.4.2 RRPGP sub-programmes

Western Australia Remote Area Power Supply (RAPS) sub-programme

This sub-programme has a budget of 18 million AUD (approximately 13,7 MUSD) in RRPGP funds to provide rebates of 55 % (50 % RRPGP rebates, 5 % Western Australian rebates) of the initial capital costs of renewable generation equipment installed in remote areas of Western Australia. Eligible target groups are:

- indigenous communities;
- isolated households; and
- commercial operations, including pastoral properties, tourist operations and mining operations.

General eligibility criteria

General eligibility criteria are as follows:

- Funding is available for renewable energy systems that partly or fully replace diesel generation, or where diesel would otherwise have been used.
- Facilities that will be served by renewable energy systems must not be connected to the South-West or Pilbara grids.
- In general, expenditure on renewable energy-generating equipment must be at least 30 % of total expenditure on which a rebate is sought.
- Renewable energy technologies must be appropriate for given applications and must meet relevant standards and regulations.
- Sufficient resources must be available for the effective implementation of projects.

Rebates

The Renewable Remote Power-Generation Programme provides rebates of up to 50 % of the capital cost of renewable energy systems, including:

- renewable energy-generating equipment, such as photovoltaic arrays and wind turbines;
- essential enabling equipment, such as power converters and energy storage systems;
- essential non-equipment expenditure, such as design and installation.

Large projects

Large renewable energy-based generation projects, with a rebate value greater than 500 000 AUD (approximately 380 000 USD), could be implemented in Western Australia through the RRRGP. These include wind, photovoltaic, solar thermal, biomass and hydro systems.

The best opportunities for large projects are in isolated towns because the cost of supplying electricity using diesel power stations is high and there is substantial demand for electricity. Opportunities may also exist in large communities and tourist operations in remote areas.

Funding applications for large projects have to be individually assessed and approved by both the state and commonwealth governments. Funding may also be available for scoping studies that assess the viability of potential renewable energy projects.

5.4.3 Private investment, superannuation funds and ethical trust funds

Despite encouraging growth forecasts and a fair amount of enthusiasm, private investment in the renewable energy sector is currently very modest and is largely limited to energy-sector specialists. The industry still heavily depends upon government and public-sector funding, and favourable political policies and framework conditions are still essential for the commercialization of renewable energy.

Studies are showing that a growing number of investors are interested in integrating social and environmental criteria within their investments. A 2002 study published by the United Nations Environment Programme (UNEP) Finance Initiatives, and supported by leading financial and insurance institutions, made recommendations to the global financial community and policy-makers, aimed at promoting stronger clean technology markets and more commercially attractive greenhouse gas reduction markets. Pension funds are under more and more pressure to integrate social and environmental criteria within their investment decisions, and, as has been mentioned, there is some private equity and venture capital looking at sustainable energy businesses. CleanTech Venture based in the US organizes project developers, and investors meet twice a year.

The *Sustainable Energy Finance Directory* is an international directory of lenders and investors who provide finance to the renewable energy and energy efficiency sectors. Produced jointly by UNEP Energy, UNEP Finance Initiatives and (Basel Agency for Sustainable Energy (BASE), the directory is a practical search tool designed to assist project developers in seeking capital for sustainable energy projects.

The CVC Renewable Energy Equity Fund (REEF) is a 26,5 AUD million (approximately 20,1 MUSD) venture capital fund established by the federal government of Australia to increase investment in renewable energy technologies through providing equity finance. It is funded up to 18 million AUD (approximately 13,7 MUSD) by the Australian government's REEF programme and 8,5 million AUD (approximately 6,5 MUSD) comes from private sources. The fund invests in emerging high-growth Australian companies with domestic and global market potential in the renewable energy industry. However, there is no such fund in Australia that is able to support the installation of very large solar PV projects.

Australian Ethical Investments Ltd (AEI) is a well-known Australian ethical fund that has invested in a number of solar energy manufacturing companies. Australian Ethical Investments was established in 1986 for the purpose of environmental and socially responsible investment. AEI manages four unit trusts, with around 100 individual investments. In addition, all trusts are available for superannuation. However, there is no evidence that they have yet participated in any syndication of funds for developing renewable energy-based power-generation projects.

5.4.4 Ownership and management issues

The project development and management of VLS-PV projects in the Perenjori region have to be considered carefully since the local utility, Western Power, which owns the electricity Network or Synergy, which retails the power in the region, may not be interested in setting up and operating these systems. Private project developers have to be asked to participate in the setting up of such projects on a build, operate and transfer (BOT) basis. If project developers come forward to set up these projects at a later stage, such as after ten years or so, the operation and management of the system can be transferred to local end users. Alternatively, local mining companies who will use the power from these projects could be persuaded to extend their interest in these projects on a build, operate and own (BOO) basis.

5.4.5 Power purchase agreements with potential end users

The cost of power generation from the pilot project of 10 MW after availing of a 50 % subsidy from the government will be approximately 0,14 AUD/kWh (0,11 USD/kWh), after deducting 0,04 AUD/kWh (0,03 USD/kWh) for trading a Renewable Energy Certificate under the Mandatory Renewable Energy Target (MRET) of the federal government of Australia, which is very much comparable with the cost of power generation from a diesel power project. At a price of 0,36 AUD/litre (0,27 USD/litre) for diesel fuel available to the mining companies in Perenjori, the cost of power generation from diesel power projects comes to about 0,12 AUD/kWh (0,09 USD/kWh). Therefore, the mining companies would be interested in purchasing power from the proposed pilot project of 10 MW.

As a back-up power source, it has been suggested that the diesel-based power project will be most suitable at Perenjori. The mining companies themselves can finance installing the diesel power project.

Prior to setting up the pilot project, arrangements have to be made by the project developers to sell the power to the mining operators. The power purchase agreements (PPAs) should be signed for the whole lifetime of the project, which is 30 years. The Shire of Perenjori and Mid West Development Commission would play an important role in negotiating the terms and conditions of the PPAs. The assistance offered by these two interests would be necessary to attract the developers for this project, as well as for other projects to be installed later on.

5.5 EXPECTED EFFECTS INDUCED BY VLS-PV AT PERENJORI

5.5.1 Greenhouse gas emissions abatement

It has been calculated that a total 21 815 tonnes of CO_2 could be eliminated from the environment and over 10 000 tonnes of coal can be saved per year by installing the 10 MW PV system at Perenjori. The cumulative saving over 30 years of the project's lifespan could be as high as 655 000 tonnes of CO_2 and 300 000 tonnes of coal.

Figure 5.9 shows the saving of CO_2 and coal every year.

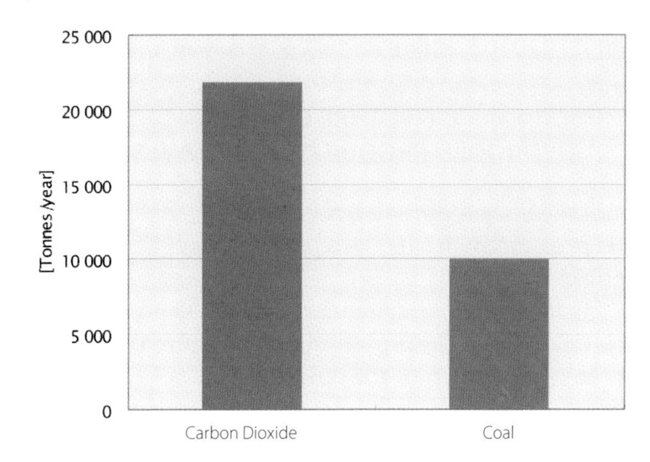

Figure 5.9 CO_2 and coal saving by 10 MW PV plants

5.5.2 Economical and social development of Perenjori region

Solar power creates local jobs, increases the tax base and increases the economic multiplier, which is the number of times each dollar is re-spent in the local community.

Solar industries directly employ people and support a number of jobs in diverse areas such as steel moulding, electrical and plumbing contracting, architecture and system design, and battery and electrical equipment.

5.5.3 Local industry development

This project presents the potential to set up a solar module manufacturing facility either at Perenjori or at a nearby location, and power generated from the project will become a base from which to attract industries in the region. Mining industries would be the major industry to bring multi-billion dollars of investment to the region.

The potential industries can be:

- solar cells and modules manufacturing industries;
- mining operations;
- dairy industries;

- flour milling;
- food processing;
- farm machinery manufacturing and repairing industries;
- water bottling; and
- ecotourism.

5.5.4 Future investment potential in the region

As mentioned earlier, mining operations would be the major industries to be attracted to the Perenjori region. Mining alone can bring billion of dollars to the region. As soon as industrial development begins, a number of other industries will follow, bringing in new investments.

5.6 FUTURE OPTION: HYDROGEN PRODUCTION FROM VLS-PV

Since there is no provision for unused energy storage in the project, hydrogen production may be a viable option.

Hydrogen, the simplest element, is composed of one proton and one electron. It makes up more than 90 % of the composition of the universe. It is the third most abundant element in the Earth's surface and is found mostly in water. Under ordinary (Earthly) conditions, hydrogen is a colourless, odourless, tasteless and non-poisonous gas composed of diatomic molecules (H_2).

The use of hydrogen:

- offers an elegant way to store and regenerate electricity;
- does not produce carbon dioxide or any other greenhouse gas;
- avoids all the costs of producing and using fossil fuels;
- is clean – in the cycle of storing and releasing energy, it uses water as both the source and end product;
- fits with our existing electrical grid infrastructure and is an excellent way to power a vehicle or a house;
- could solve our electrical grid's load management issues;
- addresses the finite nature of the Earth's fossil fuel reserves;
- avoids the hidden ecological, health, aesthetic and property damage costs of fossil fuel use; and
- does not produce acid rain or deplete the ozone layer.

Hydrogen's potential use in fuel and energy applications includes powering vehicles, running turbines or fuel cells to produce electricity, and generating heat and electricity for buildings. The current focus is on hydrogen's use in fuel cells.

A fuel cell works like a battery but does not run down or need recharging. It will produce electricity and heat as long as fuel (hydrogen) is supplied, and consists of two electrodes – a positive electrode (or anode) and a negative electrode (or cathode) – sandwiched around an electrolyte. Hydrogen is fed to the anode and oxygen is fed to the cathode. When activated by a catalyst, hydrogen atoms separate into protons and electrons, which take different paths to the cathode. The electrons go through an external circuit, creating a flow of electricity. The protons migrate through the electrolyte to the cathode, where they reunite with the oxygen and electrons to produce water and heat.

No economic viability of hydrogen production from solar PV has yet been established, and so a careful step has to be taken to introduce such a concept.

5.7 RECOMMENDATIONS

In this study, Perenjori was selected as a potential site for setting up a VLS-PV project.

From the perspective of regional development, providing electricity to mining operations, global warming and other environmental issues, it is apparent that VLS-PV systems can contribute substantially to local energy needs, can play a considerable role in the environment and socio-economic development of the region, and can become economically and technologically feasible.

To establish a VLS-PV project at Perenjori, the following is recommended:

- A systematic route as proposed in this study has to be adopted for the development of VLS–PV.
- Rainwater harvesting should be tried out to utilize the facility effectively without incurring much cost, thus compensating for the high cost of electricity generation from solar PV projects.
- Hydrogen production from the unused electricity should be tried out; but the economic cost of such a new concept should be carefully worked out.
- Local industry should be promoted for the manufacturing of solar cells, modules and silicon wafers for

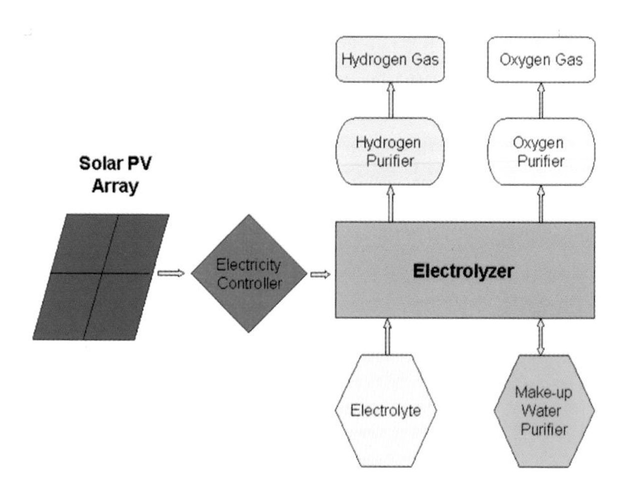

Figure 5.10 Schematic flowchart of the hydrogen production from PV system

supplying the modules to VLS–PV projects, as well as for the export market.

- A sustainable funding arrangement by the government of Western Australia and the federal government of Australia has to be made to promote and set up VLS–PV projects.
- The Research Institute for Sustainable Energy (RISE) can undertake a range of sustainable energy activities, including systems design, testing and training in operation and management.
- Since there are sufficient funds available for AGO/SEDO to provide a 50 % subsidy, immediate efforts have to be made to apply for the subsidy before these funds lapse.
- A funding of 50 000 AUD (approximately 38 000 USD) should be made available to take the project to its next stage, including a subsidy for setting up the proposed 10 MW project. A funding of 300 000 AUD (approximately 228 000 USD) would also be required to carry out a feasibility study and set up the specifications for the tender to be floated.
- A consultant or an officer of the Mid West Development Commission or Shire of Perenjori should be responsible for driving this project further, as well as for pursuing the required financing.

REFERENCES

1 Kurokawa, K., Takashima, T., Hirasawa, T., Kichimi, T., Imura, T., Nishioka, T., Iitsuka, H. and Tashiro, N. (1996) 'Case studies of large-scale PV systems distributed around desert area of the world', *PVSEC-9*, Miyazaki, November, pA-IV-4

2 Kurokawa, K., Takashima, T., Hirasawa, T., Kichimi, T., Imura, T., Nishioka, T., Iitsuka, H. and Tashiro, N. (1997) 'Case studies of large-scale PV systems distributed around desert areas of the world', *Solar Energy Materials and Solar Cells*, vol 47, nos 1–4, pp189–196

3 IEA PVPS Task VI/Subtask 50 (1999) *A Preliminary Analysis of Very Large-Scale Photovoltaic Power Generation (VLS-PV) Systems*, IEA-PVPS VI-5 1999:1

4 Kurokawa, K. (ed) (2003) *Energy from the Desert: Feasibility of Very Large Scale Photovoltaic Power Generation (VLS-PV) Systems*, James and James, London

5 IEA PVPS Task 8 meeting held in Perth in February 2004

6 Australian Greenhouse Office (AGO), Canberra

7 Sustainable Energy Development Office of Western Australia (SEDO), Perth

8 Shire of Perenjori

9 Mid West Development Commission, Geraldton

10 Advanced Energy Systems Ltd, Perth

11 Prime Solar Limited, Perth

12 RISE Lab, Perth

13 Shell Solar, Germany

14 Australian Greenhouse Office (AGO) *Energy Saving Calculator*, AGO, Canberra

15 Pyle, W., Healy, J. and Cortez, R. (1994) 'Solar hydrogen production by electrolysis', *Home Power Magazine*, February–March, pp32–38

Desert Region Community Development

6.1 INTRODUCTION

In this chapter, the following issues are researched and investigated for developing and constructing a new concept of integrated renewable energy technology and photovoltaic (PV) generation, installed in arid regions or deserts with agricultural technology:

- the possibility of utilizing VLS-PV in desert areas;
- a sustainable scenario for installing VLS-PV systems;
- modelling of a sustainable society with solar photovoltaic generation and greening in the desert;
- specific problems of very large scale photovoltaic power generation (VLS-PV) systems in desert areas;
- the use of solar PV generation with regard to elemental technology/systems for sustainable agriculture;
- sustainable agriculture utilizing other regional renewable energy options;
- desertification issues;
- issues regarding irrigation; and
- the effects of VLS-PV on the location environment.

Voluntary advisers and students assembled and formed the Desert Working Group at the Tokyo University of Agriculture and Technology's (TUAT's) 21st Century Centre of Excellence (COE) Programme, Evolution and Survival of Technology-Based Civilization. The working group built a network featuring local and international research activities, which aims to establish large scale photovoltaic systems in desert areas.

6.2 APPROACHES TO ESTABLISHING A DESERT COMMUNITY

6.2.1 Sustainable PV stations

Research of VLS-PV design

In this case study, we design a system as the preceding step to the actual verification test, assuming that the VLS-PV will be installed in a desert area with a large amount of solar radiation and that the generated electricity will be transmitted to the main power system. An image of the system is shown in Figure 6.1.

Here, our objective is to evaluate the system economically and environmentally. In this study, the detailed design of a 100 MW VLS-PV system supposed to be installed in a desert area was examined. Two types of mountings, fixed flat plate and tracking, were also evaluated. Regarding the module type, poly-Si PV and a-Si PV modules were evaluated. These designs were evaluated using life-cycle analysis (LCA) methodology. As a result, the generation costs will be economical when the price of PV modules decreases to 2,0 USD/W and will then be able to compete against existing electrical power. In addition, installations of VLS-PV in developing countries have the potential to create employment. The commerce and industry of countries that are economically growing may benefit by producing the mountings, foundations and other components of the system. Energy payback time is 1,8 years, and life-cycle carbon dioxide (CO_2) emission intensity is 12 g-C/kWh. From these results, the impact of VLS-PV on the environment can be minimized. As mentioned earlier, the VLS-PV system is promising for solving global environment problems from environmental and economic perspectives.

Analysis of solar energy resources

We develop a remote sensing model using satellite imagery and select a good location for constructing a VLS-PV system in a desert area. Based on this model, we propose an efficient ground truth method in the research zone.

This study reports, in particular, on the results of analysing PV resource potential in the Gobi Desert of Mongolia and the Thar Desert of India. Figure 6.2 shows the estimated result for a suitable area in the Gobi Desert in Mongolia. The details are provided in

Figure 6.1 Image of VLS-PV system

Appendix A, section A.2 'Resource analysis of solar energy by using satellite images'.

Regional characteristics of the photovoltaic generation system

Sun-dried brick is sustainable on unused ground, such as a desert, and can be made in simplified production facilities at low temperatures (790 to 1 000° C). Transportation and equipment costs will be reduced by using sun-dried bricks instead of steel for the mountings for PV modules. In forested lands, it is possible to use wood for these mountings.

Desalination through PV systems

The salt ratio in groundwater in desert areas is high. For people and agriculture, desalination technology is important in arid areas. This chapter discusses reverse osmosis and electro-dialysis processes for desalination. Both technologies are already in use, but there are few case examples. Figure 6.3 provides an example of an electro-dialysis desalination system using PV.

6.2.2 Regional society

Essentially, almost all desert areas are harsh environments in which to live. Therefore, the population density in desert regions is low, and almost all areas are undeveloped and generate low income. In addition,

Figure 6.2 Estimated results for a suitable area within the Gobi Desert in Mongolia

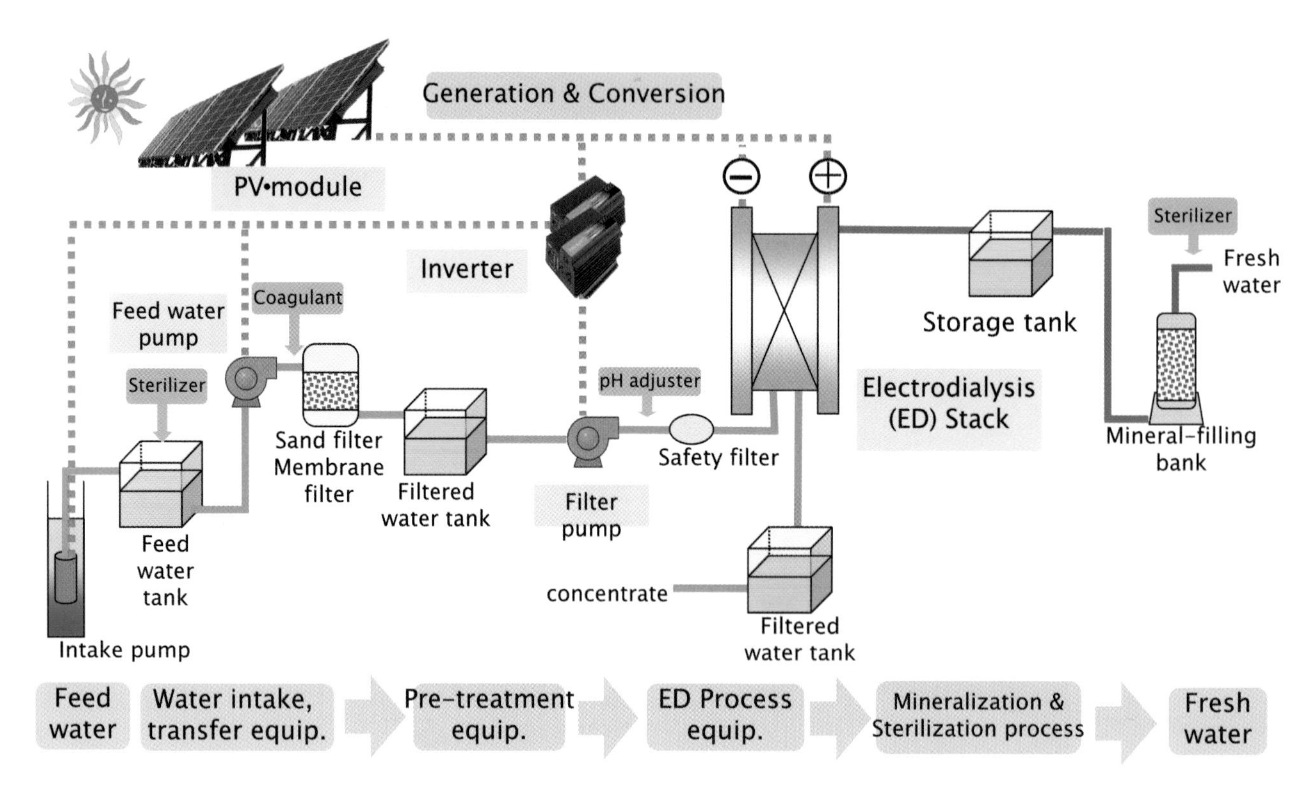

Figure 6.3 A flow diagram of an electro-dialysis desalination system using PV

desert region ecosystems are very sensitive. If people create too much pasturage, fell too many trees or poorly manage water resources, this easily causes desertification. People then go to other areas, resulting in increasing pressures on desert lands and an increase in desertification.

It is therefore important to research a regional society model that can supply a sufficient level of energy for living, as well as sustainable economic activity, by using solar resources with developed technology.

Figure 6.4 shows the basic economy of a village in a desert area. An analysis of the configuration of the economy depicts a suitable community in an arid land area.

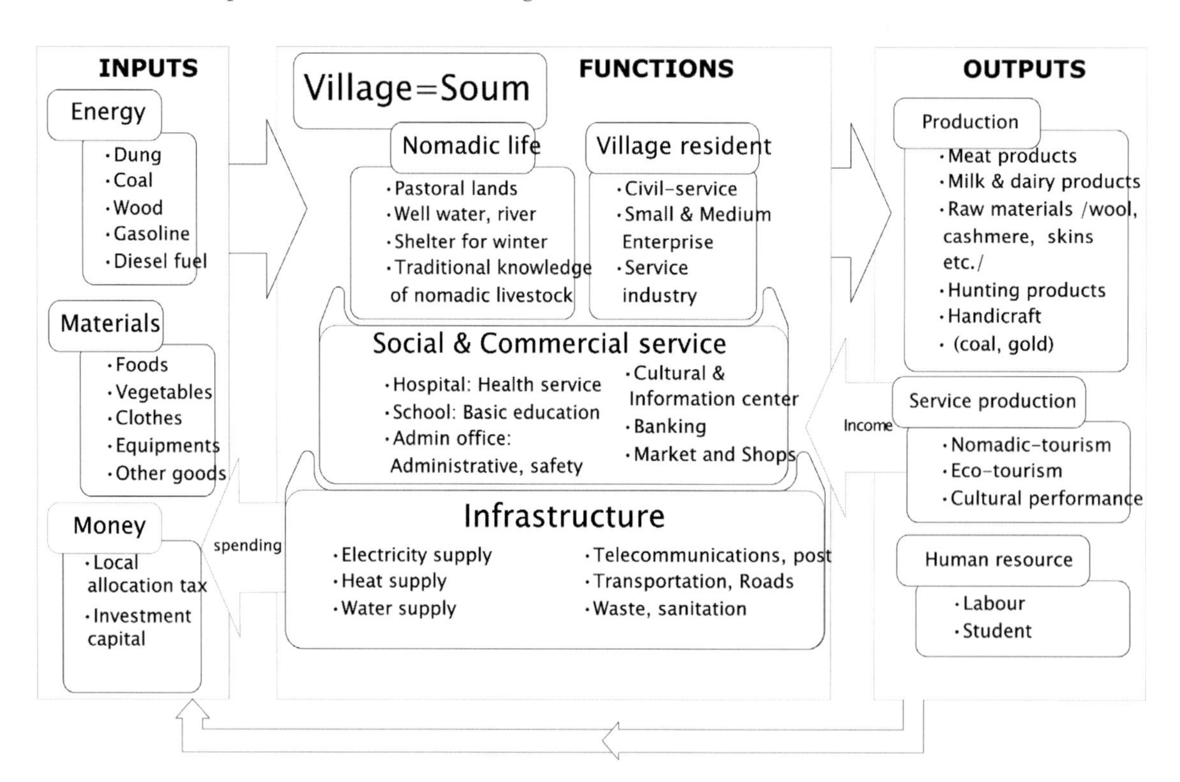

Figure 6.4 Example of a desert area village's material, energy and currency flow

6.2.3 Suggestion for development modelling of solar photovoltaic generation and greening model society with stability

This research aims to achieve the self-reliant sustainable development of an area with a driving force consisting of a combined VLS-PV system installed on an ongoing basis on unused ground, such as a desert, combined with the development of greening and agriculture that utilize the abundant electrical power produced by the system. As a result, it is important not to introduce too large a PV system, and agricultural development must not result in abandoned land afterwards. Self-reliance and stability in the social economy of a district must also be maintained.

We propose a new development model where abundant electrical power, supplied from a PV system in a desert area, is utilized for the greening agricultural development of the neighbourhood.

Community proposal

This section discusses what kind of community should be established and what kind of technologies are required. We focus on semi-arid land areas, which are diverse; as a result, various ideas are elaborated upon:

- *Collaboration on agriculture and technology*: renewable energy is utilized, to a large extent, for people and agriculture. Electricity derived from renewable energy is stored for use at night. Extra energy generated during the day is distributed to other cities. Renewable energy is utilized for agriculture and tree planting. Salt damage is prevented by using desalination equipment.
- *Economically sustainable community*: many kinds of renewable energy equipment are installed, and power generation is distributed to other cities. Grid electricity is used at night.
- *Self-supporting community*: middle-class equipment and power storage are installed in the community for people's daily use. The community is not connected to a grid.
- *Electrification*: accelerate electrification in villages. By using electricity derived from renewable energy, the community's standard of living can be improved. If a power grid exists near the village, it is possible to sell the electricity.
- *Using salt-damaged land*: salt-damaged land is suitable for PV power plants because the land is difficult to use for other purposes, such as agriculture.
- *Farm and power plant*: a specialized community for agriculture with a PV power plant can reduce battery storage. Farms can use enough electricity.
- *Portable system*: this targets nomads, who move by season. It is not that small, but it can be moved and is a simple system.

Table 6.1 summarize these ideas. We focus on the community's ability to collaborate on agriculture and technology.

Plan of collaboration on agriculture and technology

The desert working group aims to achieve an ideal community. Agriculture and tree planting can be facilitated with plentiful renewable energy, and electricity is used by neighbouring cities. The community must have the potential to appeal to other people through the following themes:

- *Sustainable PV stations: sustainable energy production*. Here, the VLS-PV system is the main feature, using sunshine, along with wind power and other renewable energy. At night, electricity comes from battery storage instead of from the grid.
- *Forest, grassland and farmland: sustainable farming*. Utilizing soil and water conservation technology, we will conserve and rehabilitate the landscape. This may not be difficult to achieve with PV support

Table 6.1 Summary of community proposed

	Collaborating on agriculture and technology	Economically sustainable community	Self-supporting community	Electrification	Farm and power plant	Portable system
Renewable energy	A	A	A	A	A	A
Battery	A	n.a.	A or n.a.	n.a.	(night time)	A (small size)
Agriculture; greening	A	A	A	n.a.	A	Seasonal
Grid connection	A	A	A	A	A	n.a.
Stable electricity	A	A	n.a.	n.a.	n.a.	n.a.
Salt-damaged land	n.a.	n.a.	n.a.	A	n.a.	n.a.
Groundwater	A	A	A	n.a.	A	Depends upon people
Living people	A	A	A	n.a.	a little	A
Electricity demand	Community; city	City; community	Community (city)	(City)	Agriculture; city	Small village; ger

Notes: A = applicable; n.a. = not applicable.

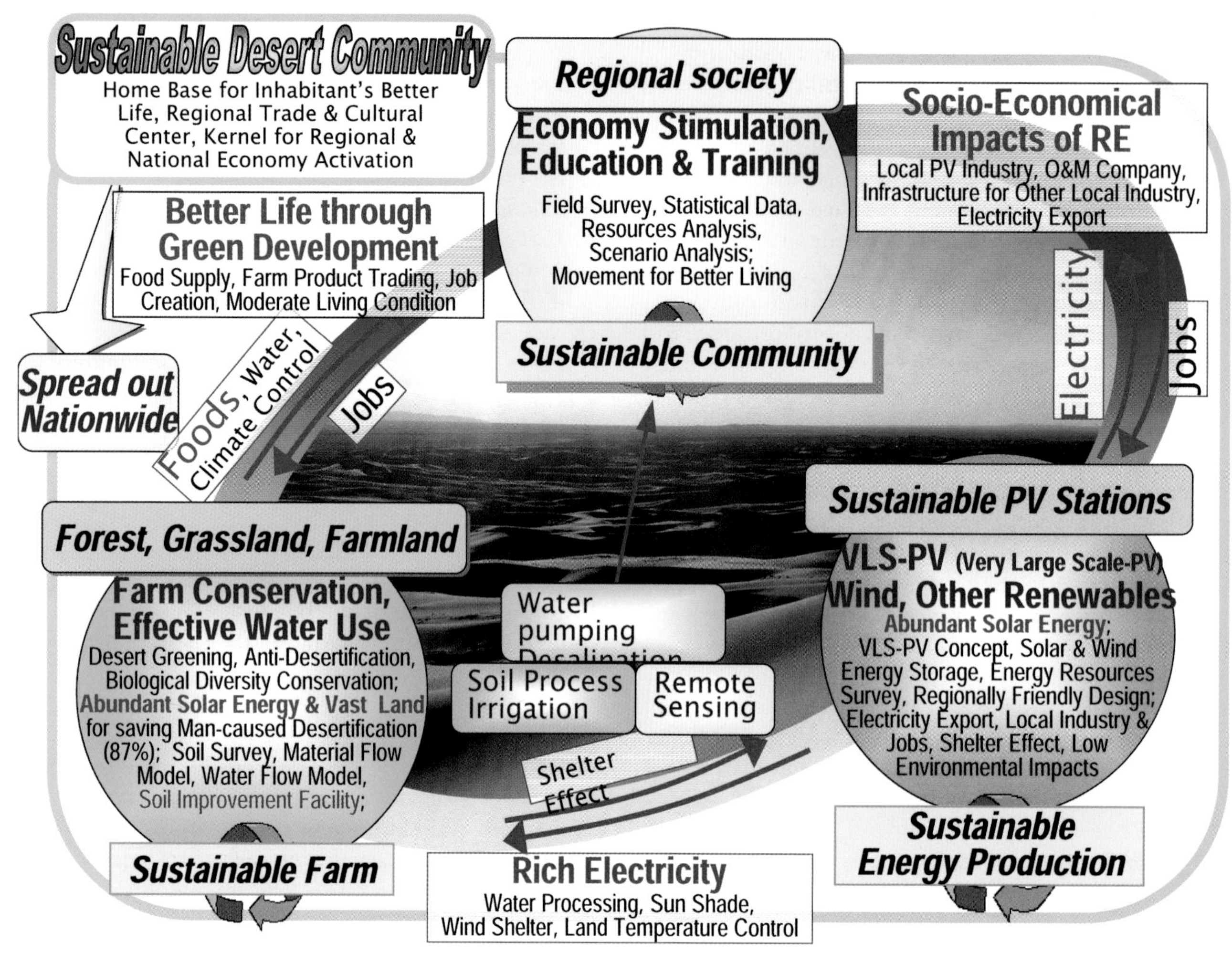

Figure 6.5 Framework of desert community development and research topic

because the main reason for desertification is human activity relating to energy needs. Conservation and rehabilitation will take plant and animal ecology into account.

- *Regional society: sustainable community.* Statistical and scenario analyses are used to develop an ideal community. In order to sustain regional society, in addition to the facilities and technology needed, education and training are provided.
- *Remote sensing.* Remote sensing technology has the potential to find suitable places in which to implement VLS-PV and wind power systems. It can generate data on soil and water required for sustainable agricultural production.
- *Desalination.* Renewable energy power can operate a desalination system, which will supply drinking and irrigation water.
- *Collaboration between renewable station and forest–grassland–farmland.* Proper operation of a water pump and desalination system can provide good quality water and remove salt from the groundwater. This can enhance crop yields and reduce the use of fuelwood. Renewable energy-driven greenhouse agriculture can produce high quality and high product yields. These systems can reduce greenhouse gas (GHG) emissions.
- *Effect of collaboration between PV station and regional society.* PV stations can go beyond supplying electricity to local communities to producing more employment in the region. This would also increase local incomes through selling electricity. PV structures can also protect houses from strong winds.
- *Effect of collaboration between forest–grassland–farmland and regional society.* Agriculture can supply food to communities and leads to wind protection effects and employment. People in local communities can obtain more income by selling products.
- *Technology.* Implementing a subsurface drainage system may save groundwater quality, and desalination equipment protects soil from damage due to salinization.

This community is theoretically ideal, but it requires groundwater and a power grid nearby.

6.3 SUSTAINABLE AGRICULTURE DEVELOPMENT USING PV SYSTEMS

6.3.1 Basic characteristics of desert soil and countermeasures against over-cropping

Chemical properties of desert soil

The disintegration of rock, both chemically and physically, leads to grain refining and the soil formation process. The rate of soil formation is estimated at about 0,1 mm of soil layer per year. It depends upon climatic conditions such as temperature and precipitation. Generally, the rate of soil formation is high in humid and moderate regions, while it is slow in cool and dry regions due to the low chemical reaction rate and lack of water, which is essential for chemical disintegration. In the physical weathering process, repeating heating and cooling causes the parent material to turn into small fragments. In the chemical weathering process, minerals dissolved in water may react and partly precipitate to form clay particles.

The major cations in the Earth's crust, in order of susceptibility to weathering, are:

1 calcium (Ca), magnesium (Mg) and sodium (Na);
2 potassium (K);
3 silicon (Si);
4 aluminium (Al) and iron (Fe).

Therefore, soils in tropical and subtropical regions, where there is high precipitation and warm temperatures, and thus higher rates of chemical reactions, tend to show a red colour due to the abundance of less weatherable Si, Fe and Al oxides. In cold or dry areas, where slow weathering is expected, soils tend to have more alkaline elements, such as Na, Ca and Mg. Soils in arid and semi-arid regions tend to contain more Na and Ca due to the soil formation process mentioned earlier. These elements affect soil productivity; therefore, sustainable agricultural development needs to consider soil formation history.

Climate of arid and semi-arid regions

Generally, arid and semi-arid regions are described according to their annual precipitation. Strictly speaking, however, it is better to think of a balance of precipitation and evapotranspiration in the area. Annual precipitation in semi-arid areas is less than half of evapotranspiration. In this region, it is necessary to introduce irrigation for agricultural production. Figure 6.6 indicates the variation within the Sahara Desert's precipitation levels and areal expansion for the 1980s. Precipitation trends coincide with changes in desert area. When precipitation shifts to less than the average, the area of the Sahara Desert, shown in red and in the bar chart, increases. In this way, the desert area may change with changes in the natural condition. When discussing the desertification of different land-use areas shown in Table 6.2, it is important to remember that fluctuation of natural conditions may change the area of desert land.

The year 1984 was the end of a drought period, and the same year showed the maximum area of desertification. Precipitation recovered during 1984 to 1990 and draught started again in 1991. It can be suggested that desertification increased after 1991. The fact that the area of pasture land and rain-fed fields are larger in 1991 than in 1984 does not imply that countermeasures against desertification were successful during this period. Furthermore, the desertification rate for irrigated land did not decrease, indicating that farmers still employed inappropriate irrigation practices.

Major reasons for desertification of agricultural land

Overgrazing, shifting cultivation, erosion of farmland, high groundwater level due to excess irrigation water-

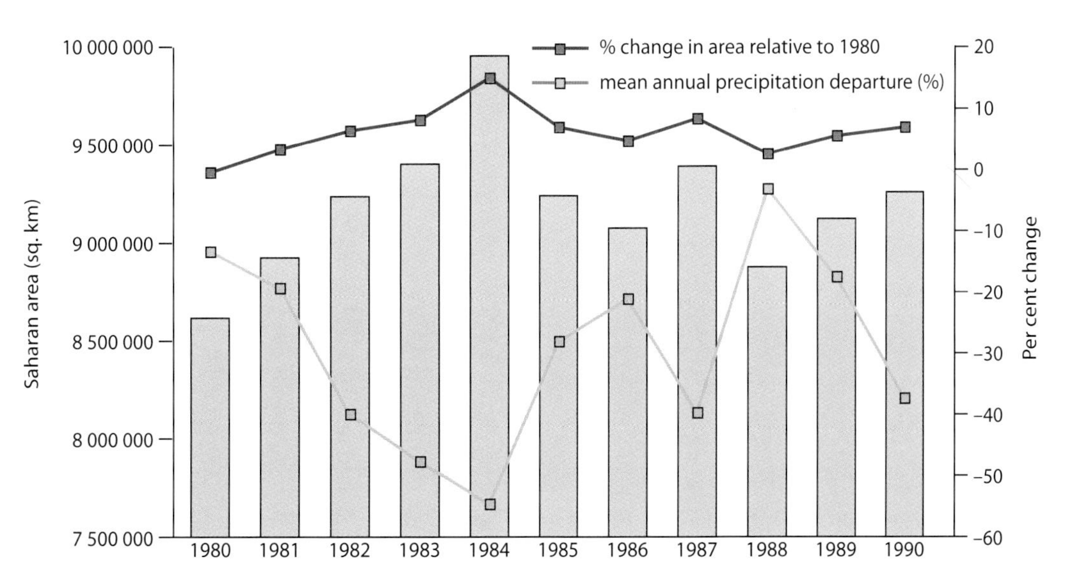

Figure 6.6 Expansion of Sahara Desert[1]

Table 6.2 Area of major arid regions[2]

	Utilizable arid region (10 000 km²)	Hyper desert (10 000 km²)	Total arid regions (10 000 km²)
Algeria	45,7	190,1	235,8
Burkina Faso	25,6		25,6
Egypt	5,1	94,9	100,0
Ethiopia	80,7	1,4	82,1
Mali	84,6	31,5	116,1
Mauritania	59,4	43,7	103,1
Namibia	73,2	9,0	82,2
Niger	41,6		41,6
Somalia	61,7	1,3	63,0
Sudan	151,1	68,7	219,8
Western Sahara	14,1	12,5	26,6
P. R. China	496,5	37,9	534,4
India	161,8		161,8
Mongolia	129,0		129,0
Pakistan	80,2		80,2
USSR	677,1		677,1
Australia	702,4		702,4
Mexico	153,0	1,7	154,7
USA	385,5	1,3	386,8
Argentina	194,3		194,3
Brazil	80,8		80,8
TOTAL	3703,4	494,0	4197,4
World	5281,7	917,8	6199,5

logging, and salinization following inappropriate irrigation practices are the major factors that induce desertification. Seepage from open channel canals for irrigation does not concern local people, but this causes the serious problem of rising groundwater levels. Although details of over-harvesting are not discussed here, we should reduce the over-harvesting of wood by introducing alternative energy sources into people's daily lives, since large quantities of cut wood are for fuelwood.

As mentioned earlier, since chemical weathering of parent rock does not occur in arid and semi-arid regions, soils in these regions tend to contain more alkali-metal and alkaline-earth elements. These minerals dissolve into irrigation water and are mobilized.[3] Mobilized salts such as Na and Ca chlorides move towards the surface soil and concentrate there due to the high rate of evaporation. This causes salt accumulation in the surface soil and deteriorates soil structure, reducing soil productivity. Using saline water for surface irrigation is another cause of salinization. This leads to surface input of salt into agricultural lands. As the salt accumulates at the land surface, increasing the concentration of minerals near the surface layer drastically decreases the land yield. For example, the Syr Darya River near the Aral Sea of Central Asia, which is well known for water quality deterioration, showed electrical conductivity (EC) and total dissolved salt (TDS) of 0,18 dS/m and 1,3g/L. Assuming irrigation water contains 1g/L of salts, 500 mm of irrigation for a cropping season may induce 5 kg/m² of salt into the irrigated land. Research in Punjab, Pakistan, revealed that 570 kg/ha of salts accumulated in agricultural land in a year as a result of using saline water for irrigation.[4]

The Food and Agriculture Organization (FAO)[5] proposed an equation to predict the yield reduction due to salty water irrigation.

$$Y_r = 100 - b(ECe - a) \qquad [6.1]$$

where Y_r = is relative yield = (yield under saline water irrigation)/(yield under good water irrigation) × 100 (%),
a = threshold concentration of salt for no reduction of yield equivalent to electrolyte conductivity dS/m,
b = ratio of reduction of yield to increasing salt concentration, and
ECe= salt concentration of saturated extract solution from soil (equivalent to electrolyte conductivity dS/m).

Examples of data on crop tolerances to salinity and some specific ions and elements are given in Table 6.3.[5] Irrigated land, 16 % of the world's arable land, contributes 40 % of world food production. About 60 % of the irrigated lands belong to arid or semi-arid regions and are affected by degradation processes such as waterlogging and salinization. Consequently, nearly 1 million ha of irrigated lands are abandoned because of productivity reduction. During the late 1990s, 25 % of the world's irrigated lands – 60 million to 70 million ha – was salt affected to some extent.[6] For example, in some areas in the Indus River basin in Pakistan, the

Table 6.3 Salt tolerance of plants

Crop	*ECe** (dS/m)	Slope* (%/dS/m)	EC of 50 % reduction of yield (ds/m)	*ECe* of 50 % germination (dS/m)
Barley	8,0	5,0	18	16~24
Cotton	7,7	5,2	17	15
Wheat	6,0	7,1	13	14~16
Alfalfa	2,0	7,3	8,9	8~13
Tomatoes	0,9	9,0	7,6	7,6
Cabbage	1,0	14,0	7,0	13
Corn	1,7	12,0	5,9	21~24
Paddy rice	3,0**	12,0**	3,6	18

Note: * a, b in Equation 6.1; ** paddy rice is under pond water during cropping season

groundwater level (previously more than 15 m) is now less than 1,5 m below ground following the introduction of widespread irrigation in the region. The area of salt-affected lands was 1,22 million ha (12 % of the cultivated land) and 1,19 million ha (20 % of the cultivated land) in the Punjab and Sind states in 1979. It has increased to 1,47 and 1,60 million ha.[7]

The rise of the groundwater level due to irrigation is affected by the amount of irrigation water used, irrigation practices and water resource management. Even in Japan, the efficiency of surface irrigation is less than 70 %. This means that more than 30 % of irrigated water is lost through deep percolation and evaporation. In larger fields, more water loss is expected. Irrigated agriculture in arid and semi-arid regions usually relies upon water supplied by open channels, as shown in Figure 6.7. Leakage from the channel bed becomes a significant loss of irrigation water. This leakage also induces increasing percolation and raises groundwater levels. In large and flat plains, additional water head difference is required to distribute water to the whole area. Attaining the head difference requires additional water consumption for raising the water level of the irrigation pond. This additional water will finally irrigate the fields and contribute to increased groundwater levels. Alternatively, sprinkler irrigation, such as pivot irrigation, is employed in industrialized countries (see Figure 6.8).

Figure 6.7 Irrigation canal during the dry season, Gansu Province, China

Figure 6.8 Pivot irrigation

Effect of salt on soil behaviour

In arid and semi-arid regions, chemical weathering can be delayed due to lack of water. Thus, minerals that are easily weathered, such as Na and Ca, still remain in the soil without leaching. When irrigation is introduced in such soil, these minerals dissolve into irrigated water and move upwards due to evapotranspiration, finally accumulating at the ground surface. This especially occurs when appropriate drainage or saline water is used for irrigation. Generally, leaching is implemented to improve salt-accumulated soil. The efficiency of leaching depends upon soil structure and permeability.

The difficulty in improving salt-accumulated soil is that during salt accumulation sodification occurs. Sodification is the process of becoming Na-rich, following adsorption of much Na to the soil. Sodic soil has an unstable structure and often has low permeability. Especially under low electrolyte concentration, which may typically happen with the leaching of salts, sodic soil disperses and most water-percolating pores in the sodic soil are clogged. Once clogging occurs, it is impossible to achieve further leaching of salts since almost no water flows through the soil. More dispersion and unstable soil structure may be a factor in enhanced water erosion since low permeability causes more runoff, and more dispersion makes the soil prone to becoming eroded. The composition and species of salts in soil water, the ions adsorbed on the soil's surface and the type of clay minerals all affect soil structure.[8]

Salinization of soils in arid regions often involves soil sodification. The formation of sodic soil largely depends upon soil colloidal properties. The amount of charge in the soil depends on the type of colloid. This means that the soil is a large ion-exchange material. Since soil particles usually have a negative charge, excess cations distribute in the vicinity of soil particles. This counterbalances the effect of the negative charge of the soil particles and maintains electric neutrality. Thermodynamic equilibrium is also attained between adsorbed ions and ions in soil water.

When there is a change in the composition of soil solution due to an increase in concentration following evaporation or an input of additional ions from irrigation, ion exchange occurs, reaching an alternative equilibrium condition. When the soil contains the two major cations, Na and Ca, an Na–Ca ion exchange reaction in the soil is expressed as follows:

$$\text{Soil–Ca} + 2\text{Na}^+ \leftrightarrow \text{Soil–2Na} + \text{Ca}^{2+} \qquad [6.2]$$

Here, 'soil' in the equation denotes exchange sites of the soil. The equation is called the Gapon convention. It is an empirical equation used widely to predict cation exchange equilibrium between mono-valent Na^+ and divalent Ca^{2+}:

$$\frac{Q_{Na}}{Q_{Ca}} = K_G\left(\frac{C_{Na}}{\sqrt{\frac{C_{Ca}}{2}}}\right) \qquad [6.3]$$

where C is the concentration of cations in soil water [mol_c/L or eq/L], Q is adsorbed cation per dry mass of soil [mol_c/kg] and the subscript represents the variety of ion. K_G is the Gapon constant, and when we follow the custom as C_{Na} and $C_{Ca}/2$ expressed by charge equivalent [$mmol_c$/L], K_G becomes approximately 0,5 [$(mmol_c/L)^{-0,5}$] for much of the soil. In addition, when $mmol_c$/L (= meq/L) is used as the unit of concentration, the bracket on the right-hand side of Equation 6.3 is called the sodium adsorption ratio (SAR).

Considering the total concentration of a bi-component system (Na^+ and Ca^{2+}) as C_o (= $C_{Na} + C_{Ca}$), the ratio of Na^+ concentration to the total concentration (= C_{Na}/C_o) as f_{Na} and the ratio of adsorbed Na^+ to the cation exchange capacity (CEC) as N_{Na} (=Q_{Na}/CEC), then Equation 6.3 becomes:

$$\frac{N_{Na}}{1 - N_{Na}} = K_G\frac{f_{Na}\sqrt{C_o}}{\sqrt{\frac{(1 - f_{Na})}{2}}} \qquad [6.4]$$

This equation implies that the even ratio of the Na solution is constant, and a higher total concentration causes more Na adsorption. This also suggests that as evaporation proceeds and the concentration of the salt in soil water becomes higher, the ratio of adsorbed Na increases – thus sodification occurs. Exchangeable sodium percentage (ESP), which is the degree of saturation of soil exchange sites by sodium, is widely used as a sodification index. Cation exchange capacity is more laborious to measure than cation concentration in solution; but ESP can be estimated by a solution composition based on the Gapon convention. Since selectivity of divalent Mg adsorption is often similar to that of Ca, the Gapon convention is also applied for those divalent cations, and now SAR may be calculated by the following formula:

$$\text{SAR} = \frac{C_{Na}}{\sqrt{\frac{(C_{Ca} + C_{Na})}{2}}} \qquad [6.5]$$

Substituting Equation 6.5 for Equation 6.3 (Gapon convention) with the correct dimension, we have a relation between ESP and SAR:

$$\frac{\text{ESP}}{100 - \text{ESP}} = \frac{K_G}{\sqrt{1\,000}}\text{SAR} \approx 0{,}015 \times \text{SAR} \qquad [6.6]$$

Equation 6.6 is widely used for practical purposes to predict the extent of sodification from the quality of soil water or irrigation water. Here, we show a result of Russo and Bresler.[9] They investigated the relation between soil permeability as a function of electrolyte concentration and SAR. The soil sample was loamy soil with 20 % (by weight) of clay fraction. Unsaturated hydraulic conductivity of Ca-saturated soil (SAR = 0) was determined as a unique function of volumetric water content.

In contrast, when SAR and the ratio of adsorbed Na was increased (sodificated), the soil showed a different electrolyte dependence, with a different volumetric water content. It can be expected that the flow pattern of solutions percolating through a soil may differ based on the electrolyte concentration and the flux density of the applied solution. This suggests that we should employ complicated water management for higher water efficiency and that the amount of irrigation

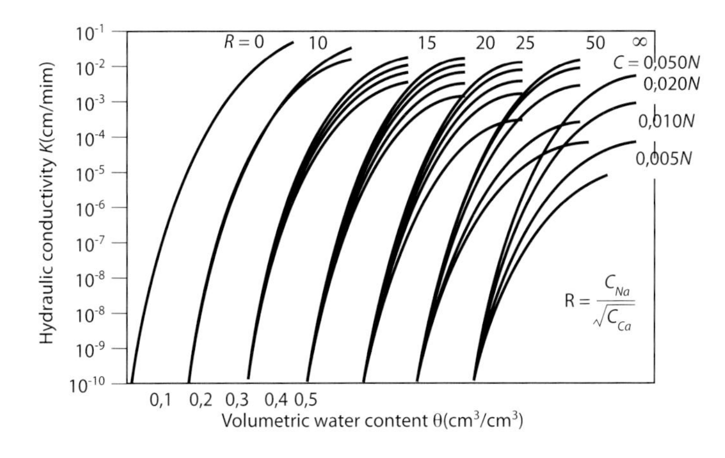

Figure 6.9 Sodium adsorption ratio (SAR), concentration, moisture content and soil permeability[9]

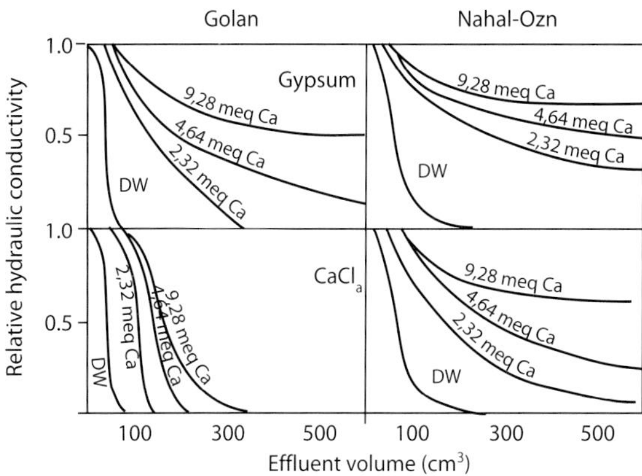

Figure 6.10 Electrolyte concentration and hydraulic conductivity[10]

should be adjusted to adapt to the electrolyte concentration. In addition, with a higher value of SAR, the concentration dependence of hydraulic conductivity was prominent even at low water content, as shown in Figure 6.9.

Figure 6.10 shows changes in hydraulic conductivity of sodic soils with SAR of 20 when distilled water was applied following surface application of $CaCl_2$ and gypsum ($CaSO_4$).[10] The numbers in Figure 6.10 (2,32, 4,64 and 9,28 milli-mole charge equivalent (meq) of Ca) are the amounts of calcium cation applied to the soil as either $CaCl_2$ or $CaSO_4$ in meq of Ca, and correspond to 30 %, 60 %, and 120 % of the amount of exchangeable sodium contained in the 10 cm-thick surface soil layer. In the case of applying Ca equivalent to 30 % of exchangeable sodium for the layer, the SAR of the surface 3 cm-deep layer decreased while, applying the Ca equivalent to 120 % of the exchangeable sodium of the surface soil layer, most of the adsorbed sodium of the 10 cm-thick soil layer was exchanged by Ca. When distilled water was applied to sodic soil of SAR 20 (denoted by DW in Figure 6.10), the permeability decreased immediately after the infiltration experiment began. This is because of clogging within the soil's pore space as a result of percolation of distilled water, which has a lower electrolyte concentration than critical coagulation concentration (CCC).

Natural rainfall induces similar soil dispersion because it contains few electrolytes like distilled water. This is why soil crust was easily formed on sodic soil under natural rainfall. In addition, when low electrolyte concentration water is applied to the soil to amend salt accumulation of sodic soil, the same result may emerge. When calcium salt was applied, decreasing permeability was observed irrespective of salt variety.

The efficiency of maintaining permeability became prominent with an increasing amount of applied calcium. This indicates that applying more calcium may enhance replacing exchangeable Na with Ca, and the soil may become less dispersive. However, in an actual field, it is unrealistic to apply huge amounts of calcium salt because of the cost involved. Generally, when the soil is ESP less than 15, soil dispersion and deterioration from irrigation and leaching will not occur. We can determine the amount of calcium needed to reduce the ESP to less than 15 by using Equations 6.5 and 6.6. The tendency for a change in permeability to occur during leaching is different for applications of $CaCl_2$ or $CaSO_4$ solutions because of the difference in electrolyte concentration of the leaching water. Highly soluble $CaCl_2$ dissolves quickly into leaching water and gives a higher electrolyte concentration of leaching solution. Soon after, however, most of the $CaCl_2$ dissolves and electrolyte concentrations in the leaching water decline. Therefore, when $CaCl_2$ is applied, electrolyte concentration quickly decreases with increasing percolation and soil dispersion, and a decrease in permeability subsequently occurs. The electric conductivity of the soil water when permeability decreased was 0,08 dS/m, and this corresponds to approximately 1 mmolc/L of electrolyte concentration. In the case of gypsum ($CaSO_4$), which has a solubility lower than $CaCl_2$, applied $CaSO_4$ dissolves into leaching water gradually, and the lowest electric conductivity is 0,5 dS/m. This concentration is higher than CCC and dispersion and clogging occur in the soil.

Plant salt tolerance

Semi-desert regions, 20 % of the land territory, extend around deserts. These regions have deteriorated to desert due to human activity; however, since there used to be an abundance of vegetation, this suggests the possibility of producing food again. In semi-desert regions, water scarcity is the most serious limitation for crop production. At the same time, salinity stress also restricts crop production (see Table 6.4). Salinity stress occurs with increasing water stress. There are complex problems in dealing with the direct effects of Na, Ca, excess Ca and Na, and plant nutrient deficiency of Mg, Fe and other micro-nutrients. Although some salt tolerance substances, such as glycine betaine, are reported,

Table 6.4 Salt content of soils and reduction of yield[11]

Type	Species	Yield reduction 0 %	10 %	25 %	50 %
		ECw (dS/m)			
Fruit	Date palm	2,7	4,5	7,3	12,0
	Orange	1,1	1,6	2,2	3,2
	Apple	1,0	1,6	2,2	3,2
	Grape	1,0	1,7	2,7	4,5
	Avocado	0,9	1,2	1,7	2,4
Field crops	Barley	5,3	6,7	8,7	12,0
	Cotton	5,1	6,4	8,4	12,0
	Sugar beet	4,7	5,8	7,5	10,0
	Wheat	4,0	4,9	6,4	8,7
	Soybean	3,3	3,7	4,2	5,0
Forage crops	Bermuda grass	4,6	5,7	7,2	9,8
	Perennial ryegrass	3,7	4,6	5,9	8,1
	Tall fescue	2,6	3,9	5,7	8,9
	Alfalfa	1,3	2,2	3,6	5,9
	Orchard grass	1,0	2,1	3,7	6,4

the mechanism of salt tolerance is not yet fully understood. Consequently, when planning farming in semi-arid regions, cultivated species are selected by referring to data on plant salt tolerance from previous studies.

In order to maintain better physical soil condition, such as water content, permeability or water retention, Ca and Mg salts are frequently applied. Salinity tolerance of the cultivated crop should be considered when those salts are applied to a field. In most cases, calcium sulphate is thought to be a sodification-preventing material without reducing the yield because of its moderate solubility.

6.3.2 Electricity needs for sustainable irrigated agriculture

Countermeasures for salt-affected soils

Electricity can act as a countermeasure to salinity in agricultural land. Countermeasures to salinity problems attributed to rising groundwater levels are as follows:

- decreasing the amount of seepage water; and
- reducing the groundwater table by pumping up groundwater.

Figure 6.11 shows the edge of the Loess Plateau in China. Excess irrigation water quickly percolates downward through the loess because of its high permeability. Excess irrigation water flows laterally along the surface of the bedrock and seeps out from the edge of the plateau. Under these circumstances, the water flow on farmland above the plateau essentially moves downward and salt accumulation does not occur. Constructing drainage pipes below farmlands, therefore, and draining excess irrigated water to prevent rising groundwater levels seems to be an effective way of preventing salt accumulation following excess irrigation.

In Egypt, efforts were made to reduce salinization by combining irrigation and drainage practices. Consequently, as a ten-year average, yield increased by 27 to 37 % for wheat, 21 to 35 % for cotton, and 21 to 38 % for maize. The use of concrete channels or pipelines for distributing irrigation water is an effective method of preventing leakage from open channels. To reduce groundwater levels, it is also beneficial to construct many wells and to pump up the groundwater. There are some places where the groundwater level decreases by 0,9 to 1,8 m after ten years of pumping. It is not realistic for local residents in developing countries to pump up water to reduce groundwater levels when the rise of the groundwater level is in excess of several metres, as in Punjab State, Pakistan. Here, farming is possible when the groundwater level is around 2,5 m. For a 100 ha farmland, however, we have to pump up 100 million m^3 of water to reduce the groundwater

Figure 6.11 Downward leaching of salt at the Loess Plateau in Gansu Province, China

level by 1 m. It is difficult for farmers to utilize huge amounts of electricity to drive pumps. In addition, pumped water cannot always be used directly for irrigation since it usually contains salt. Some regions have the following regulations:

- Pumped water can be used directly for irrigation when the salt concentration is lower than 1 000 parts per million (ppm).
- Combine with low saline stream water to dilute pumped water when the salt concentration is 1 000 to 3 000 ppm.
- Drain away water with salt concentrations of higher than 3 000 ppm.

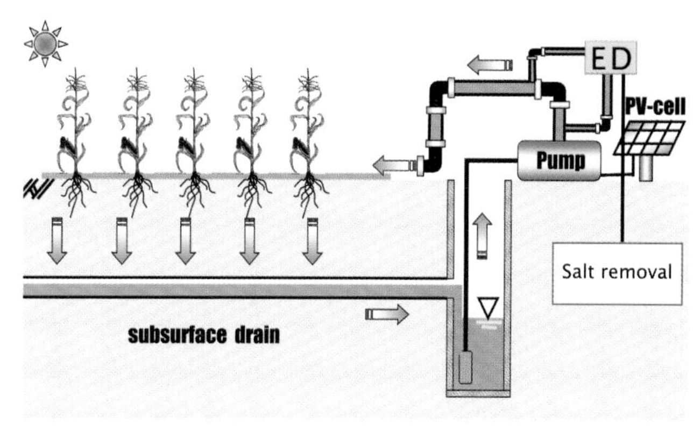

Figure 6.12 Irrigation–drainage combined system

Arid and semi-arid regions are better suited to agriculture that uses solar irradiation. Sometimes sunshine is excessive; but too much sunshine can be easily reduced. Rain-fed agriculture is limited by low precipitation. Although irrigated agriculture has been maintained in some places, it often causes soil degradation. The main causes of soil degradation are improper water distribution and inadequate irrigation practices and quality.

In figuring out the appropriate amount and quality of irrigation water – including groundwater behaviour and required amount of water – quality control of irrigation water and control of the amount of irrigation using pumps are required. While these points are easy to achieve in industrialized countries, arid and semi-arid regions still have problems with the cost and energy resources to improve water distribution when using pumps. We suggest realistic proposals that assume the use of PV systems for supplying energy to control the quality and quantity of irrigation water.

Introducing PV systems in arid and semi-arid regions has the potential to introduce electricity as an alternative energy source to fuelwood. This has implications for forest conservation by reducing reliance on fuelwood, and it can help to conserve water courses by reusing desalinated groundwater. In the past, water was desalinated through distillation by using solar heat, but this requires a huge area to desalinate salty water because the distillation rate of solar heat is quite low. A considerable amount of low saline water is available when we use a PV system to generate electricity to drive a desalination system, such as reverse osmosis (RO) or electro-dialysis (ED).

Complex system of irrigation and drainage

As mentioned earlier, one method of saving irrigation water and ensuring sustainably irrigated agriculture is through the complex irrigation and drainage system that we can see in Egypt. However, high salt-concentrated water cannot be used and drained in a conventional system. As a result, the performance of this system is limited in areas of severe salinization. Here, we propose a method that uses an electric pump and a desalination system driven by PV to reuse excess irrigated water (see Figure 6.12).

Since it is difficult to supply water for cereal crops by drip irrigation, we propose to prevent the rise of groundwater levels by subsurface drainage and by reusing drained water. Using existing sprinklers is also a possible alternative to surface irrigation. Although desalinated pumped water is directly supplied to surface irrigation, it is better to mix it with river water or other groundwater in order to control electrolyte concentrations to prevent the degradation of soil structure.

Subsurface irrigation

Evaporation at the ground surface is significant in arid and semi-arid regions, especially in inland areas. When surface irrigation is implemented, most irrigated water is lost through evaporation. In this case study, the application of an underground irrigation system may be beneficial. Underground irrigation is dominated by the soil's water retentiveness, soil permeability and root distribution of the crop. Occasionally, not enough water can be supplied (see Figure 6.13).

Drip irrigation

Drip irrigation is well known as a form of agriculture in arid regions. Researchers in Israel and the US are contributing to developing drip irrigation systems, and have accumulated a huge volume of technical knowledge and data. Drip irrigation is not adequate for field

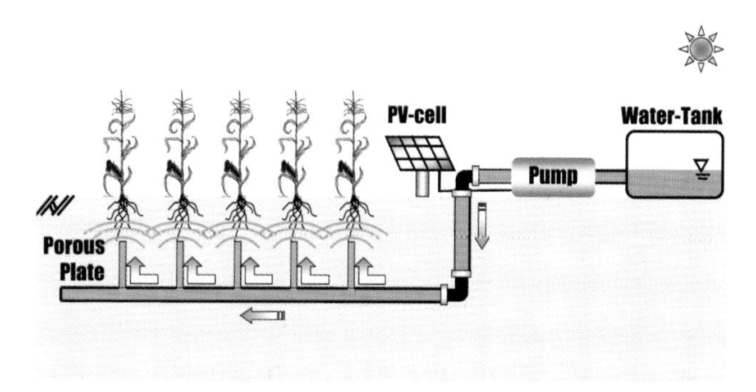

Figure 6.13 Subsurface irrigation

crops, such as cereals, because the emitters limit the water supplied. However, since running and plant costs are high for cash crops, such as fruits, olives and nuts, they are generally appropriate for drip irrigation. Drip irrigation rarely results in rising groundwater levels or salt accumulation as long as there is appropriate information about crop water requirements, soil texture, water retentiveness and permeability. There are still problems that we have to resolve, such as the amount of water-soluble salt in irrigation water, which influences efficiency, and securing the energy resources for moving the drip irrigation system. Figure 6.14 shows a drip irrigation system with a desalination system using PV as an energy resource.

Energy-free irrigation–drainage systems that exist in developing countries need to be improved. Although they can adapt to native natural conditions, the productivity of the conventional energy-free irrigation systems is still low, and they will not supply enough food for the growing population. In addition, as mentioned earlier, the irrigation of salty water degrades both the physical and chemical condition of agricultural soil. Electricity is required to improve this. Fuelwood is the only energy resource possible, other than electricity, to control the salt concentration of water through distillation; however, this would cause a huge depletion of woodland and enhance desertification. While pump-driven irrigation drastically increases crop yield, too much irrigation often causes groundwater levels to rise and salt accumulation in the topsoil. This then reduces the yield within a few decades of excess irrigation.

Irrigation–drainage and/or drainage–subsurface irrigation systems are effective ways of establishing sustainable irrigated agriculture. They assume PV use for pumping the irrigated and drained water, as well as for water quality control. Further studies should clarify their technical and economic feasibility, as well as their role in rural sociology. For example, in the process of technical transfer, the sociological aspect of a rural community is an important issue. In addition, the cost and life-cycle assessment of introducing electro-dialysis and reverse osmosis water treatment systems for agricultural purposes to semi-arid regions are also important, as is discussing how to establish the operating community. Education and training of staff are crucial to implementing the system sustainably in a region. In previous cases, trained staff sometimes moved to other jobs, attracted by higher incomes. This kind of problem can be prevented by higher payment; but the impact of high incomes on local society should also be considered. Higher payments to technical workers may cause inflation of local economies or collapse of the society. This is not favourable to introducing new technology within a rural area.

Life-cycle assessment (LCA) is also crucial in attaining sustainable development and efficient agricultural production. The costs of achieving sustainability, reducing CO_2 emissions and energy consumption should be

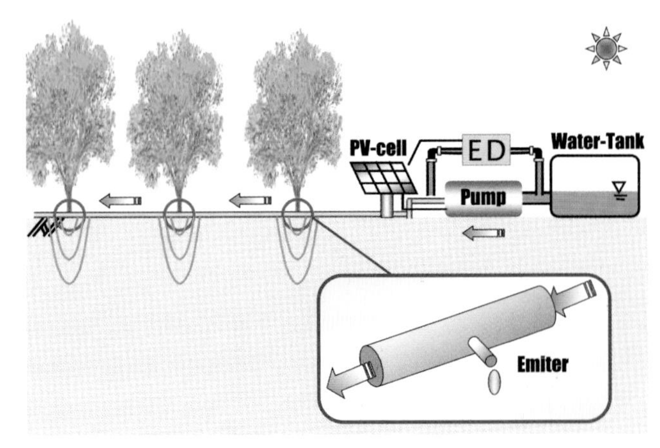

Figure 6.14 Desaliniatd drip irrigation system

issues for discussion. A further issue to consider is that the cost of introducing a PV system for irrigation and desalination is not always affordable based on the income from trading agricultural products. Moreover, soils in arid and semi-arid regions are quite fragile, and even today it is almost impossible to produce soil artificially. This means that when introducing new technology to semi-arid regions for the purposes of agriculture, care should be taken not to harm the soils in the region. As a result, informed discussions using theoretical, numerical and experimental information are required otherwise mistakes will be made in introducing new irrigation systems to semi-arid regions. Trial and error practices *in situ* should never occur.

6.3.3 Effects of VLS-PV on the local environment

Factors to consider

During substance exchange between soils and the Earth's atmosphere, the functions of the atmospheric boundary layer, which is formed on the ground surface, play a considerable role. The atmospheric boundary layer is affected by wind velocity, as well as by the terrestrial and geomorphic features, and physical structures of the plant community.[12] According to these factors, if VLS-PV systems are installed widely in the area, the micro-meteorology or the surroundings of the PV array will possibly be affected.

Ground surface is regarded as the boundary between the underground (soil water and groundwater) and the atmosphere from the viewpoint of the hydrological cycle. Different conditions of the micro-meteorology above the ground surface significantly affect the ecosystem – that is, the vegetation of the region. Evaporation and transpiration at the ground surface involve a complicated interrelation of atmospheric environmental factors, water regimes in soils and factors affecting changes in the plant environment (humidity, temperature, etc.), as shown in Figure 6.15. For example, although a plant's physiological activity, starting with photosynthesis, is often regarded as biological behav-

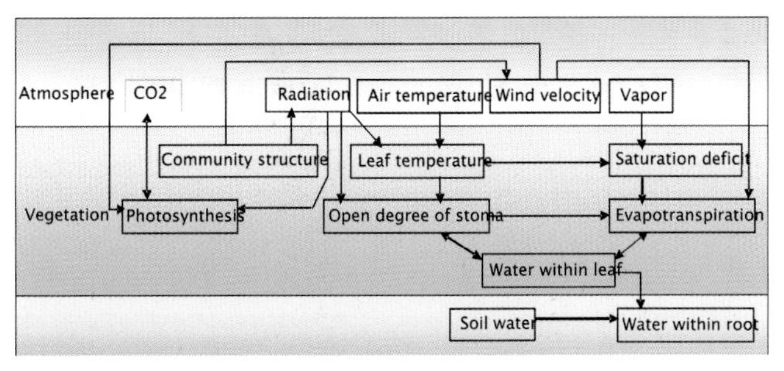

Figure 6.15 Interrelationship among atmosphere, vegetation and soil in the water cycle[13]

iour, the driving force of a photosynthesis reaction is the change in entropy by evaporation. The rate and direction of evaporation itself is determined mostly by the chemical potential (or the energy state) of water in the soil surrounding the plant roots and of the vapour in the atmosphere around leaf stomata.

Plants take in CO_2 from the atmosphere and release oxygen (O_2) produced by photosynthesis in exchange. In this process, the transport of CO_2 via the stomata of the leaf surface is related to the difference in CO_2 concentration between open air (C) and inside the stoma (Cs), and the wind velocity (U). Using these factors, CO_2 exchange ratio (κ_C) and leaf area density (s) can be expressed as follows:

$$CO_2 \text{ amount by exchange} = \kappa_C \cdot sU(H - Hs) \quad [6.7]$$

Using moisture (vapour) density in open air (H) and in stoma (Hs), and vapour exchange ratio (κ_H), evaporation is derived as follows:

$$\text{Moisture amount by exchange} = \kappa_H \cdot sU(H - H_s) \quad [6.8]$$

Added to this are the exchange ratios κ_C and κ_H, which are expressed as $\kappa_H, \kappa_C \sim U^{-0,5} \cdot O_P^{0,66}$, depending upon the wind velocity and the degree of openness of the stomata.[14]

Evaporation from the ground surface depends upon the difference between absolute humidity (vapour pressure difference) in the atmosphere and the soil; the heat supply at the ground surface (consumed as latent heat in evaporation); the water content and the speed of upward water movement in the soil; the supply of water for evaporation; and the wind velocity in the open air. Especially when the water supply is not a limiting factor, the available evapotranspiration rate at the ground surface, based on heat balance, is expressed as follows:

$$\ell E_S = \frac{\frac{\Delta}{0.5}(S - G) + \ell E_a}{\frac{\Delta}{0.5} + 1}$$

where

$$\ell E_a = 0.35(e_a - e_d)\left(0.5 + \frac{u_2}{100}\right) \quad [6.9]$$

where ℓE_S is the amount of available evapotranspiration, e_a is the saturated vapour pressure at mean temperature, e_d is the mean vapour pressure in the air, u_2 is the mean wind velocity at a point 2 m above ground, S is the net amount of radiation, G is the amount of heat conduction, e_0 is the saturated vapour pressure at ground surface temperature (T_0), and $\Delta = \frac{e_0 - e_a}{T_0 - T_a}$ (Penman equation).[12]

As described so far, the evapotranspiration rate is affected by meteorological environmental factors, such as vapour pressure in the atmosphere, wind velocity, temperature and radiation levels. In this regard, in arid or semi-arid regions, water supply to the ground surface that causes the movement of unsaturated soil water may affect salt accumulation at the soil surface, which is the reduction of the chemical potential of water at the soil surface fringe associated with the accumulation (that is, depression of boiling point) and control of evaporation. In this case, analysing the transport of unsaturated water and substances in the soil is also important, as well as monitoring micro-meteorology on the ground surface.

If PV panels are installed in an arid or semi-arid area, the following physio-environmental changes can be predicted:

- a decrease in the amount of solar radiation in the shaded area beneath the PV module; and
- a deformation of the structure of the atmospheric boundary layer.

These factors affect evapotranspiration, the soil's hydrological cycle and available water resources. It has been reported that when the distance travelled by the wind becomes greater than 100 m, the shape of the ground surface's atmospheric boundary layer is affected by the morphology of the ground surface. This also leads to the movement of water in the soil; thus, evaporation under large scale PV arrays may be affected by the VLS-PV panels. Consequently, it is important to evaluate the effect of the VLS-PV installation on micro-meteorology and plant ecosystems.

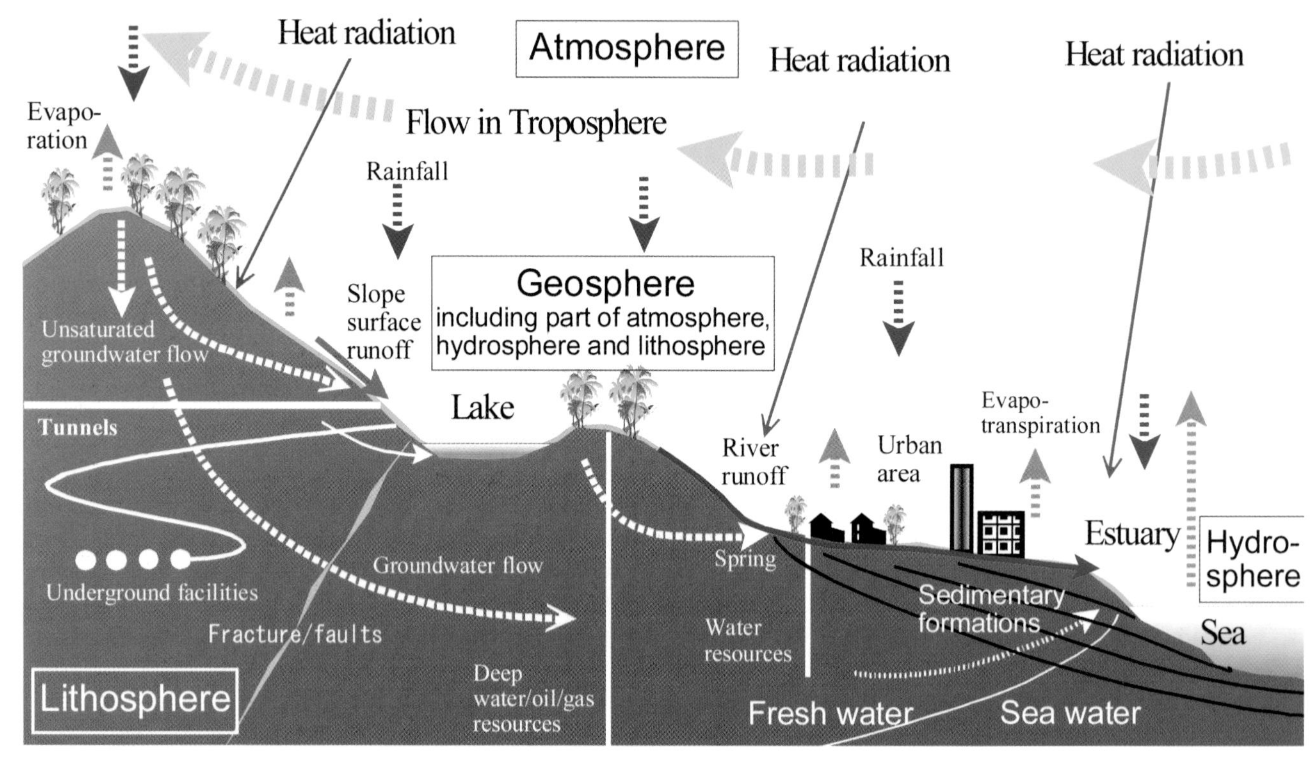

Figure 6.16 Regional water circulation
Source: courtesy of Mr Tosaka

These ideas take field-scale to small watershed-scale hydrological phenomena into account. It is important to consider broader regional water circulation when discussing harmonization between sustainable production, the surrounding environment and local ecosystems. Numerical simulation using physical-based models is often required when analysing hydrological phenomena quantitatively, using available data such as groundwater levels, wind direction and velocity, humidity, radiation evapotranspiration, and precipitation. Figure 6.16 provides a conceptual outline of hydrological circulation over seashore to mountainous areas. Compared to the field-scale simulation (although individual processes are rough), large scale simulation includes many more elements.

Hydrologic circulation change prediction following VLS-PV installation

More than 30 general circulation models (GCMs) are running at research institutes all over the world; but the results of these simulations do not agree well. One reason for this conflict is that we still do not have enough data for world water circulation. To improve this situation, a multi-institutional project (the Global Energy and Water Cycle Experiment – Asian Monsoon Experiment, GEWEX/GAME project) aims to predict future water resources and flood disasters by developing a high-resolution hydrologic circulation model in monsoon Asia from fundamental hydrologic data. Data acquisition is the most important mission of the early stage of the project. The project also aims to bring together the output of each research body, and to collect observation data to develop a Japanese model in order to contribute to the fourth evaluation report by the Intergovernmental Panel on Climate Change (IPCC).

Our understanding of natural physical processes and their impacts has lagged considerably behind the pace of development. However, recent advancements in the field of environmental science have made it possible to assess, mitigate and project the environmental impacts of man-made changes with a fair degree of accuracy. Today, it is well known that the capacity of the Earth and its ecosystems to annul the effects of man-made changes has been declining ever since the onset of civilization. Land surface properties, too, have changed markedly during the past several centuries, decisively affecting local and global climate. The genesis of climatological changes in temperate latitudes, for example, is traced to the large scale conversion of forests and grassland into highly productive cropland and pasture. This influenced, beyond a doubt, the interaction of the Earth's energy and water balance with the terrestrial biosphere and the atmosphere. An increase in the number and intensity of thunderstorms over the years, therefore, is regarded as a direct consequence of an increase in thermal energy derived from sensible heating at the Earth's surface and condensation/freezing of water vapour.

Weather forecasters use a variety of parameters derived from the vertical profile of thermodynamic variables, such as local temperature, wind circulation, available potential energy, evaporation and transpiration, to assess the potential for rainfall. Several studies were undertaken to precisely establish various correlations between different physical processes in the environment, particularly interactions between vegetation cover and climate. These, by and large, indicate that initial changes in the landscape had a cooling effect owing to a gradual increase in albedo–vegetation morphology. However, a turning point in the global climate has been observed, and further land cover changes are likely to have a warming effect, governed by physiological mechanisms rather than vegetation morphological mechanisms, especially in the tropics and sub-tropics. Therefore, the environmental impact of large scale land-use change must be ascertained before embarking upon technologically driven mega-projects. Nevertheless, a proper mix of strategies has the potential to mitigate the adverse impact on the environment and to reverse negative consequences.

Early investigations in this field revealed evidence of strong linkages between land surface, albedo and atmospheric circulations.[15,16] Another land surface property – surface roughness – was also found to play an important role in energy and water fluxes to the atmosphere.[17–19] These developments opened up new approaches and opportunities to an understanding of land–atmosphere interactions. A plethora of scientific activities followed. These created immense interest in allied investigations, such as linking leaf area index to climate changes. Chase et al (1996) even used the leaf area index to show that human disturbance of land cover can change atmospheric circulation and can affect climate in areas distant from the actual disturbance.[20] Newly developed techniques have been further assimilated into boundary layer modelling. These developments have provided an insight into the linkages between the landscape and deep cumulus convection.[21–24] Initial forays were primarily on land-use change-induced global climate changes.[25] They were followed by investigations on regional-scale or meso-scale changes. Researchers invariably examined the effects of extra-tropical land cover on temperature and hydrology and of land cover change on climate.[26–30] However, most recent efforts have undoubtedly concentrated more on the regional effects of tropical deforestation and regional desertification because of the vulnerability of these ecosystems and their importance to human populations.[2,31,32] Gradually, it is being accepted that anthropogenic land cover changes of the type already observed can significantly influence global climate. Chase et al (1996) made concerted efforts at simulating and analysing the effects of realistic global green leaf area changes to explain winter climate anomalies.[20] They correlated the phenomenon with tropical deforestation affecting low-latitude convection and global-scale circulations. Zheng and Eltahir (1997) further explored and simulated mid-latitude tele-connections between complete tropical deforestation and a linear wave model.[33] They concluded that appropriate conditions existed for the propagation of tropical waves into the extra-tropics. Other significant developments included simulation of isolated extra-tropical effects due to physical or physiological changes in tropical vegetation, as well as understanding biosphere–precipitation relationships.[18] The effort of Narisma and Pitman (2003) to simulate the impact of 200 years of land cover changes on the Australian near-surface climate is a significant step forward.[34]

Changes in land surface albedo can result from land-use alterations and may thus be tied to an anthropogenic cause. Hansen and Nazarenko (2003) estimated that a forcing of –0,4 W/m^2 has resulted so far,[35] approximately half of which occurred in the industrial era only. The largest effect is estimated to be at the high latitudes, where snow-covered forests that had a lower albedo have been replaced by snow-covered deforested areas. Some researchers have pointed out that the albedo of a cultivated field is affected more by a given snowfall than the albedo of an evergreen forest. These simulations were based upon pre-industrial vegetation replaced by current land-use patterns. It was also found that global mean forcing could be –0,21 W/m^2, with the largest contributions coming from deforested areas in Eurasia and North America. Betts et al (1997, 1998) estimated an instantaneous radiative forcing of –0,20 W/m^2 by surface albedo change replacing natural vegetation by present-day land use.[36,37] Kalnay and Cai (2003a) reported an evaluation of a second influence of land use – an influence other than urbanization – on climate – namely, the role of agriculture and irrigation.[38] They used the reanalysed upper-air data to examine trends over the last 50 years, a method independent of surface-based measurements, to separate urbanization from non-urbanization effects. They concluded, in their follow-up report of corrected measurements, that the trend in daily mean temperature due to land-use changes is an increase of 0,35°C per century.[39]

Thus, landscape pattern and its average condition exert a major control on weather and climate[40,41] at the meso-scale as well as the global scale. The direct surface heating anomalies and subsequent changes in circulation associated with land-use change are permanent. A tele-connection pattern is observed since land-use changes are found to be a major factor, capable of governing the temperature and precipitation distributions worldwide. Therefore, mega-technological projects such as very large scale photovoltaic power generation (VLS-PV) installation in deserts[42] are capable of changing the entire landscape. An initial investigation using green area modelling and meso- as well as global-scale modelling indicates that such projects are likely to increase convective rainfall under certain conditions. At this point, the land cover change is manifested as a change in surface greenness or a change in vegetation pattern, not as actual VLS-PV installation due to time constraints and data availability. The analysis, basic assumptions and projections are addressed in the following sections.

Preliminary modelling of VLS-PV installation

Changes in land cover influence climate and, more specifically, the hydrological cycle through partitioning the incoming solar radiation into turbulent, sensible and latent heat fluxes. The energy budgets at the surface can be written as follows:[40]

$$R_N = Q_G + H + L\,(E + T) \qquad [6.10]$$

$$R_N = Q_S\,(1 - A) + Q_{LW}\,(\downarrow) - Q_{LW}\,(\uparrow) \qquad [6.11]$$

$$P = E + T + RO + I \qquad [6.12]$$

where R_N represents the net radiative fluxes; P is the precipitation; E is evaporation (conversion of liquid water from the soil surface into water vapour); T is transpiration (phase conversion to water vapour through stomata of plants); Q_G is the soil heat flux; H is the turbulent sensible heat flux; L (E + T) is the turbulent latent heat flux; L is the latent heat of vaporization; RO is runoff; I is infiltration; Q_s is insolation; A is albedo; Q_{LW} ($\downarrow$) is downwelling long-wave radiation; and Q_{LW} ($\uparrow$) is upwelling long-wave radiation, which is:

$$Q_{LW}\,(\uparrow) = (1 - \varepsilon)\,Q_{LW}\,(\downarrow) + \varepsilon\,\sigma\,T_s^4 \qquad [6.12]$$

where, ε is the surface emissivity; σ is Stefan's constant and T_s is the surface temperature.

Any land-use change that alters one or more of the variables in Equations 6.10, 6.11 or 6.12 will directly affect the potential for rainfall. For example, decrease in albedo (A) would increase R_N, thus making more heat energy available for Q_G, H, E and T. If the surface is dry and bare, all of the heat energy would go into Q_G and H. The heat that goes into H increases the temperature potential. Thus, changes in the Earth's surface can induce significant disturbances in the surface energy and moisture budgets. These would influence the heat and moisture fluxes within the planetary boundary layer, the convective available potential energy (CAPE) and other fundamentals of deep cumulus cloud activity.

Choice of model

The standard version of the National Centre for Atmospheric Research (NCAR) Community Climate Model version 3.0 (CCM3), coupled with the Land Surface Model (LSM)[31], is used for simulation in the current investigation. The LSM is a GCM-scale parameterization of atmospheric–land surface exchanges and accounts for vegetation properties as functions of one of 24 basic vegetation types. The LSM includes a single-level canopy (the lake model) and calculates averaged surface fluxes due to sub grid-scale vegetation types and hydrology. Two phonological properties, leaf area index (LAI) and stem area index (SAI), are interpolated between prescribed monthly values. Albedos are calculated as a function of LAI and SAI among other factors. All other vegetation properties are considered to be seasonally constant.

The Community Land Model (CLM) was developed as a collaborative venture between land modelling groups and combines the best features of three well-documented and modular land models, the Land Surface Model (LSM) of Bonan (1996),[43] the Biosphere–Atmosphere Transfer Scheme (BATS) of Dickinson et al (1993),[44] and the 1994 version of the Chinese Academy of Sciences, Institute of Atmospheric Physics LSM (IAP94).[45–47] CLM is preferred over other land modelling schemes such as RAMS (Regional Atmospheric Modeling System),[48] VIC (Variable Infiltration Capacity) and NOAH (National Centres for Environmental Prediction/Oregon State University/Air Force/Hydrologic Research Laboratory) because CLM makes better use of a two-leaf canopy model.[46] CLM also features better soil depth parameterization and soil temperature. It simulates using a heat diffusion equation in ten soil layers, which is important as feedback and memory of the water table, while 'deep' ground processes are important to the surface water and energy budgets. It has a thinner and more realistic top soil layer (1,75 cm), which gives more accurate results for the estimation of surface soil water fluxes and the diurnal cycle of surface soil temperature. An improved annual cycle of runoff water and, consequently, evapotranspiration (because of improved soil depth parameterizations) is a significant improvement over other models.

Methodology

It is assumed that the PV system covers the land surface horizontally and not inclined at any angle for the entire area of the Sahara, Gobi and Thar deserts. The net energy flux at the ground surface is the sum of beam and diffused solar radiation and long wavelength radiation from the Earth and the atmosphere. The sensible heat flux and latent heat flux were calculated using the following equations:

$$H = \rho_{atm}\,C_p\,(\theta_{atm} - T_g)/r_{ah} \qquad [6.14]$$

where, ρ_{atm} is air density; C_p is the specific heat of the air; θ_{atm} is the potential; T_g is the ground temperature; r_{ah} is the aerodynamic resistance to sensible heat flux and water vapour transfer between atmosphere at height z_1. The potential temperature is again a function of ambient temperature, surface pressure and atmospheric pressure. The latent heat flux can be obtained from the following equation:

$$\lambda E = \rho_{atm} C_p (e_{atm} - e^*(T_g))/\gamma (r_{aw} - r_{srf}) \quad [6.15]$$

where, r_{aw}, r_{srf} are aerodynamic resistances to sensible heat flux and water vapour transfer between the atmosphere at height z_2 and z_3; e_{atm} is vapour pressure; $e^*(T_g)$ is saturation vapour pressure at the ground surface; and γ is the psychrometric constant.

A CLMl was run for one year for two cases of the assumed albedo for the whole area of the Gobi (162,8 × 10^4), Sahara (743 × 10^4 km^2) and Thar deserts (44,3 × 10^4 km^2). The following two cases were considered:

1 case A: visible range radiation = 0,20, infrared radiation = 0,20;
2 case B visible range radiation = 0,10, infrared radiation = 0,20.

The input global data were obtained from the CCM3 website (www.cgd.ucar.edu/cms/ccm3/) to run the model in off-line mode for one-year projections.

Model results and discussion

The surface albedo is of utmost importance in determining the absorption of solar energy, although large variations are possible due to the type and nature of the vegetation. Albedo also generally decreases as the surface wetness increases. The range of albedo for snow-free conditions is from about 0,1 for tropical forest to about 0,4 for some dry sandy surfaces. With incident mean daily solar fluxes typical of the tropics (300 to 400 W/m^2 in cloudless conditions), a variation of about 100 W/m^2 is possible. However, this would be an upper limit to spatial variations and is only conceivable in an area with an extreme climatic change or extensive human intervention, such as PV installation, deforestation or irrigation on highly reflective soil.

VLS-PV reduces albedo considerably compared to dry sandy desert soils. Therefore, it is invariably supposed to increase sensible heat flux (H) and surface heat flux (Q_G) substantially. However, a complex interaction of various parameters seems to undermine this hypothesis according to the results obtained from a preliminary study of VLS-PV installation in the Thar, Gobi and Sahara deserts. This is a very positive indication for VLS-PV installation. Initial results for a single-year simulation of CLM indicate a very small increase in the global sensible heat flux (H) and surface heat flux (Q_G). This implies that the installation of VLS-PV in deserts would have a positive radiative force due to surface albedo change replacing the normal desert ecosystem. The projected increase in daily global mean temperature due to this land-use change in the Sahara, Gobi and Thar deserts is approximately 0,001°C in one year. A global mean temperature rise of 0,1°C in one century can definitely be concluded without green area patches around VLS-PV.

The most significant finding of this case study is that a proper green area modelling coupled with VLS-PV would have a positive impact on convective rainfall, rather than a negative impact due to a rise in sensible and surface heat flux. These green areas would further decrease dust storms, reduce CO_2 emissions and mitigate several ill effects of fossil-fuel energy use. They would attract monsoon due to evapotranspiration characteristics and would regulate temperature. Although the current study is based upon land cover change, manifested as a change in surface greenness or a change in vegetation pattern, more realistic assumptions and meso-scale modelling with longer periods of simulation are needed to reveal the actual impact of VLS-PV on the global climate.

6.4 RECOMMENDATIONS

In order to achieve the sustainable development of desert region communities, it is important to consider technology innovation and to protect the community and its environment. Maintaining community food production is a further important issue.

Agriculture is necessary for food supply, and lack of appropriate knowledge of agriculture often leads to soil degradation and wasting water, thus worsening desertification. In this chapter, we proposed different types of irrigation systems using PV to save soil and water in arid and semi-arid areas. In addition, it is advisable to provide an impact assessment of the climatic results of introducing VLS-PV into a desert region since a PV system will affect quality of life and food production in arid and semi-arid regions.

REFERENCES

1. Tucker, C. J., Dregne, H. E. and Newcombe, W. W. (1991) 'Expansion and contraction of the Sahara desert from 1980 to 1990', *Science*, no 253, pp299–301
2. Xiu, A. and Pleim, J. E. (2001) 'Development of a land surface model. Part I: Application in a mesoscale meteorological model', *Journal of Applied Meteorology*, vol 40, pp192– 209
3. Ogino, Y. and Murashima, K. (1992) 'Salinity and waterlogging problems in Indus River basin', *Journal of Japan Society of Irrigation Drainage and Reclamation Engineering*, vol 60, no 11, pp1007–1010 (in Japanese)
4. Kitamura, Y. and Yano, T. (2002) 'Water and soil degradation caused by secondary salinization and remedial measures in irrigated lands in Central Asia', *Journal of Japan Society of Irrigation Drainage and Reclamation Engineering*, vol 70, no 7, pp605–609 (in Japanese)
5. Rhoades, J. D., Kandish, A. and Mashali, A. M. (1992) *The Use of Saline Waters for Crop Production*, FAO Irrigation and Drainage Paper No. 48

6. Hacchou, N. and Tsutsui, T. (1998) 'Salt management and its damage in the arid areas', *Journal of Japan Society of Irrigation Drainage and Reclamation Engineering*, vol 66, no 8, pp801–805 (in Japanese)
7. Kitamura, Y. (1993) 'The view and countermeasures on problems to combat desertification', *Journal of Japan Society of Irrigation Drainage and Reclamation Engineering*, vol 61, no 1, pp37–40 (in Japanese)
8. Nishimura, T. and Toride, N. (2003) 'Colloidal behaviour and hydrology', in *Colloids in Soils*, Gakkai Shuppan Centre Inc, Japan, Chapter 7 (in Japanese)
9. Russo, D. and Bresler, E. (1977) 'Effect of mixed Na-Ca solutions on the hydraulic properties of unsaturated Soils', *Soil Science Society of America Journal*, vol 41, pp713–717
10. Shainberg, I., Sumner, M. E., Miller, W. P., Ferina, M. P. W., Pavan, M. A. and Fey, M. V. (1989) 'Use of gypsum on soils: A review', *Advances in Soil Science*, vol 9, pp1–101
11. Suarez, D. L. (2003) *World Water Forum*, Kyoto, Japan
12. Maki, T. (1986) 'Global environmental resources and agro-ecosystem', in T. Nagano and K. Ohmasa (eds) *The Agricultural Meteorology and Environmentology*, Asakura-shoten Publication Co, Tokyo, pp52–62 (in Japanese)
13. Hino, M. and Kanda, M. (1992) 'The soothing effect of vegetation on climate and application for land environment', in *The Global Environment and Fluid Dynamics*, Asakura-shoten Publications Co, Tokyo, Chapter 13, pp230–247 (in Japanese)
14. Kondo, J. (1982) *The Atmospheric Boundary Layer Science,* Tokyo Publishing Co Ltd, Tokyo, p175 (in Japanese)
15. Charney, J.G. (1975) 'Dynamics of deserts and drought in the Sahara', *Quarterly Journal of the Royal Meteorological Society*, vol 101, pp193–202
16. Charney, J. G., Quirk, W. J., Chow, S. H. and Kornfield, J. (1977) 'A comparative study of the effects of albedo change on drought in semi arid regions', *Journal of Atmospheric Science*, vol 34, pp1366–1385
17. Deardorff, J. W. (1972) 'Parameterization of the planetary boundary layer for use in general circulation models', *Monthly Weather Review*, vol 100, pp83–106
18. Sud, J. C. and Smith, W. E. (1984) 'Ensemble formulation of surface fluxes and improvement in evapotranspiration and cloud parameterization in a GCM', *Boundary Layer Meteorology*, vol 29, pp185–210
19. Sud, Y. C., Lau, K. M., Walker, G.. K. and Kim, J. H. (1995) 'Understanding biosphere precipitation relationships: Theory, model simulations and logical inferences', *Mausam*, vol 46, pp1–14
20. Chase, T. N., Pielke, R. A., Kittel, T. G. F., Nemani, R. and Running, S.W (1996) 'The sensitivity of a general circulation model to large-scale vegetation changes', *Journal of Geophysical Research*, vol 101, pp7393–7408
21. Stull, R. B. (1988) *An Introduction to Boundary Layer Meteorology*, Kluwer Academic Publishers, Dordrecht
22. Kain, J. S. and Fritsch, J. M. (1993) 'Convective parameterization for mesoscale models: Kain–Fritsch scheme', in A. Emanuel and D. J. Raymond (eds) *The Representation of Cumulus Convection in Numerical Models*, American Meteorological Society, Boston, MA, pp165–170
23. Huang, X., Lyons, T. J. and Smith, R. C. G. (1995) 'Meteorological impact of replacing native perennial vegetation with annual agricultural species', in J. D. Kalma and M. Sivapalan (eds) *Scale Issues in Hydrological Modelling*, Advanstar Commun, Chichester, UK, pp401–410
24. Arya, S. P. (2001) *Introduction to Meteorology*, Academic Press, San Diego, CA
25. Mintz, Y. (1984) 'The sensitivity of numerically simulated climates to land–surface boundary conditions', in J. Houghton (ed) *The Global Climate*, Cambridge University Press, Cambridge, pp79–105
26. Bonan, G. B, Pollard, D. and Thompson, S.L. (1992) 'Effects of boreal forest vegetation on global climate', *Nature*, vol 359, p716
27. Copeland, J. H., Pielke, R. A. and Kittel, T. G. F. (1996) 'Potential climatic impacts of vegetation change: A regional modeling study', *Journal of Geophysical Research*, vol 101, pp7409–7418
28. Stohlgren, T. J., Chase, T. N., Pielke, R. A., Kittel, T. G. F. and Baron, J. (1998) 'Evidence that local land use practices influence regional climate and vegetation patterns in adjacent natural areas', *Global Change Biology*, vol 4, pp495–504
29. Liang, X. and Xie, Z. (2001) 'A new surface runoff parameterization with sub grid-scale soil heterogeneity for land surface models', *Advances in Water Resources*, vol 24, pp1173–1193
30. Lyons, T. J. (2002) 'Clouds prefer native vegetation', *Meteorology and Atmospheric Physics*, vol 80, pp131–140
31. Berg, A. A., Famiglietti, J. S., Walker, J. P. and Houser, P. R. (2003) 'Impact of bias correction to reanalysis products on simulations of North American soil moisture and hydrological fluxes', *Journal of Geophysical Research*, vol 108, no D16, pp4490
32. Xue, Y. and Shukla, J. (1993) 'The influence of land surface properties on Sahel climate. Part I: Desertification', *Journal of Climate*, vol 6, pp2232–2245
33. Zheng, X. and Eltahir, E. A. B. (1997) 'The response to deforestation and desertification in a model of West African monsoons', *Geophysics Research Letters*, vol 24, pp155–158
34. Narisma, G. T. and Pitman, A. J. (2003) 'The impact of 200 years land cover change on the Australian near-surface climate', *Journal of Hydrometeorology*, vol 4, pp424–436
35. Hansen, J. and Nazarenko, L. (2003) 'Soot climate forcing via snow and ice albedos', *Proceedings of the National Academy of Science*, vol 101, pp423–428
36. Betts, R. A., Cox, P. M., Lee, S. E. and Woodward, F. I. (1997) 'Contrasting physiological and structural vegetation feedbacks in climate change simulations', *Nature*, vol 387, pp796–799

37. Betts, A. K., Viterbo, P., Beljaars, A., Pan, H.-L., Hong, S.-Y., Goulden, M. and Wofsy, S. (1998) 'Evaluation of land–surface interaction in ECMWF and NCEP/NCAR reanalysis models over grassland (FIFE) and boreal forest (BOREAS)', *Journal of Geophysical Research*, vol 103, no D18, pp23079–23085
38. Kalnay, E. and Cai, M. (2003a) 'Impact of urbanization and land-use change on climate', *Nature*, vol 423, pp528–531
39. Kalnay, E. and Cai, M. (2003b) 'Impact of urbanization and land-use change on climate – Corrigenda', *Nature*, vol 425, p102
40. Pielke, R. A., Sr. (2001) 'Influence of the spatial distribution of vegetation and soils on the prediction of cumulus convective rainfall', *Review of Geophysics*, vol 39, pp151–177
41. Pielke, R.A., Sr. (2002) *Mesoscale Meteorological Modeling*, 2nd edition, Academic Press, San Diego, CA,
42. Kurokawa, K. (ed) (2003) *Energy from the Desert: Feasibility of Very Large Scale Photovoltaic Power Generation (VLS-PV) Systems*, James and James, London
43. Bonan, G.. B. (1996) *The NCAR Land Surface Model (LSM version 1.0) Coupled to the NCAR Community Climate Model*, Technical Report NCAR/TN-429, NCAR, Boulder, CO
44. Dickinson, R. E., Henderson-Sellers, A. and Kennedy, P. J. (1993) *Biosphere Atmosphere Transfer Scheme (BATS) Version 1e as Coupled to the NCAR Community Climate Model*, NCAR Technical Note, NCAR, Boulder, CO
45. Dai, Y. and Zeng, Q. (1997) 'A land surface model (IAP94) for climate studies. Part I: Formulation and validation in off-line experiments', *Advances in Atmospheric Science*, vol 14, pp443–460
46. Dai, Y., Dickinson, R. E. and Wang, Y. P. (2004) 'A two-big-leaf model for canopy temperature, photosynthesis and stomatal conductance', *Journal of Climate*, vol 17, no 12, pp2281–2299
47. Zeng, X., Dickinson, R. E., Barlage, M., Dai, Y., Wang, G. and Oleson, K. (2005) 'Treatment of under-canopy turbulence in land models', *Journal of Climate*, vol 18, no 23, pp 5086–5094
48. Cotton, W. R., Pielke, R. A., Sr., Walko, R. L., Liston, G. E., Tremback, C. J., Jiang, H., McAnelly, R. L., Harrington, J. Y., Nicholls, M. E., Carrió, G. G. and McFadden J. P. (2003) 'RAMS 2001: Current status and future directions', *Meteorology and Atmospheric Physics*, vol 82, pp5–29

CHAPTER SEVEN

Conclusions and recommendations

7.1 GENERAL CONCLUSIONS

The scope of this report is to examine and evaluate the potential of VLS-PV systems, which have a capacity ranging from several megawatts to gigawatts, and to develop practical project proposals for demonstrative research towards realizing VLS-PV systems in the future.

It is apparent that VLS-PV systems can contribute substantially to global energy needs, can become economically and technologically feasible, and can contribute considerably to environmental and socio-economic development. With the objective of ensuring these contributions and outlining the actions necessary for realizing VLS-PV in the future, this report has developed concrete project proposals.

When thinking about a practical project proposal for achieving VLS-PV, a common objective is to find the best sustainable solution. System capacity for suitable development is dependent upon each specific site with its own application needs and available infrastructure, especially access to long-distance power transmission, and human and financial resources

Although the expected impacts of VLS-PV will differ in each region, depending upon the local situation, and upon any options and concerns, there are several conclusions in common.

7.1.1 Background

- World energy demands will increase as world energy supplies diminish due to trends in world population and economic growth in the 21st century. In order to save conventional energy supplies while supporting growth in economic activity, especially within developing countries, new energy resources and related technologies must be developed.
- The potential amount of world solar energy that can be harnessed is sufficient for global population needs during the 21st century. It has been forecast that photovoltaic technology shows promise as a major energy resource for the future.
- Much potential exists in the world's desert areas. If appropriate approaches are found, they will provide solutions to the energy problem of those countries that are surrounded by deserts.

7.1.2 Technical aspects

- It has already been proven that various types of PV systems can be applied for VLS-PV. PV technologies are now considered to have reached the necessary performance and reliability levels for examining the feasibility of VLS-PV.
- To clarify technical reliability and sustainability, a step-by-step approach will be required to achieve GW scale. This approach is also necessary for the sustainable funding scheme.
- Technological innovation will make VLS-PV in desert areas economically feasible in the near future.
- Global energy systems, such as hydrogen production and high temperature super-conducting technology, as well as the higher conversion efficiency of PV cells and modules, will make VLS-PV projects more attractive.

7.1.3 Economic aspects

- The economic boundary conditions for VLS-PV are still unsatisfactory at current PV system prices without supporting schemes, so that lowering the investment barrier, additional support, and/or higher feed-in tariffs for large systems are still required.

- VLS-PV using flat-plate PV modules or CPV modules produced using mass production technologies are expected to be an economically feasible option for installing large central solar electric generation plants in or adjacent to desert areas in the future as a replacement for central fossil-fuelled generating plants.
- Intelligent financing schemes such as higher feed-in tariffs in developed countries, and international collaboration schemes for promoting large scale PV or renewable energy in developing countries will make VLS-PV economically feasible.
- For the sustainable development of VLS-PV, initial financial support programmes should be sufficient to completely finance the continued annual construction of VLS-PV plants and the replacement of old VLS-PV plants by new ones, eliminating the need for any further investment after the initial financing is repaid.

7.1.4 Social aspects

- Renewable energy development is a promising way of ensuring social development and is one of the most important policies in developing countries.
- VLS-PV projects can provide a pathway to setting up PV industry in the region, while the existence of a traditional and strong PV industry will provide an additional factor contributing to an economic and political environment that favours the application of PV.
- International collaboration and institutional and organizational support schemes will lead to the success of VLS-PV projects in developing countries.
- Developing VLS-PV will become an internationally attractive project, making the region and/or country a leading country/region in the world.
- When we think about developing VLS-PV, we should be careful to consider the social sustainability of the region – for instance, in terms of complementing agricultural practices and desert community development.

7.1.5 Environmental aspects

- In light of global environmental problems, there is a practical necessity for renewable energy, and it is expected to be an important option for the mid 21st century.
- A proper green area coupled with VLS-PV would positively impact upon convective rainfall, rather than produce a negative impact due to a rise in sensible and surface heat flux. It would also decrease dust storms, as well as reduce and mitigate CO_2 emissions.

7.2 REQUIREMENTS FOR A PROJECT DEVELOPMENT

VLS-PV is anticipated to be a very globally friendly energy technology, contributing to the social and economic development of the region in an environmentally sound manner.

However, concrete realization of VLS-PV projects depends upon successful negotiation between project developers and PV and electricity industries, and must have the support of government regarding its acceptance, sustainability and economic incentives.

In order to make negotiations successful and to establish a VLS-PV project, approaches required for project development are as follows:

- Clarify critical success factors on both technical and non-technical aspects.
- Demonstrate technical capability and extendibility.
- Demonstrate economic and financial aspects.
- Show local, regional, and global environmental and socio-economic effects.
- Assess, analyse and allocate project risks such as political and commercial risks.
- Find available institutional and organizational schemes.
- Provide training programmes for installation, operation and maintenance.
- Develop instructions or a guideline for these approaches.

7.3 RECOMMENDATIONS

There are strong indications that VLS-PV could directly compete with fossil fuel and with existing technology as the principal source of electricity for any country that has desert areas. This could be accomplished by finding an investment scheme and by getting institutional and organizational support for its implementation.

In addition, the technology innovations regarding PV and global energy systems in the future will make VLS-PV economically and technologically attractive and feasible.

The following recommendations are outlined to support the sustainable growth of VLS-PV in the near future:

- Discuss and evaluate future technical options for VLS-PV, including electricity network, storage and grid management issues, as well as global renewable energy systems.
- Analyse local, regional and global environmental and socio-economic effects induced by VLS-PV systems from the viewpoint of the whole life cycle.
- Clarify critical success factors for VLS-PV projects, on both technical and non-technical aspects, based on experts' experiences in the field of PV and large

scale renewable technology, including industry, project developers, investors and policy-makers.

- Develop available financial, institutional and organizational scenarios, and general instruction for practical project proposals to realize VLS-PV systems.
- The International Energy Agency (IEA) PVPS community will continue Task 8 activities. Experts from the fields of grid planning and operation, desert environments, agriculture, finance and investment should be involved.
- The IEA PVPS community welcomes non-member countries to discuss the possibility of international collaboration in IEA PVPS activities.

APPENDIX A

Case studies of VLS-PV systems in deserts

A.1 A LIFE-CYCLE ANALYSIS FOR VARIOUS VERY LARGE SCALE PHOTOVOLTAIC POWER GENERATION (VLS-PV) SYSTEMS IN WORLD DESERTS

A.1.1 Introduction

According to an International Energy Agency (IEA) report,[1] total carbon dioxide (CO_2) emissions and primary energy supply in the world in 2030 will be twice as much as in 2000. If world energy demands continue to increase, fossil fuel resources may dry up in this century. In addition, too much energy consumption causes serious environmental problems, such as global warming and acid rain. Renewable energies are expected to resolve both the energy and environmental problems. Photovoltaic power generation systems are one promising renewable power. The photovoltaic (PV) system needs no fuel, produces no emissions and requires very low maintenance at the operation stage. However, solar energy has the disadvantage of its low energy density by nature. To generate a large amount of power, such as comparable to that of a nuclear power plant, the PV system must be introduced on a very large scale.

The life-cycle analysis (LCA) methodology is an appropriate measure to evaluate the potential of VLS-PV systems in detail because the purpose of this methodology is to evaluate its input and output from cradle to grave. In this study, generation cost, energy payback time (EPT), and the CO_2 emission rate of the VLS-PV system were calculated with this method. These indices are defined by the following equations:

$$\text{EPT [year]} = \frac{\text{Total primary energy requirement of the PV system throughout its life cycle [GJ]}}{\text{Annual power generation [GJ]}}$$

$$CO_2 \text{ emission rate [g-C/kWh]} = \frac{\text{Total } CO_2 \text{ emission throughout its life cycle (g-C)}}{\text{Annual power generation [kWh/year]} \times \text{lifetime [year]}}$$

$$\text{Generation costs [cents/kWh]} = \frac{\text{Annual expense of the PV system [cents/year]}}{\text{Annual power generation [kWh/year]}}$$

EPT means years to recover primary energy consumption throughout its life cycle through its own energy production. In this study, the calculation of EPT is based on electricity generation in China. The CO_2 emission rate is a useful index to know how effective the PV system is for global warming.

A.1.2 General assumptions

A.1.2.1 General information of desert areas

Deserts consist of sand, rock, gravel or other materials. Table A.1 (page 120) shows a variety of deserts. Gravel deserts are the best for installing very large scale PV systems because they have the least sand, and sandstorms thus cause minimum damage. High irradiation and very large unutilized land areas exist in world deserts. For example, even the Gobi Desert, located at high latitudes, has higher irradiation (4,7 $kWh/m^2/d$) than Tokyo (3,5 $kWh/m^2/d$). Furthermore, the Sahara Desert has even more irradiation (7,4 $kWh/m^2/d$). Theoretically, PV systems installed in the Gobi Desert with a 50 % space factor have the potential to generate as much energy as the recent world energy supply (384 EJ in 2000).

Eight reference points in six deserts were selected for case studies of life-cycle analysis in light of their economics and environment. Figure A.1 provides geographic information about the eight regions. Almost all deserts are located at the regression line with arid

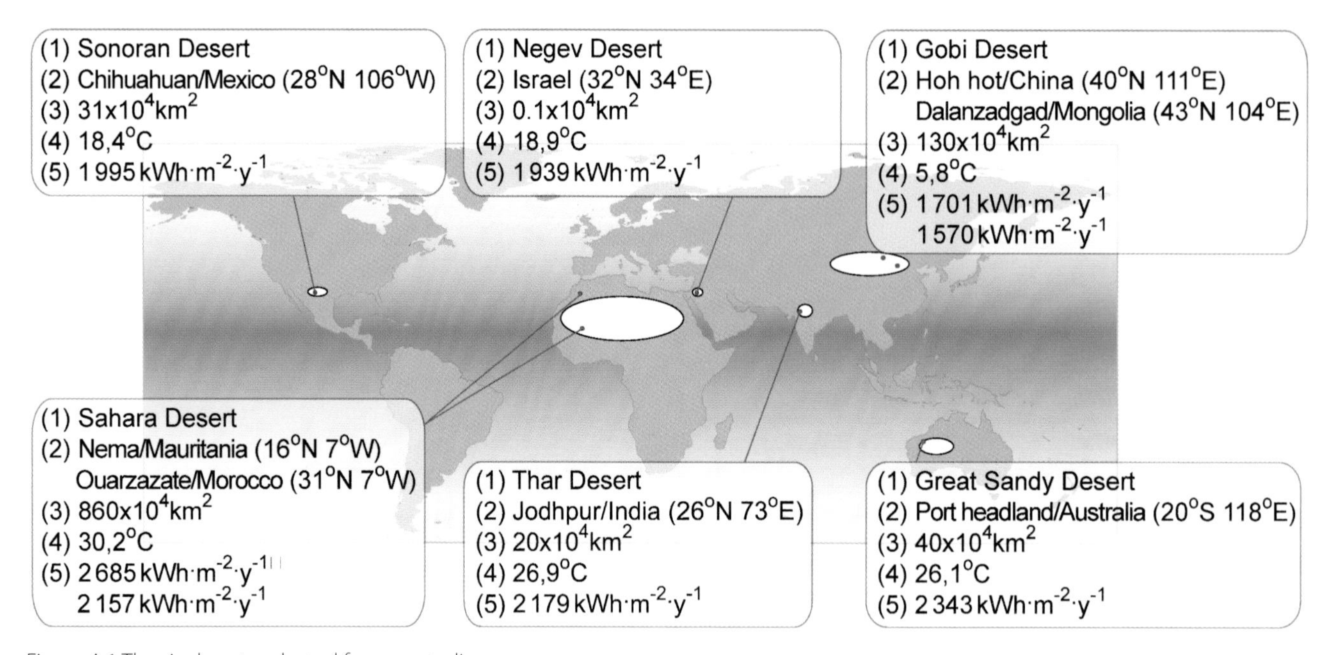

Figure A.1 The six deserts selected for case studies

Notes: (1) desert name; (2) reference point; (3) area; (4) annual average ambient temperature; (5) annual horizontal global irradiation.

climates. Only the Gobi is a characteristic desert. It is located at a high latitude and high elevation: 900 m in the east and 1 500 m in the west. It consists of rock desert, gravel desert and a small sand desert in the west. The Sahara, Thar and Great Sandy deserts, among the selected six deserts, are over 2 000 kWh/m^2/y. Irradiation in Mauritania in the Sahara Desert is especially high.

A.1.2.2 VLS-PV design and configuration

An image of the VLS-PV system installed in a desert area is shown in Figures A.2 and A.3. Figure A.2 shows an image of a basic array and a 500 kW array unit. It consists of 30 modules for the basic array, and 4 200 modules for the 500 kW array unit, which is about 100 m in length and width in the m-Si case. A 100 MW system and a conceptual image of a 1 GW VLS-PV system are shown in Figure A.3. A 100 MW system consists of 200 sets of 500 kW array units. It requires 840 000 modules, 10 1000 tonnes of steel, and 140 000 tonnes of concrete. The length of cable is 1 300 km. The system size is 1,1 km × 2,1 km for a 100 MW system. Buffer plants should be planted beside the system as a windbreak and for environmental reasons. If each 100 MW system is 1 km in width, then the total 1 GW PV system will be 15 km east–west and 2 km north–south.

VLS-PV systems are assumed to be installed in six deserts: the Sahara, Negev, Thar, Sonoran, Great Sandy and Gobi deserts. The performance ratio depends on the operating temperature, degradation, load matching factor, efficiency factor, inverter used, and other data. Figure A.4 is the thematic circuit diagram of a 100 MW VLS-PV system. 200 sets of inverters, 208 sets of 6,6 kV circuit breakers, 5 sets of 30 MVA transformers, 18 sets of 110 kV GIS, 10 sets of 110 kV disconnecting switches, 2 sets of static var compensators (SVC) and a common power board are installed in a 100 MW unit. Table A.3 (page 120) provides information about the balance of systems (BOS) for a 100 MW system. Figure A.4 depicts a thematic circuit diagram of a 100 MW VLS-PV system.

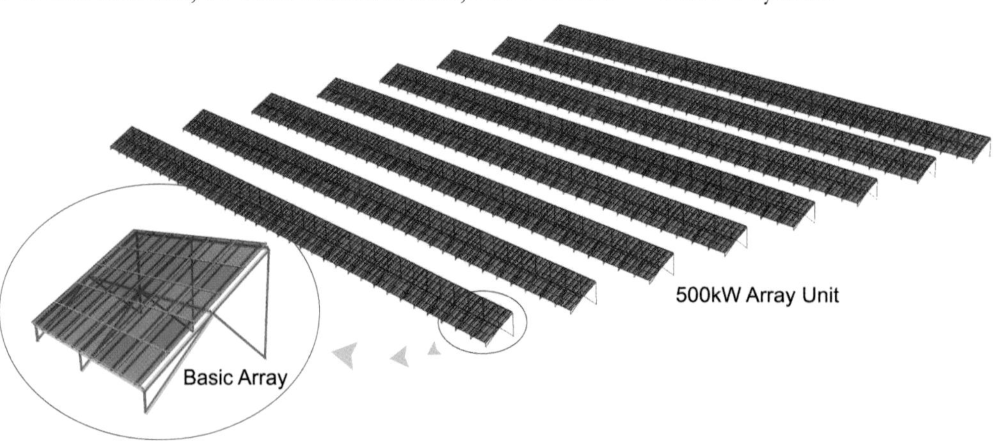

Figure A.2 Image of basic array and 500 kW array unit

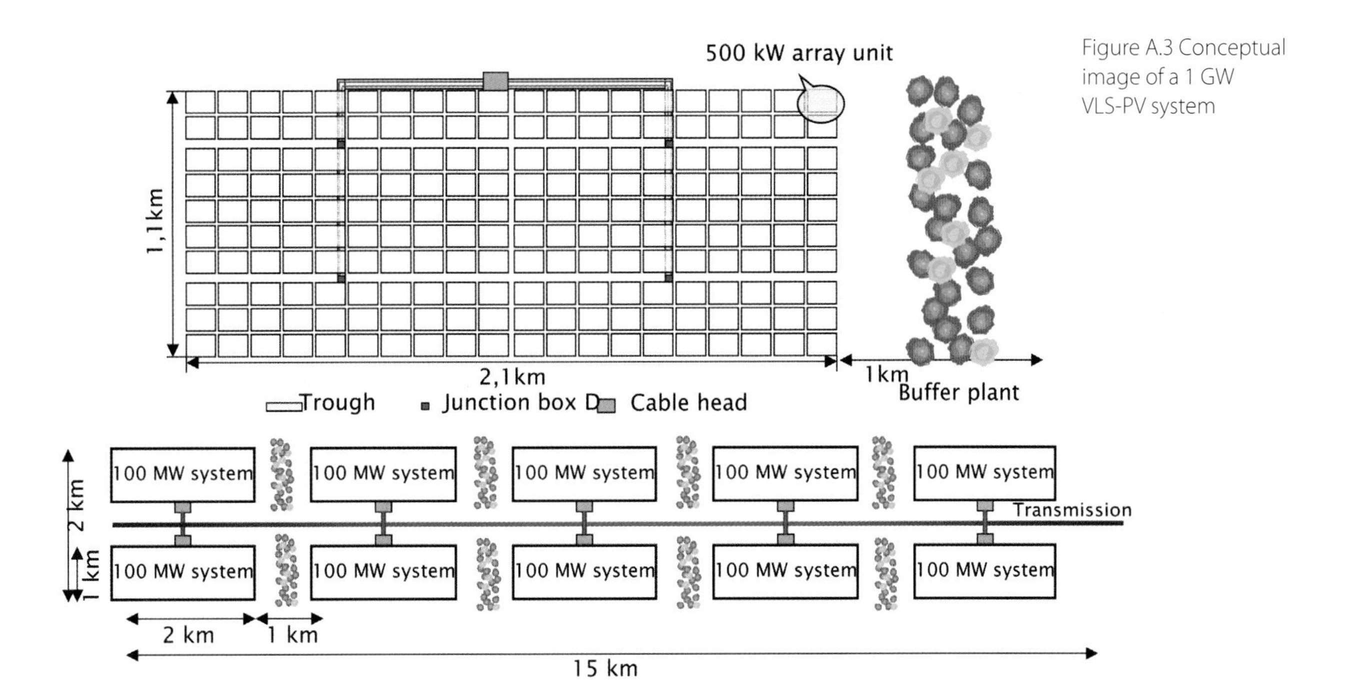

Figure A.3 Conceptual image of a 1 GW VLS-PV system

A.1.2.3 Variety of case studies

This research is divided in two parts, a world desert case study and a comparative study. The world desert case study assumes a fixed flat-plate system with m-Si PV modules. Its cost will be discussed for eight desert regions. The comparative study is focused on an environmental comparison. Fixed flat-plate systems installed with m-Si, a-Si and cadmium telluride (CdTe) PV modules will be compared for cost, EPT and CO_2. A sun-tracking system with m-Si PV modules will also be discussed concerning the three indices. A comparative study is assumed in the Gobi Desert in China only; but an economic study is assumed in eight desert areas. The case study list is shown in Table A.4 (page 120). For this calculation, PV module energy and CO_2 data were obtained from a New Energy and Industrial Technology Development Organization (NEDO) report.[2] Table A.5 (page 120) shows the module specifications, which came from Photon International.[3]

A.1.2.4 Data preparation for this case study

For life-cycle analysis of VLS-PV systems, cost data in each desert area, energy data and CO_2 emissions data in China and Japan were collected. Table A.6 (page 120) shows the price of steel and concrete in local prices for

Figure A.4 Thematic circuit diagram of a 100 MW VLS-PV

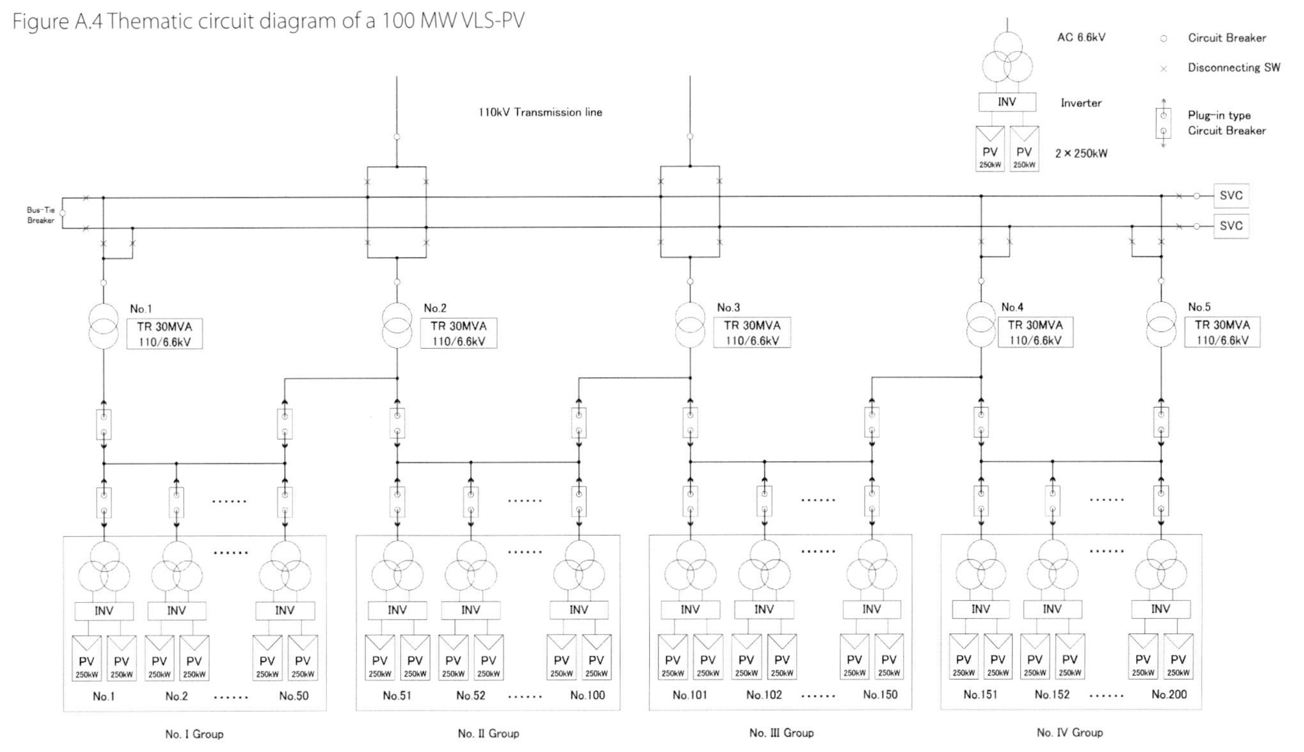

constructing a PV array and transmission tower. Steel and concrete are assumed to be domestic material. Cables, inverters and common equipment are imported from advanced countries. Their prices are shown in Table A.7 (page 120). It is assumed that inverter price is related to module price, which is a parameter, because the VLS-PV system is assumed to be installed in the near future and the module price will decrease by that time.

The details of operation and maintenance are described in Section A.1.3.5. Maintenance costs are also calculated based on actual results of a US PV project. In this case, the costs for repair parts were 0,084 % per year of the total construction costs, and labour for maintenance involved one person per year. Table A.8 (page 121) shows the results for annual operation and maintenance costs.

Table A.9 (page 121) shows the energy and CO_2 emission data of an existing electric power plant in order to calculate materials that require electricity in their production. Basic materials used in the VLS-PV system emit energy and CO_2 in their production. The energy and CO_2 contents are shown in Table A.10 (page 121). These values are taken from NEDO's report[2] in Japan. The energy requirement of an a-Si PV module and a CdTe PV module is one half and one third of an m-Si.

Basic transport data is shown in Table A.11 (page 121). Land transport is assumed to be 600 km and marine transport is assumed to be 1 000 km. The comparative study, which is a Chinese desert case, contains land transport calculations; but the world desert case study does not contain land transport.

The construction period is assumed to be one year, or 240 working days, as shown in Table A.12 (page 121). Land cost is not included in this case study because of the low value of desert land. In addition, the environmentally friendly system might be supported by the government. Table A.13 (page 121) presents economic data for the VLS-PV system. The currency exchange rates are assumed to be 13 Japanese yen/Chinese yuan and 120 Japanese yen/USD. The interest rate is 3 %, and the depreciation years and system lifetime is 30 years.

A.1.3 VLS-PV system design

A.1.3.1 Irradiation

Both irradiation and ambient temperature data were taken from the *World Irradiation Data Book*[4] used for system designs, as shown in Figure A.1. If the installation sites have no direct and diffuse irradiation data, they are estimated from global irradiation data by using the Iqbal model.[5] If no data about duration of sunlight is available, the Liu-Jordan model[6] is applied (see Table A.14, page 121). In-plain irradiation data is calculated by using the RB, Hey,[7] and isotropic models.[8] The irradiation of the tracking system is obtained to calculate a method, referred to as JSES,[8] which changes part of this method.

The simple one-axis sun-tracking PV systems consist of PV modules mounted on a horizontal axis rotating from east to west in synchronization with the sun's position in the sky. The irradiation of the sun-tracking system is calculated for each angle's irradiance, and these figures are totalled. Results of irradiation estimates are shown in Table A.15 (page 122).

A.1.3.2 Transmission loss

Large parts of arid or semi-arid land areas may be far from existing power grid lines. Therefore, if VLS-PV systems are installed, transmission costs have to be factored in. This section describes transmission losses when MW- or GW-scale VLS-PV systems are installed in desert areas. We also investigate the feasibility of these systems.

System configuration

Basic system configurations are described in Section A.1.2.2 and cable and transmission line length in Table A.18 (page 122). For calculations of transmission loss, efficiency of transformer and static var compensator (SVC) are assumed to be 99,21 % and 99,5 %, respectively, and the power factor is assumed to be 0,9, as shown in Table A.16 (page 122).

Approach

Irradiation and power current

Essentially, transmission loss is calculated by Equation A.1. For photovoltaic systems, irradiation changes during the day, so the power current is also changing. Therefore, we assume the power current is in proportion to irradiance, and we estimate the power current from irradiation data. There is strong and low irradiation during the day, such as at noon and in the morning. Since the transmission loss is calculated by Equation A.1, the daily average is not suitable, but root mean square (RMS) should be applied. Differences of monthly averages are also considered. In a desert area, it is difficult to obtain detailed data. Therefore, the daily irradiation curve is calculated by the Berlage model from the monthly average irradiation data of the Japan Weather Association (JWA):[4]

$$\text{Transmission loss} = I^2 \cdot R \qquad \text{[A.1]}$$

For calculating transmission losses, we follow the steps below:

1 Calculate the daily irradiation curve for 12 months with the Berlage model[9] (Equation A.2), which is used for making a theoretical clear day's curve:

$$E_{gth} = \tau^{AM} \cdot I_{on} \cos\theta_z + 0.5 \frac{(1 - \tau^{AM}) \cdot I_{on} \cdot \cos\theta_z}{1 - 1.4\ln(\tau)} \quad [A.2]$$

E_{gth} :global horizontal irradiance [kW/m^2]
τ :atmospheric transmissivity
AM :air mass
θ_z :zenith angle [deg]
I_{on} :direct component of horizontal irradiance (AM = 0) [kW/m^2].

2 Divide the sunny day and cloudy day by using the rate of sunshine (Equation A.3). The sunny day's curve is assumed to be the same as the theoretical curve. The cloudy day's curve is assumed to be the condensed curve of the Berlage model, and the total irradiation in the month is the same as the original. Therefore, the average irradiation from JWA and calculated irradiation by RMS are different. For detailed calculations, in-plain irradiations are also calculated by rb, Hey and isotropic models:

$$\sqrt{\frac{G_s^2 \times d_n \times r + G_c^2 \times d_n \times (1 - r)}{d_n}}$$

$$= \sqrt{Gs^2 r + Gc^2(1 - r)} = G_{ave\,n} \quad [A.3]$$

G_s : sunny day's irradiation [kWh/day]
G_c : cloudy day's irradiation [kWh/day]
$G_{ave\,n}$: RMS of monthly irradiation [kWh/day]
r : rate of sunshine
n : number of month.

3 Calculate the RMS of the day's irradiation from the RMS of each month. This is the final step for calculating RMS:
RMS of monthly day's irradiation=

$$\sqrt{\frac{G_{ave1}^2 \times 31 + G_{ave2}^2 \times 28{,}25 + \ldots + G_{ave12}^2 \times 31}{365{,}25}} \quad [A.4]$$

4 Calculate a day's curve from the Berlage model and scale it up to be the same irradiation as the RMS that is calculated in step 3.
5 Finally, the cable and transmission losses of the VLS-PV system are calculated.

Resistances of cable and transmission line

In hot Sahara and cold Gobi desert areas, the temperature characteristics of the resistance of cable and transmission lines should be considered when installing VLS-PV systems. Basic resistances of electric cables/lines are taken from the Fujikura electric wire manual:[10]

1 Calculate the daily average irradiance by using the above method, and calculate the current produced by the irradiance.
2 Estimate the rise in temperature due to current flow. Equation A.5 is used to calculate the permissible current. It is assumed that transmission loss is the same as heat dissipation because of convection and wind:

$$I^2 R = \pi dlKt \quad [A.5]$$

I : permitted current [A]
d : outside diameter [cm]
l : length of transmission line [cm]
K : ratio of heat dissipation [W/(°C·cm^2)]
t : rise in temperature [°C]
R : resistance of final temperature and length [Ω]

$$K = \left(h_r - \frac{W}{\pi t}\right)\eta + h_w \quad [A.6]$$

W : irradiance [W/cm^2]
η : radiative coefficient of full black
h_r : ratio of heat dissipation by heat emission

$$h_r = 0{,}000567 \frac{\left(\frac{273 + T + t}{100}\right)^4 - \left(\frac{273 + T}{100}\right)^4}{t} \quad [A.7]$$

$$h_w = \frac{0{,}00572}{\left(273 + T + \frac{t}{2}\right)^{0{,}123}} \cdot \sqrt{\frac{v}{d}} \quad [A.8]$$

T : ambient temperature [°C]
v : wind speed [m/s].

3 Ambient temperature is assumed to be average ambient temperature plus 10°C. This is because it occurs during the daytime, when the system is generating.
4 Considering the permitted current, 200 MW per line is assumed.
5 Calculate resistance by using the ambient temperature and the rise in temperature. This resistance is used for estimating cable and transmission loss.

Cable loss and transmission loss

Cable and transmission resistance are calculated with the above method. Equation A.9 is applied for calculating short distances. Equation A.10 is suitable for estimating long-distance transmission loss. It considers the charging current of the transmission line:

$$P_L = 3I^2 Rl \quad [A.9]$$

P_L : total resistance of three-phase 3 line [W]
I : power current [A]
R : resistance of line [Ω/km]
l : length of transmission line [km]

$$P_L = 3Rl\left(I^2 - II_C\cos\phi + \frac{1}{3}I_C^2\right) \quad [A.10]$$

I_C : charging current at sending end [A]
$\cos\phi$: power factor (assumed to be 0,9)

$$I_C = 2\pi fClE \times 10^{-6} \quad [A.11]$$

C : capacitance of one line [μF/km]
f : frequency [Hz]
E : voltage at sending end [V]

$$C = \frac{0{,}02413}{\log_{10}\frac{D_e}{r}} \quad [A.12]$$

D_e : distance between two lines.

E is calculated by line voltage multiplied by 1/√3. Equation A.12 assumes three-phase 2 channels with a wire. D_e is the distance between two lines when lines are fully coordinated.

Results

Figure A.5 shows the annual average day's loss curve, which is estimated by the above three steps, which consider the RMS of power current, resistance-dependent temperature rise, and cable and transmission loss. Transmission loss considers the charging current. The transmission loss is the first major of loss, which is half of the total losses. Cable and transmission losses in the other deserts are shown in Figure A.6. Strong irradiation and high temperature cause great loss in Nema in the Sahara Desert. The loss ratio in Nema is 8,4 %, which is the highest in the six deserts, as shown in Table A.17 (page 122). The lowest cable and transmission losses obtained are in the Gobi Desert in China; they are 5 to 6 % of power generation.

The corrector loss with a-Si and CdTe in China is smaller than with the m-Si because the voltage of a-Si and CdTe modules are higher than m-Si modules. This causes effective transmission. However, because of PV module efficiency, the 100 MW system size is about double the m-Si system size. Therefore, the transmission loss at junction box D to the cable head is high. The total transmission loss is effective if it is compared with m-Si systems.

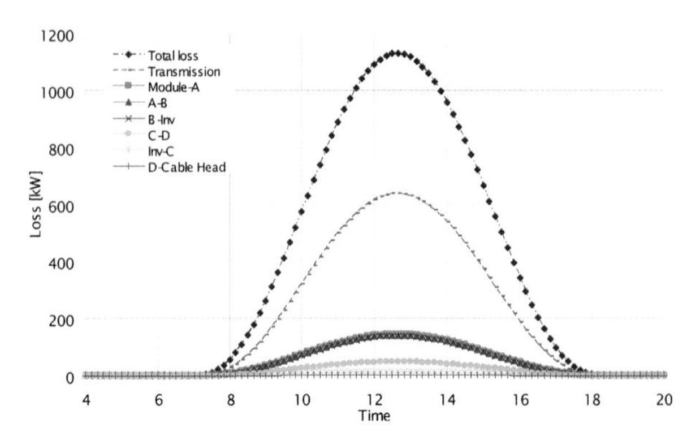

Figure A.5 Annual average estimated day's cable and transmission loss curve with 30° tilt angle in the Sahara Desert

Sensitivity analysis

Transmission loss depends upon the resistance of transmission lines. The transmission loss for TACSR 240 sq and 680 sq are also estimated, as shown in Figure A.7.

Transmission line loss of TACSR 410 sq in Nema in Sahara is 160 % of TACSR 680 sq, and 60 % of TACSR 240 sq. Transmission line cost will depend upon the installation site.

Transmission and cable losses have been obtained. The greatest loss is from the power transmission line and is much larger than other cable losses. A VLS-PV system in the Sahara Desert generates much power, but losses are still significant: results indicate that power generated amounts to approximately 16 GWh/year and 8,4 % of inverter output (annual power generation without transmission loss is 188 GWh/year with a 30° tilt angle). In the Gobi Desert (Mongolia) case, a VLS-PV system generates 155 GWh/year, which is the lowest in the six deserts; however, the transmission loss is also the lowest at 5 GWh/year and 5,7 %. The Gobi Desert produces energy effectively. Transmission loss is higher in strong irradiation areas and lower in weaker irradiation areas. Power generation is decreased by about 5 to 8 % by 100 km TACSR/AC 410 sq transmission, cable transformer and SVC. If the power transmission line is changed to TACSR/AC 680 sq, transmission loss is reduced by 40 %. However, its effect varies in different areas.

A.1.3.3 Array design

Array support structure and foundation

Figure A.8 shows the basic structure of an array support for a 30° tilt angle. The top of the foundation is 0,1 m from the ground and the lowest part of the module 0,2 m from the ground. It is assumed that the array support is made of zinc-plated stainless steel (SS 400), a thickness of chosen according to stress analysis assuming that the wind velocity is 42 m/s (based upon the design standard of structure steel[11] by the Japanese Society of Architecture). It is assumed that a worm gear pair is used to follow the sun, as shown in Figure A.10. In addition, this structure is stabilized by the worm gear pair.

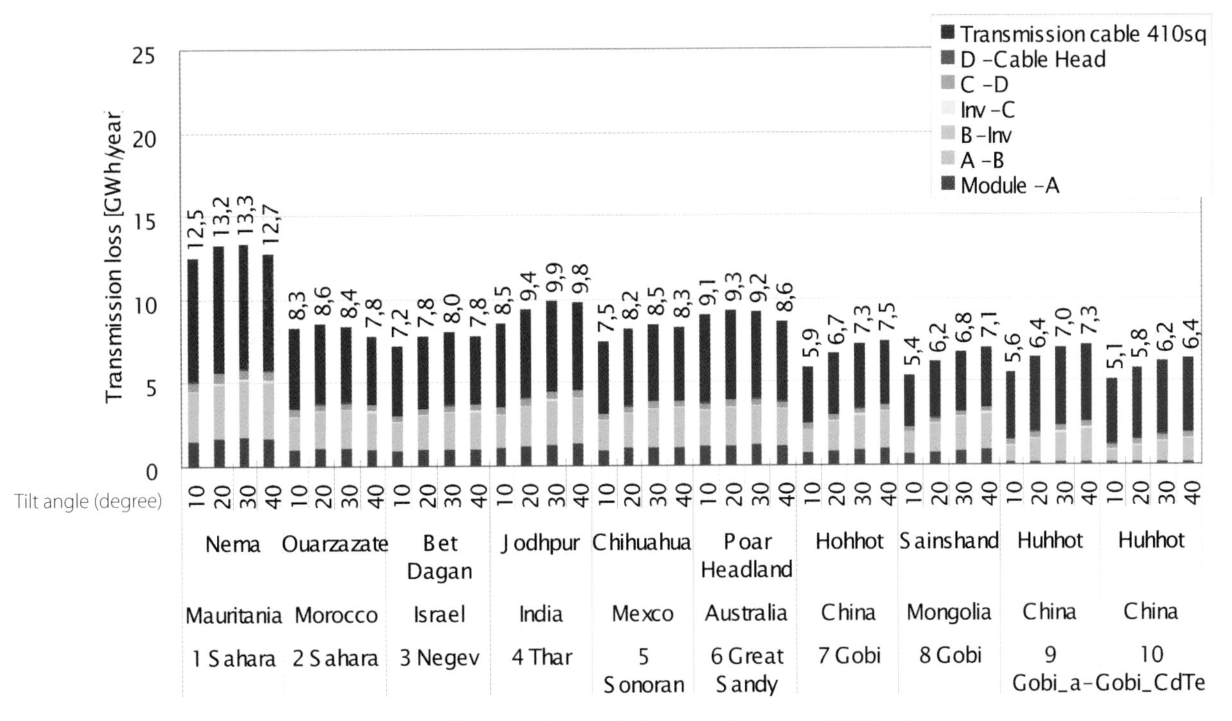

Figure A.6 Cable and transmission losses in world deserts (not including transformer and SVC loss)

Cubic foundations made of concrete are applied. They are about 1,0 m in each dimension, considering the design standard of support structure for power transmission by the Institute of Electrical Engineering in Japan. The concrete's material composition is determined in order to obtain 240 kg/cm^2 of concrete strength: 347 kg/m^3 cement, 603 kg/m^3 sand, 11 180kg/m^3 gravel and 170 l/m^3 water.

Wiring

The short and simple wiring is designed in order to prevent mis-wiring. The current capacity of cable is selected to make the voltage drop less than 4 %. It is determined from Japan Industrial Standards (JIS).

Transmission

The electric transmission system is assumed to be 100 km, two channels and 110 kV for connecting to existing transmission. It consists of steel towers, foundations, cables and ground wires. Wind velocity is considered to be 42 m/s. After calculations, the selected cables and ground wires are TACSR 410 sq and AC 70 sq; 22,0 tonne steel towers and 22,1 m^3 foundations are required with 334 towers, with foundations for 100 km transmission. Table A.18 (page 122) provides a summary of required materials for a 100 MW VLS-PV system.

The land requirements of an a-Si PV module case study in the Gobi Desert are roughly 3 to 5 km^2. This is about twice as much as the m-Si case studies because of the efficiency of a-Si PV modules, which are almost half.

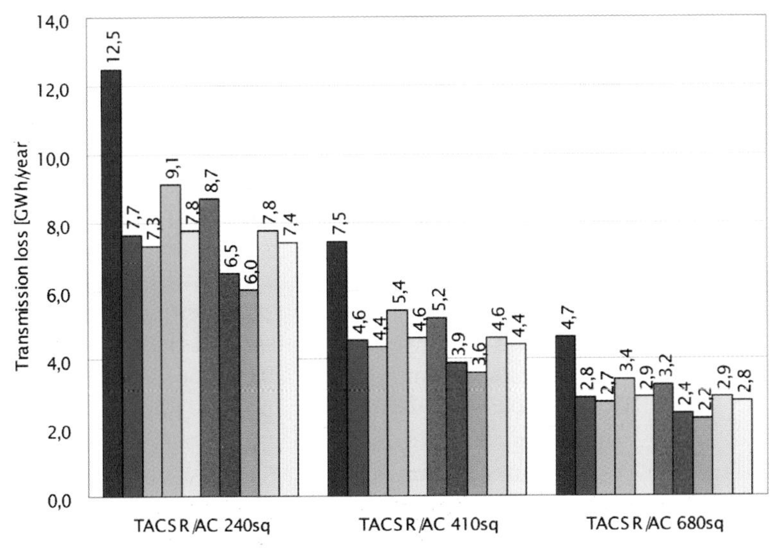

Figure A.7 Transmission losses of TACSR 240 sq, 410 sq and 680 sq

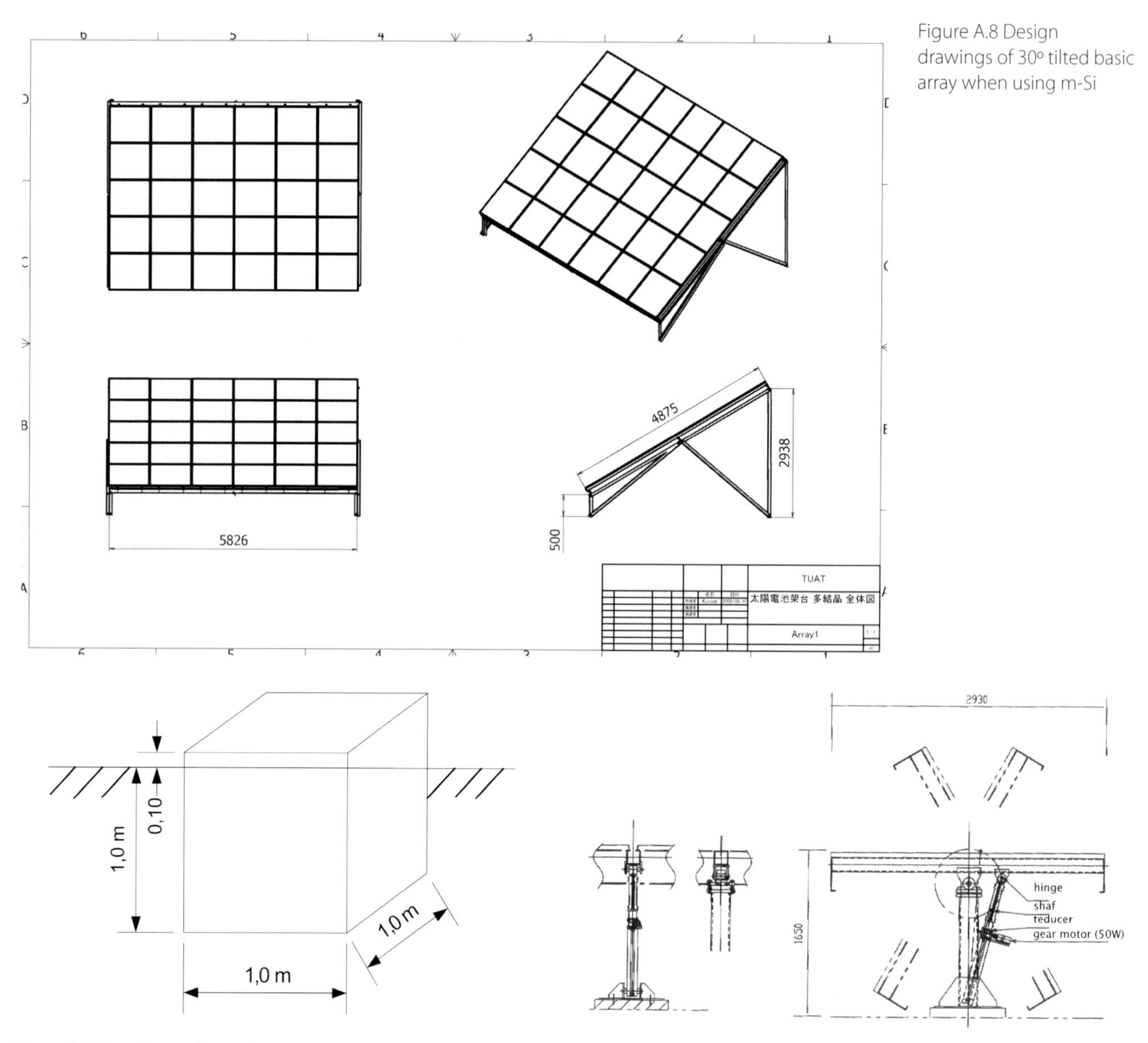

Figure A.8 Design drawings of 30° tilted basic array when using m-Si

Figure A.9 Foundation for tracking system

Figure A.10 Tracking equipment design

For the CdTe PV module case study, it is 1,5 times the m-Si case studies. Therefore, there is a relationship between land requirements and module efficiency. The array support requirement is 10 000 tonnes of steel, and the foundation requires 60 000 tonnes of concrete. The land requirement considers to spacing between PV arrays. Thin-film modules, a-Si PV and CdTe PV modules require about twice the steel and foundation. Lower efficiency results in the need for a large number of PV arrays. As a result steel and foundation requirements increase.

A.1.3.4 Transportation

Array support and foundations are produced in the country in which the VLS-PV system is installed, and other system components, such as modules, cables and inverters, are manufactured in advanced countries. All of the components are transported to the installation site by marine and land transport, as shown in Figure A.12.

A.1.3.5 Operation and maintenance of a VLS-PV system

Operation and maintenance (O&M) are calculated in light of the experience of a real PV system model, the PV-US project.[12] Three operator teams work in the 100 MW PV station. One team works in maintenance, and the other teams operate in alternate shifts. Concerning labour cost, different labour requirements for system construction were estimated by considering the varying conditions in different countries, and the unit labour cost was taken from International Labour Organization (ILO) statistics[13] and other databases. Furthermore, a supervisory labour cost is added to the cost for the installation of some apparatus. The decommissioning stage is not included in this study.

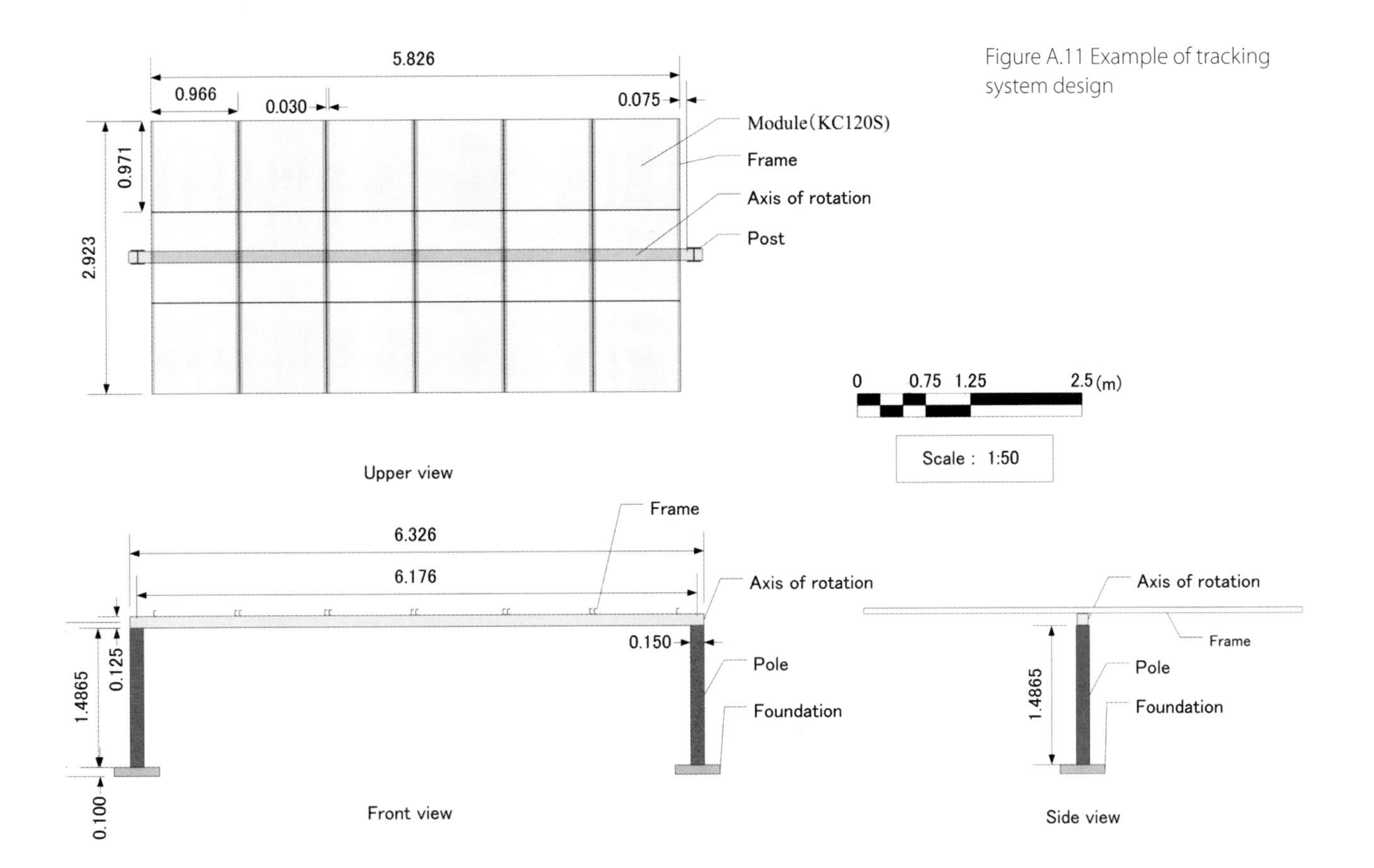

Figure A.11 Example of tracking system design

A.1.4 Evaluation results

Based on general assumptions and VLS-PV system design, results from an economic perspective have been produced for eight arid land areas and, from an environmental perspective, for the Gobi Desert in China.

A.1.4.1 World desert case studies

Initial cost

Initial costs are calculated in order to estimate generation costs of VLS-PV systems installed in the eight desert regions. These costs consist of system components, transport, transmission and construction costs. Land costs and land transport costs are not included in these studies. Almost all component costs are the same level in each desert region. However, construction costs differ between deserts because of the varying wages of the countries. For example, the wage in the Great Sandy Desert in Australia is assumed at 30 747 USD/man year, while it is only 545 USD/man year in the Gobi Desert in China. The wages are taken from the ILO *Yearbook of Labour Statistics*.[13] World desert wages are shown in Table A.8 (page 121).

Annual cost

100 MW VLS-PV systems are assumed to have 30 years of life and a 30-year depreciation period. A 3 % interest ratio, 10 % salvage value rate, 1,4 % annual property tax, and 5 % overhead expenses, operation and maintenance costs are accounted for in the annual cost. Tables A.19 to A.26 (pages 123–126) break down this annual cost. The major element to consider is PV modules in all regions. Balance of system (BOS) and property tax are second and third for all deserts. Even if the module price is reduced to 1 USD/W, it is still half of the annual cost. There are a variety of construction costs and operation and maintenance costs. 100 MW VLS-PV systems installed in the eight desert regions of the world are designed and their potential is evaluated from an economic perspective. The lowest generation costs are obtained at a 20° tilt angle in the Sahara, Negev, Thar, Sonoran and Great Sandy deserts, and at a 30° tilt angle in the Gobi Desert. Assuming a PV module price of 1 USD/W and a 3 % annual interest rate, the generation cost of the VLS-PV system with a 100 km transmission

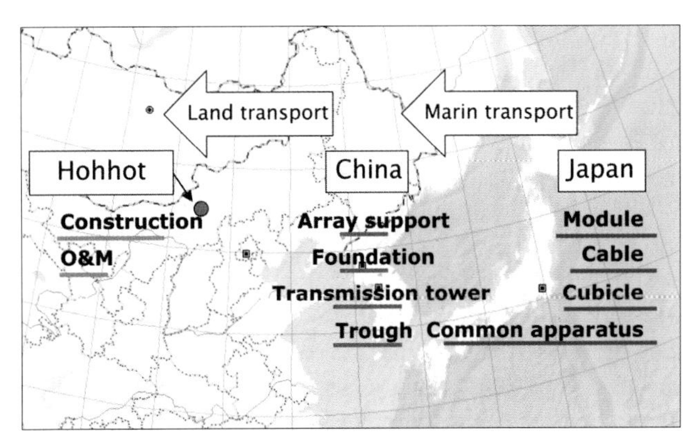

Figure A.12 Transportation method for China

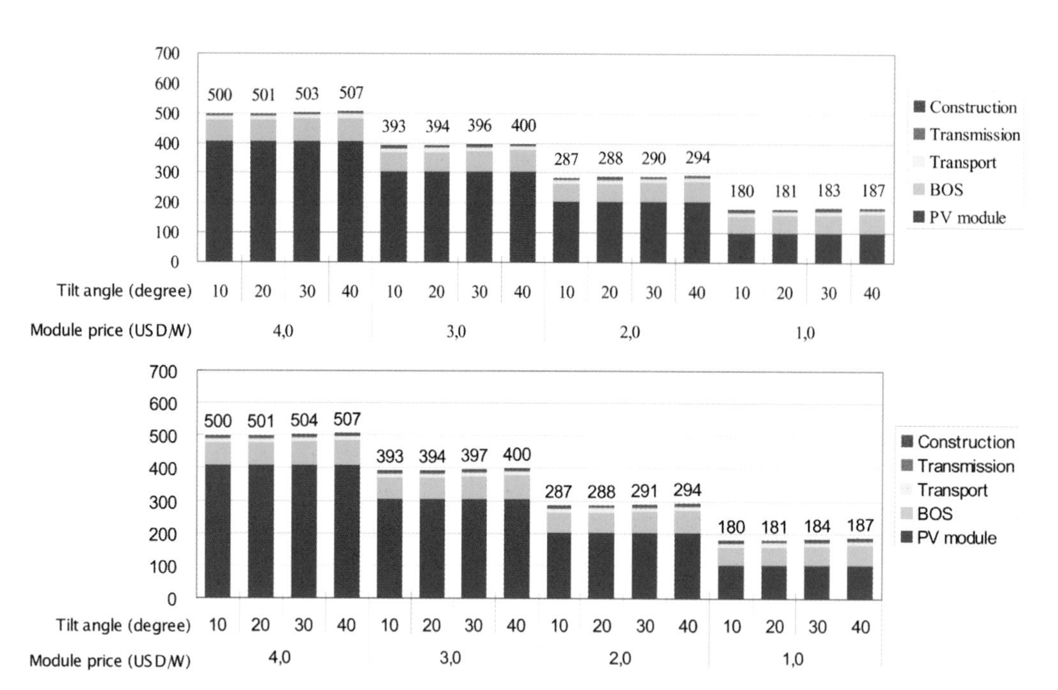

Figure A.13 Initial cost of a 100 MW VLS-PV system in the Sahara Desert (Nema)

Figure A.14 Initial cost of a 100 MW VLS-PV system in the Sahara Desert (Ouarzazate)

line is estimated at 5,8 US cents/kWh in the Sahara Desert, and 6 to 9 US cents/kWh in all desert areas, as shown in Table A.27 (page 127). Figure A.21 is a summary of the generation costs of VLS-PV in these deserts and suggests that the VLS-PV system is economically feasible for all of the sites if the module price is reduced to 1 USD or 2 USD/W.

A.1.4.2 Comparative study

A.1.4.1 discussed eight desert regions from an economic perspective. In this section, we focus on comparative studies between PV modules, which are the m-Si, a-Si and CdTe PV modules, and array structures, which are the fixed flat-plate array and one-axis flat-plate tracking array. These studies obtained cost and environmental information, such as energy payback time and CO_2 emission rate. For a detailed comparison, we focused on only the Gobi Desert case studies, but 600 km land transport is also considered in this section.

System components

The 100 MW VLS-PV systems using m-Si, a-Si and CdTe PV modules in the Gobi Desert are designed on the basis of the above assumptions. Table A.18 (page 122) shows details and Table A.28 (page 122) provides a summary of required materials for a 100 MW-scale PV system. A 100 MW VLS-PV system with a-Si PV modules requires a 4,4 km^2 land area, which is twice as much as for m-Si modules. The character of the CdTe module system is moderate, which is between the m-Si and a-Si. In the a-Si system, the array support requirement is 19 000 tonnes of steel and the foundation needed is 110 000 tonnes of concrete. The land requirement considers spacing between PV arrays. The tracking VLS-PV system is almost the same as the fixed flat-plate system with the m-Si PV module.

Cost estimation

In this study, both investment cost and operation and maintenance costs of 100 MW PV systems for each installation system are estimated to obtain generation cost. The total investment cost includes labour costs for system construction, as well as system component costs. Figures A.22 to A.25 show the initial cost, and Tables A.29 to A.32 (pages 128–129) show the annual cost for m-Si, a-Si,and CdTe PV modules, and the tracking system with m-Si PV modules. The largest of both the initial and annual cost is the PV module, and the second largest is BOS. Even if the module price is reduced to 1 USD/W, the module cost is still over 50 %. There are differences of BOS and transportation between the systems because higher efficiency of the PV module reduces required arrays, foundations and other items. The amorphous module system is a little higher than the multi-crystalline system, but the difference is very small.

Generation cost

Figures A.26 to A.29 show the results of the generation cost of the 100 MW fixed and tracking VLS-PV systems in the Gobi Desert in China. These results were obtained by dividing annual cost by power generation. Power generation at a 40° tilt angle is the highest, but wind pressure is also the high. Thick steel is required for the array and a large foundation. Therefore, the minimum generation cost is obtained at a 30° array tilt angle in each PV module system. The generation cost is around 20 US cents/kWh at a 4 USD/W PV module price. If the module price is reduced to 1 USD or 2 USD/W, generation cost reaches 7 or 11 US cents/kWh.

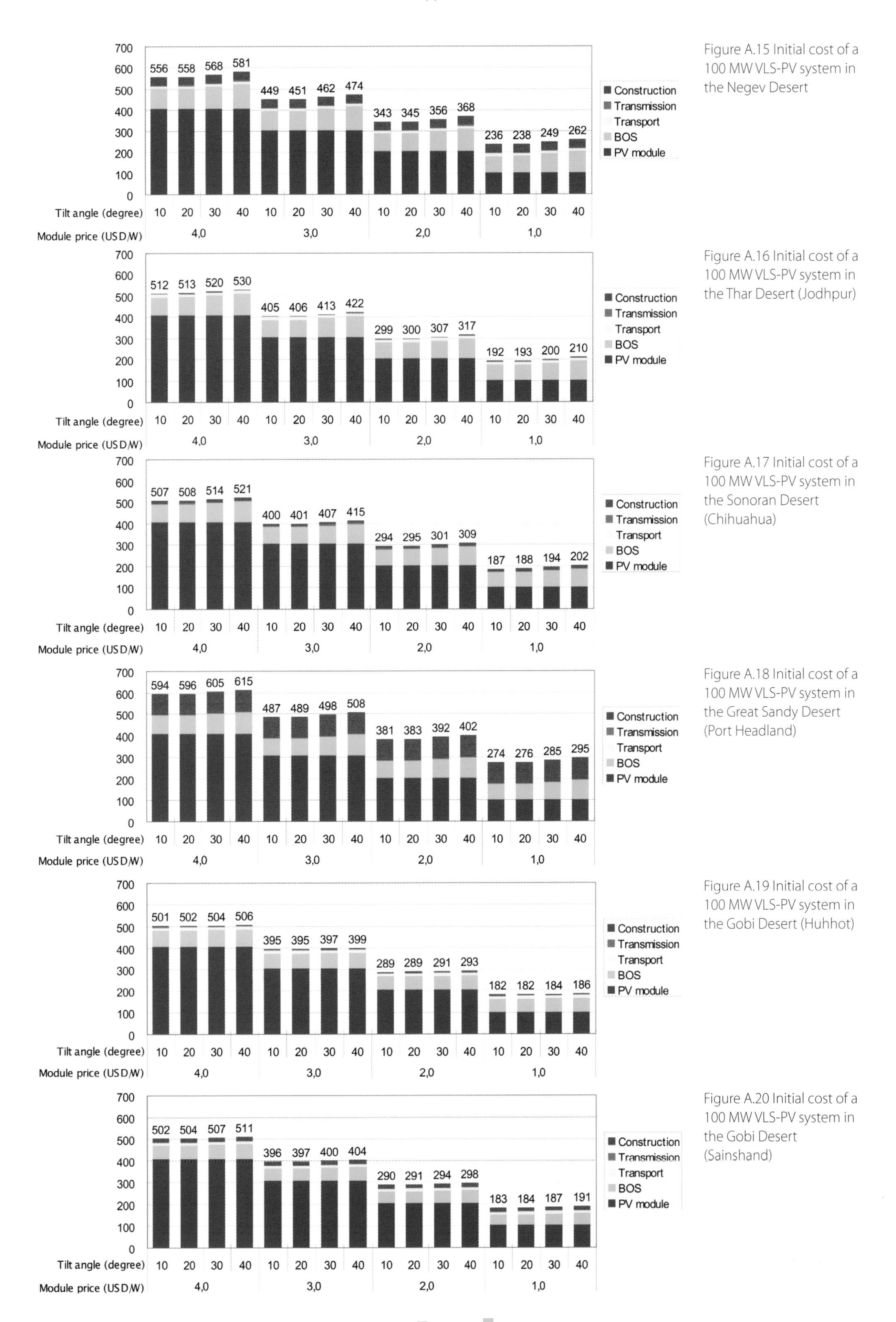

Figure A.15 Initial cost of a 100 MW VLS-PV system in the Negev Desert

Figure A.16 Initial cost of a 100 MW VLS-PV system in the Thar Desert (Jodhpur)

Figure A.17 Initial cost of a 100 MW VLS-PV system in the Sonoran Desert (Chihuahua)

Figure A.18 Initial cost of a 100 MW VLS-PV system in the Great Sandy Desert (Port Headland)

Figure A.19 Initial cost of a 100 MW VLS-PV system in the Gobi Desert (Huhhot)

Figure A.20 Initial cost of a 100 MW VLS-PV system in the Gobi Desert (Sainshand)

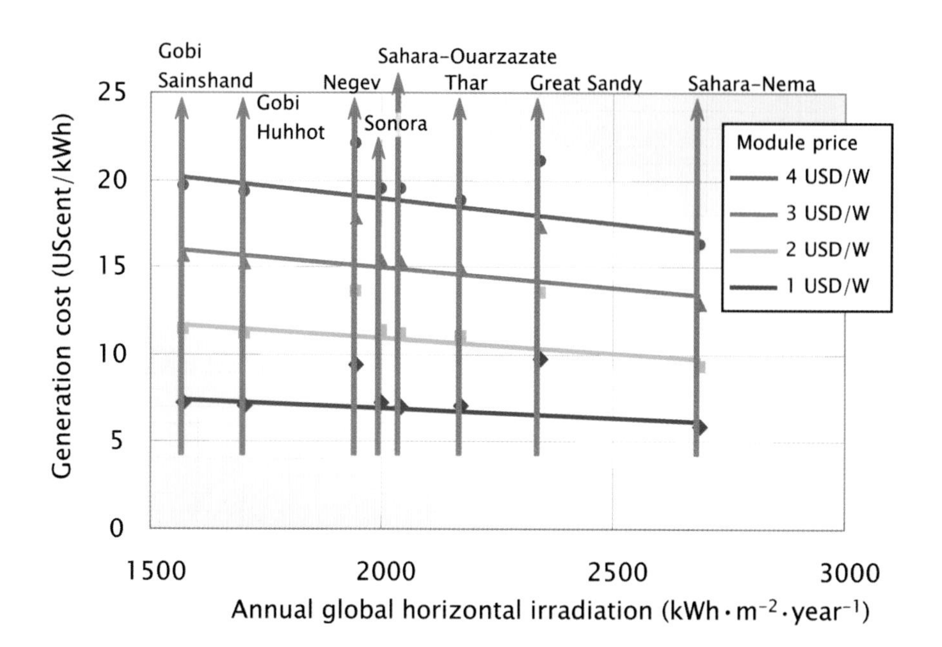

Figure A.21 Generation cost of 100 MW VLS-PV systems in eight desert regions at optimal tilt angle as a function of annual global horizontal irradiation

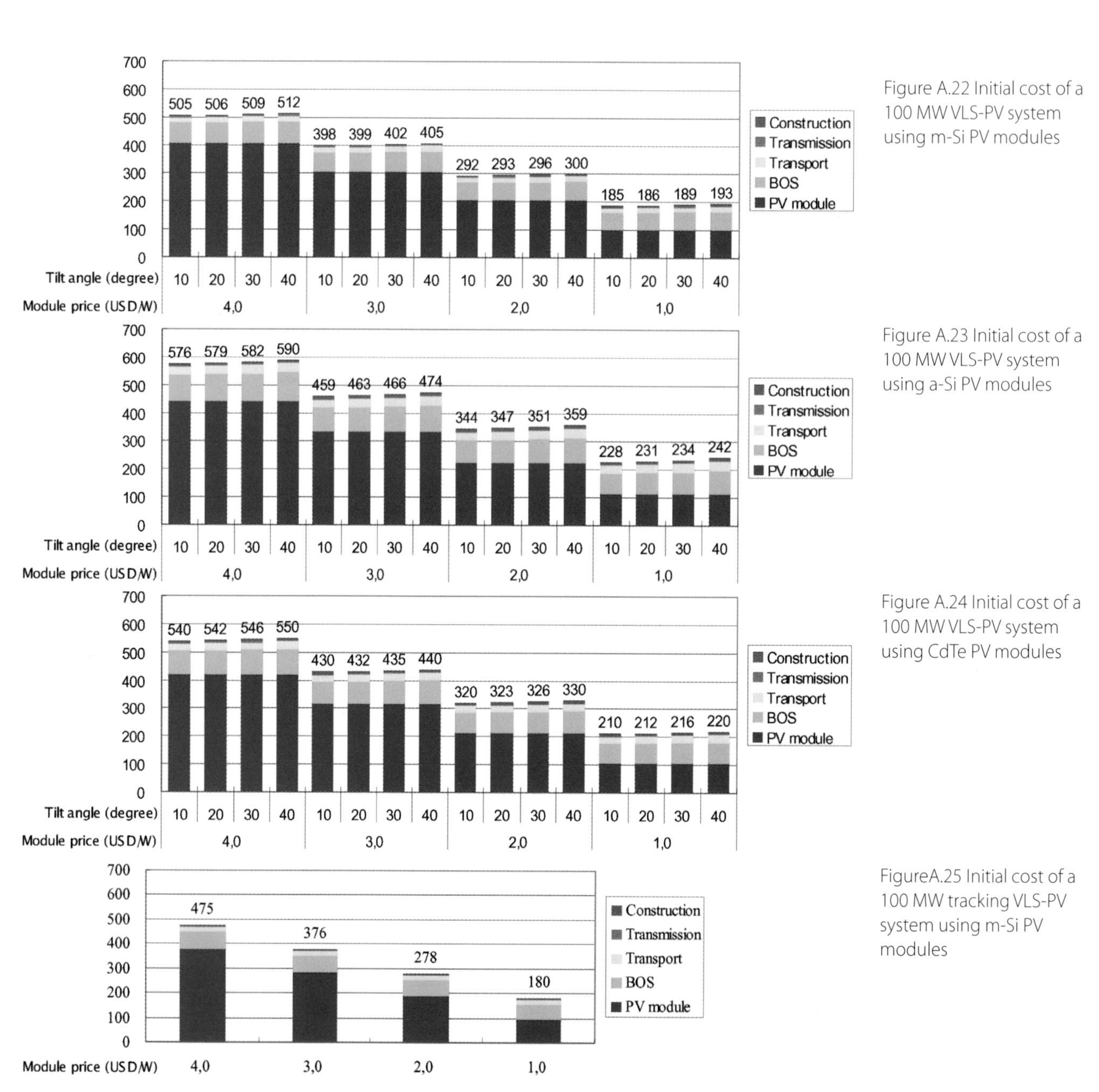

Figure A.22 Initial cost of a 100 MW VLS-PV system using m-Si PV modules

Figure A.23 Initial cost of a 100 MW VLS-PV system using a-Si PV modules

Figure A.24 Initial cost of a 100 MW VLS-PV system using CdTe PV modules

FigureA.25 Initial cost of a 100 MW tracking VLS-PV system using m-Si PV modules

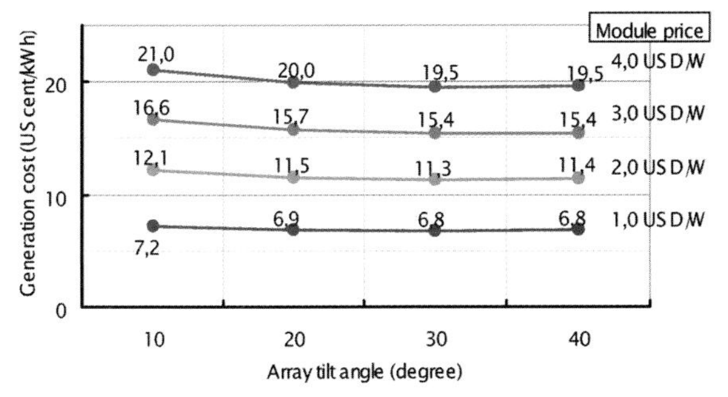

Figure A.26 Generation cost with m-Si PV modules

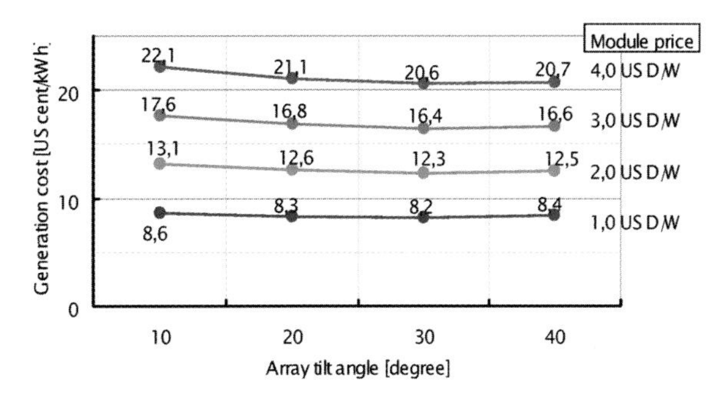

Figure A.27 Generation cost with a-Si PV modules

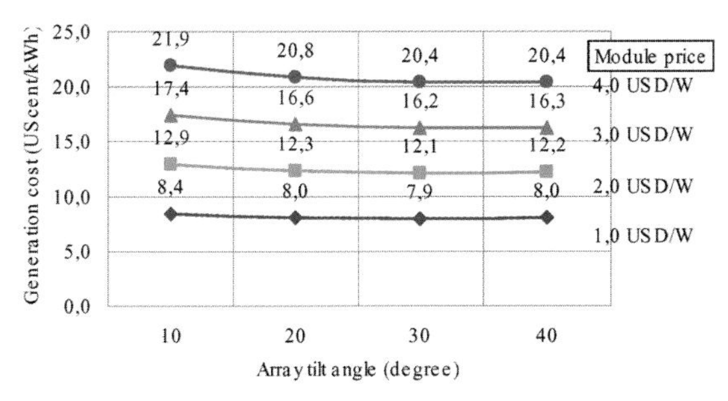

Figure A.28 Generation cost with CdTe PV modules

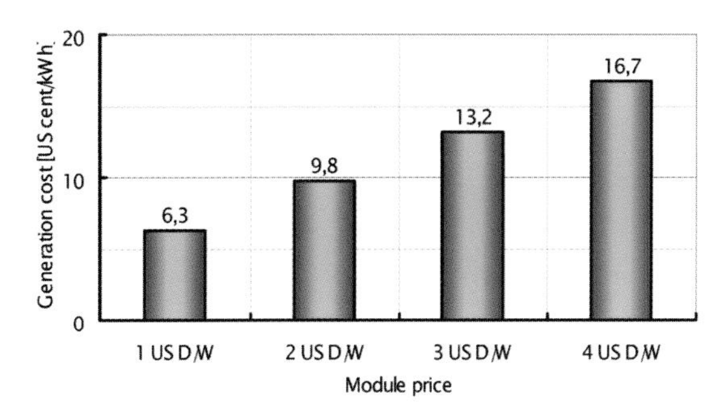

Figure A.29 Generation cost with sun-tracking systems

Tracking system transmission loss is now tentatively assumed at a 20° array tilt angle in the Great Sandy Desert. It is assumed that the same power generation produces almost the same transmission loss. Electricity is produced effectively through the tracking equipment system. It can reduce the generation cost by 10 % to 20 %, and generation costs are 16,7 US cents/kWh at a 4 USD/W PV module price and 6,3 US cents/kWh at a 1 USD/W PV module price.

Energy and CO_2 emission analysis in the Gobi Desert

Figures A.30 to A.32 and Table A.34 (page 129) show the required energy and energy payback time of each system. Energy payback times (EPTs) for each system are estimated by using LCA. With multi-crystalline silicon solar modules, the EPT is 2,2 years; 3,0 years EPT is obtained for amorphous silicon solar modules, and 2,3 years EPT for CdTe PV modules. With the sun-tracking PV system, the EPT is 1,9 years. However, if these EPTs are compared over their life time, their values are very small. These systems can produce much more energy than required in their life cycle.

Three kinds of modules are assumed for 100 MW VLS-PV systems; their estimated CO_2 emissions are shown in Figures A.33 to A.35 and in Table A.35 (page 129). The results are 13,5 g-C/kWh for multi-crystalline silicon, 19,9 for amorphous silicon and 16,9 for CdTe module technology. The majority of all systems feature array support. With a-Si, CO_2 emissions of the array support are very high if compared to a VLS-PV system using multi-crystalline modules. The sun-tracking system reduces this to a 13,3 g-C/kWh CO_2 emission rate. It is lower than fixed flat-plate systems.

Figure A.36 shows a summary of energy payback time and CO_2 emission rates. High module efficiency can reduce energy payback time and CO_2 emissions rates because it can reduce array support structures and foundations that require much energy in their production. If the PV module is installed on a rooftop or slope, there is not a significant energy difference because a small array support structure can be installed.

A.1.5 Conclusions

The 100 MW very large scale power generation systems installed in the eight desert regions were designed and their potential was evaluated according to economic and environmental perspectives. This comparative study considered all equipment, transport, operation, maintenance and transmission losses. Assuming 4,0 USD/W for the m-Si PV module price and a 3 % annual interest rate, the generation cost was estimated at 19,3 US cents/kWh, and 19,5 US cents/kWh with land transport in Huhhot in the Gobi Desert. If the module price is reduced to 1 USD/W, the generation cost is reduced to US 6 to 8 cents/kWh. From an environmental perspective, energy payback time was obtained as two to three years, and the CO_2 emissions rate was obtained as 17 to 23 g-C/kWh. Amorphous silicon solar modules and CdTe modules require less energy than crystalline modules to produce. For VLS-PV systems for desert areas, however, the total energy requirement of thin-film modules was higher than crystalline silicon modules because of its efficiency. Nevertheless, thin-film modules are now developing well and will be more sophisticated in the near future.

The feasibility of a very large scale tracking system installed in the Gobi Desert in China is evaluated in

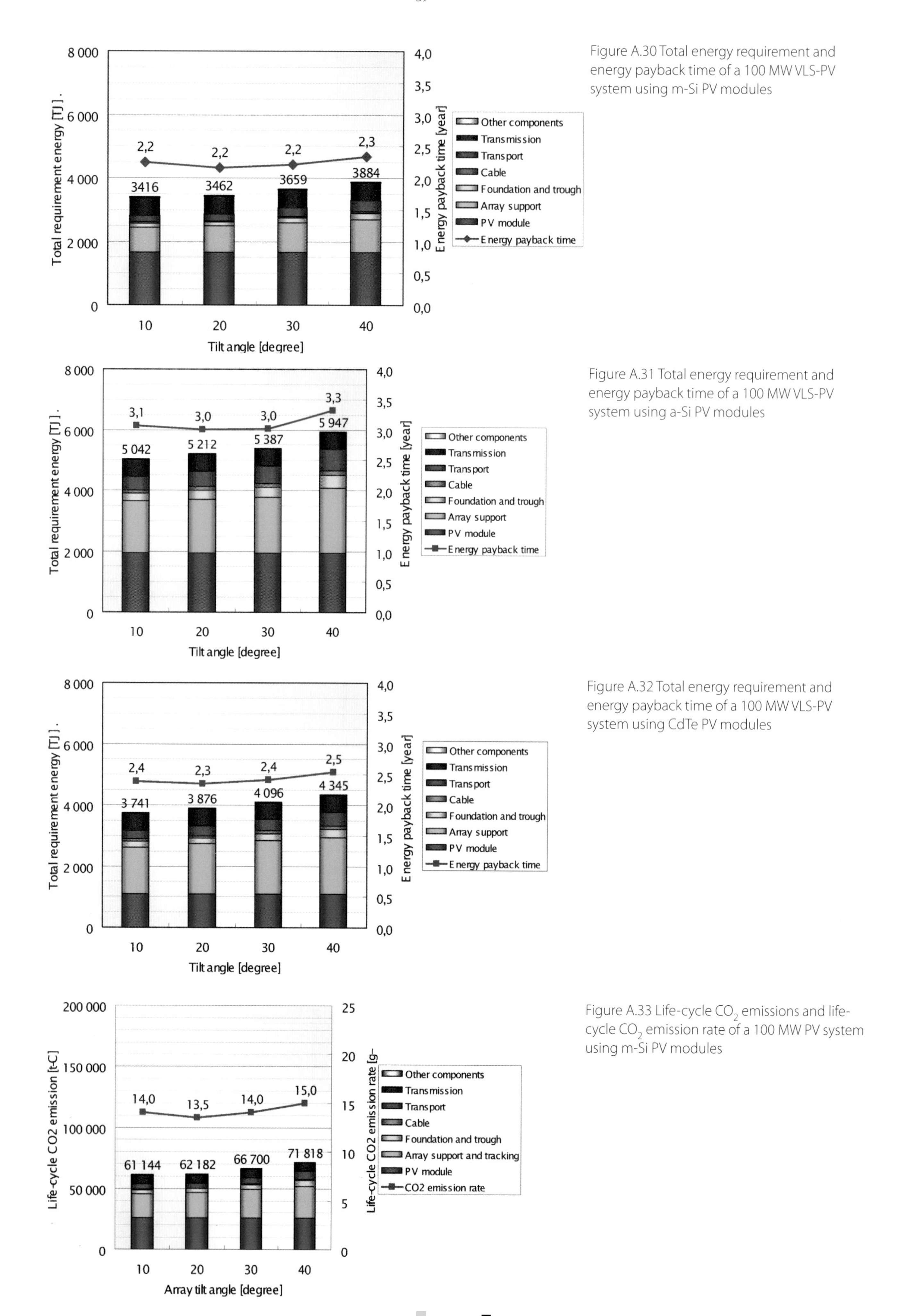

Figure A.30 Total energy requirement and energy payback time of a 100 MW VLS-PV system using m-Si PV modules

Figure A.31 Total energy requirement and energy payback time of a 100 MW VLS-PV system using a-Si PV modules

Figure A.32 Total energy requirement and energy payback time of a 100 MW VLS-PV system using CdTe PV modules

Figure A.33 Life-cycle CO_2 emissions and life-cycle CO_2 emission rate of a 100 MW PV system using m-Si PV modules

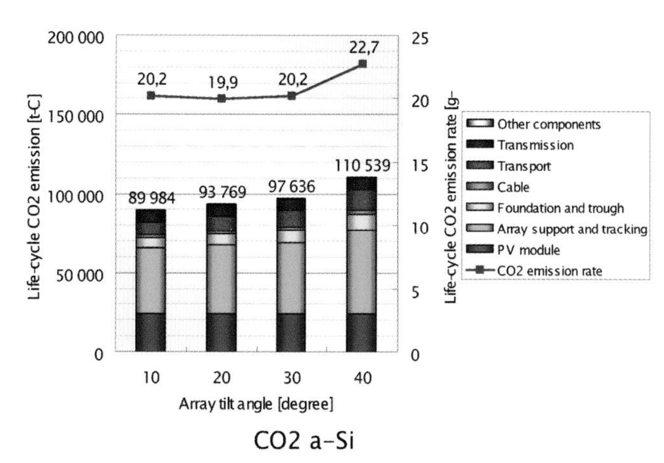

Figure A.34 Life-cycle CO_2 emissions and life-cycle CO_2 emission rate of a 100 MW PV system using a-Si PV modules

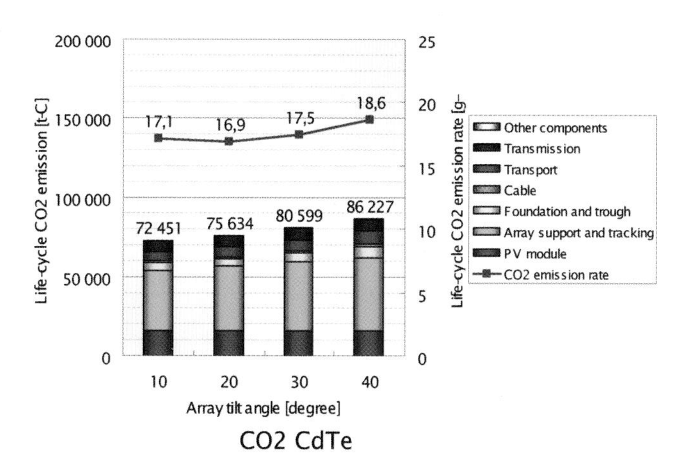

Figure A.35 Life-cycle CO_2 emissions and life-cycle CO_2 emission rate of a 100 MW PV system using CdTe PV modules

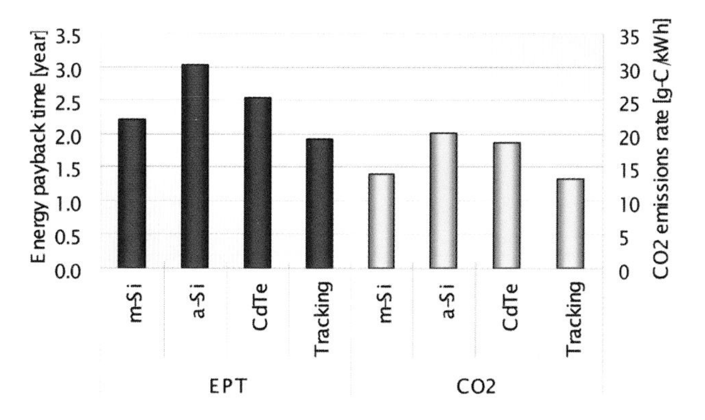

Figure A.36 Comparison of VLS-PV systems of m-Si, a-Si and CdTe modules

depth from a life-cycle viewpoint. This study suggests that the total energy requirement throughout the life cycle of the PV system, considering production and transportation of system components, system construction, operation and maintenance, can be recovered in a much shorter period than its lifetime. Therefore, the VLS-PV system promises to contribute significantly to conserving the world's energy resources. The much lower CO_2 emission rate of VLS-PV than that of existing coal-fired power plants means that it is a very effective energy technology for preventing global warming. The same conclusion must be extended to the other desert areas.

REFERENCES

1. International Energy Agency (IEA) *World Energy Outlook 2002*, IEA, Paris France
2. NEDO (2000) *Development of Technology Commercializing Photovoltaic Power Generation System, Research and Development of Photovoltaic Power Generation Application System and Peripheral Technologies, Survey and Research on the Evaluation of Photovoltaic Power*, NEDO, Japan (in Japanese)
3. *Photon International* (2005) *Photon International*, February, Solar Verlag, Aachen, Germany
4. Japan Weather Association (1991) *World Irradiation Data Book*, FY1991 NEDO Contract Report
5. Iqbal, M. (1979) 'A study of Canadian diffuse and total solar radiation data, I. Monthly average daily horizontal radiation', *Solar Energy*, vol 22, no 1, pp81–86
6. Liu, B. Y. and Jordan, R. C. (1960) 'The interrelationship and characteristic distribution of direct, diffuse and total solar radiation', *Solar Energy*, vol 4, no 3, pp1–19
7. Hay, J. E. (1979) *A Study of Shortwave Radiation on Non-Horizontal Surfaces*, Report No 79–12, Atmospheric Environment Service, Downsview, Ontario, Canada
8. Japan Solar Energy Society (2000) *Japan Solar Energy Utilization Handbook*, Japan Solar Energy Society, Tokyo, Japan
9. Berlage, H. P. (1928) *Meteor. Zeitung*, vol 45, no 5, p174
10. Fujikura Co Ltd (no date) *Electric Wire Manual*, Fujikura Co Ltd
11. Denkishoin (1980) *Design Standard on Structures for Transmissions, JEC-127: Standards of the Japanese Electrotechnical Committee*, Denkishoin, Tokyo, Japan
12. Jennings, C., Reyes, A. B. and Brien, K. P. O. (1994) *PV USA Utility-Scale System Capital and Maintenance Costs*, WCPEC-1, 5–9 December, IEEE, Hawaii
13. International Labour Organization (ILO) (1999) *Year Book of Labour Statistics*, ILO, Switzerland

APPENDIX A.1 TABLES

Table A.1 Variety of deserts

Rock desert	Few animals; low grass; a few pines grow in wetter areas
Gravel desert	Consists of small rocks; a few plants exist
Dirt desert	The grain is very small; rainwater doesn't seep into the ground; sometimes called dubbed yellow ochre
Sand desert	Consists of sand (and sometimes sand dunes) created by the wind
Salt desert	Too much irrigation causes salt damage

Table A.2 Global irradiation in the world's deserts

Major deserts	Country	Global horizontal irradiation (kWh/m²/day)
Sahara	Mauritania	7,36
Negev	Israel	5,31
Thar	India	5,96
Sonoran	Mexico	5,47
Great Sandy	Australia	8,92
Gobi	China	4,67
Tokyo (reference)	Japan	3,47

Table A.3 Balance of systems (BOS) equipment for a 100 MW system

Item	Unit	
Inverter with transformer (including spare)	Set	202
6,6 kV circuit breaker	Set	208
110 kV/6,6 kV transformer	Set	5
110 kV disconnecting SW	Set	18
110 kV GIS	Set	10
SVC	Set	2
Common power board	Set	1

Table A.4 Case studies in this research

	World desert case study	Comparative study (Gobi Desert only)
PV module type	m-Si	m-Si, a-Si, CdTe
Array type	Fixed flat-plate system	Fixed flat-plate system; tracking system
Index	Cost	Cost; EPT; CO_2 emission rate
Desert area	Eight areas in six deserts: Sahara, Negev, Thar, Sonoran, Great Sandy and Gobi deserts	Gobi Desert

Table A.5 PV module specifications

Cell type	Polycrystalline Si	Amorphous Si	CdTe (cadmium telluride)
Nominal power	120 W	58 W	65 W
Efficiency of module	12,8 %	6,9 %	9,03 %
Length, width	966 mm, 971 mm	920 mm, 920 mm	1200 mm, 600 mm
Weight	11,9 kg	12,5 kg	11,4 kg
Voltage MPP (maximum power point)	16,9 V	63,0 V	67,0 V
Current MPP	7,1 A	0,92 A	1,0 A
Voltage open circuit	21,4 V	85,0 V	91 V
Current short circuit	7,45 A	1,12 A	1,15 A
Coefficient of voltage	−82 mV/°C	−243,0 mV/°C	−0,29 %/°C
Coefficient of current	+6,0 mA/°C	+0,80 mA/°C	+0,04 %/°C
Coefficient of power	−0,5 %/°C	−0,22 %/°C	−0,25 %/°C

Table A.6 Local unit prices of steel and concrete (USD/tonne)

	Sahara		Negev	Thar	Sonoran	Great Sandy	Gobi	
	Nema	Ouarzazate					Huhhot	Sainshand
Steel	638	638	370	463	697	485	780	490
Concrete	37	37	189	141	103	134	15	42

Table A.7 Unit prices of cable and components

	Price		Price
Cable (USD/m)		Inverter (USD/W)	
600 V CV 2,0 sq single core	0,25	Module price = 4 USD/W	0,41
600 V CV 8,0 sq double core	1,13	Module price = 3 USD/W	0,36
600 V CV 60 sq single core	2,43	Module price = 2 USD/W	0,32
6,6 kV CV-T 22 sq single core	8,25	Module price = 1 USD/W	0,27
6,6 kV CV 200 sq single core	10,04	6,6 kV circuit breaker (MUSD/unit)	0,0014
110 kV CV 150 sq single core	79,17	110 kV/6,6 kV transformer (MUSD/unit)	0,62
TACSR/AC 410 sq single core	5,58	110 kV disconnecting SW (MUSD/unit)	0,57
AC 70 sq single core	1,75	110 kV GIS (MUSD/unit)	0,57
SVC (MUSD/unit)	0,34	Common power board (MUSD/unit)	0,27

Table A.8 Annual operation and maintenance costs

	Sahara		Negev	Thar	Sonoran	Great Sandy	Gobi	
	Nema	Ouarzazate					Huhhot	Sainshand
Wages (USD/man year)	1 102	1 102	15 227	403	2 187	30 747	545	1 099
Operation cost (MUSD/year)	0,02	0,02	0,25	0,01	0,04	0,51	0,01	0,02
Maintenance cost (MUSD/year)								
Module = 4 USD/W	0,42	0,42	0,49	0,43	0,43	0,54	0,42	0,42
Module = 3 USD/W	0,33	0,33	0,4	0,34	0,34	0,45	0,33	0,34
Module = 2 USD/W	0,24	0,24	0,31	0,25	0,25	0,36	0,24	0,25
Module = 1 USD/W	0,15	0,15	0,22	0,16	0,16	0,27	0,15	0,16
Total O&M cost (MUSD/year)								
Module = 4 USD/W	0,44	0,44	0,74	0,44	0,47	1,05	0,43	0,44
Module = 3 USD/W	0,35	0,35	0,65	0,35	0,38	0,96	0,34	0,36
Module = 2 USD/W	0,26	0,26	0,56	0,26	0,29	0,87	0,25	0,27
Module = 1 USD/W	0,17	0,17	0,47	0,17	0,20	0,78	0,16	0,18

Table A.9 Energy data for Japan and China

Item		Unit	Value
Japan			
Primary energy consumption for power generation		MJ/kWh	10,38
Average CO_2 emission rate of utility		g-C/kWh	114
China			
Primary energy consumption for power generation		MJ/kWh	12,01
Calorific value of Chinese standard coal		MJ/SCE-t	29 302
Retail price of diesel oil		Yuan/L	1,95
Common			
Coal	Calorific value	MJ/t	25 643
	CO_2 emission rate	g-C/MJ	24,7
Oil	Calorific value	MJ/t	45 000
	CO_2 emission rate	g-C/MJ	19,2
Diesel oil	Calorific value	MJ/L	38,5
	CO_2 emission rate	g-C/MJ	18,7
Heavy oil	Calorific value	MJ/L	40,6
	CO_2 emission rate	g-C/MJ	19,4
LNG (liquid natural gas)	Calorific value	MJ/t	54 418
	CO_2 emission rate	g-C/MJ	13,5

Table A.10 Energy and CO_2 contents of products used in this study

Product		Energy contents (GJ)	CO_2 contents (t-C)
Japan	m-Si PV module (piece)	1,96	0,03
	a-Si module (piece)	1,02	0,01
	CdTe PV module (piece)	0,66	0,01
	Silicon steel (tonne)	13,4	0,321
	Aluminium (tonne)	227	2,13
	Copper (tonne)	46,5	0,771
	HDPE (tonne)	15,8	0,264
	PVC (tonne)	29,4	0,373
	Epoxy resin (tonne)	40,5	0,754
	FRP (tonne)	81,6	2,74
	Ceramics (tonne)	0,8	0,02
China	Steel (tonne)	47,3	1,17
	Zinc-plated steel (tonne)	94,6	2,34
	Cement (tonne)	5,07	0,126

Table A.11 Data for transportation of system components

	Item	Unit	Value
Land transport in China	Transport distance	km	600
	Unit cost (truck)	yuan/(t.km)	0,3
	Fuel consumption by truck	l-diesel oil/(t.km)	0,048
Marine transport from Japan to China	Transport distance	km	1000
	Fuel consumption by cargo	kg-heavy oil/ (kt.mile)	7,99

Table A.12 Data for system construction in China

Item	Unit	Value
Construction period	Year	1
Annual working days	Days/year	240
Annual labour cost (inner Mongolia)		
Civil engineering	Yuan/year	5034
Electrical engineering	Yuan/year	7419
Number of supervisors	Person	2
Supervising cost	Million yen	15
Land price	Yuan/km^2	0

Table A.13 Economic data used in this study

Item	Unit	Value
Currency exchange rate		
Chinese yuan – Japanese yen	Yen/yuan	13
Japanese yen – US dollar	Yen/USD	120
Salvage value rate	–	01
Interest rate	/year	0,03
Property tax rate	/year	0,016
Overhead expense rate	/year	0,05
Depreciation years	Years	30
System lifetime	Years	30

Table A.14 Dividing beam and diffuse irradiation model

	H Gh	H Dh	DS	Model
Sahara (Nema)	O	×	O	Iqbal
Sahara (Ouarzazate)	O	×	O	Iqbal
Negev	O	×	O	Iqbal
Thar	O	O	O	–
Sonoran	O	×	O	Iqbal
Great Sandy	O	×	×	Liu-Jordan
Gobi (China)	O	O	O	–
Gobi (Mongolia)	O	×	O	Iqbal

Notes: H Gh = global horizontal irradiation; G Dh = global diffuse irradiation; DS = duration of sunlight.

Table A.15 Estimated irradiation in desert areas (kWh/m²/day)

In-plain irradiation	Sahara (Nema)	Sahara (Ouarzazate)	Negev	Thar	Sonoran	Great Sandy	Gobi (China)	Gobi (Mongolia)
0°	7,36	5,59	5,32	5,95	5,47	6,42	4,66	4,30
10°	7,53	5,91	5,59	6,30	5,75	6,62	5,06	4,75
20°	7,58	6,12	5,76	6,52	5,94	6,71	5,36	5,11
30°	7,45	6,17	5,79	6,59	5,98	6,63	5,53	5,34
40°	7,13	6,08	5,68	6,50	5,88	6,39	5,56	5,44
Tracking	10,16	7,90	7,54	8,24	7,52	9,11	6,60	6,43

Table A.16 Assumptions of efficiency

Equipment	Efficiency (%)
110 kV/6.6 kV transformer	99,21
Static var compensator (SVC)	99,5
Power factor	90

Table A.17 Loss ratios in six deserts with 30° tilt angle structure

	Corrector loss* (GWh/y)	Transmission loss (GWh/y)	Actual annual power generation (GWh/y)**	Total loss ratio (%)***
Sahara (Mauritania)	8,3	7,5	173	8,4
Sahara (Morocco)	6,0	4,6	155	6,4
Negev (Israel)	5,7	4,4	146	6,5
Thar (India)	6,6	5,4	158	7,1
Sonoran (Mexico)	5,9	4,6	151	6,6
Great Sandy (Australia)	6,2	5,2	160	6,7
Gobi (China)	5,5	3,9	149	5,9
Gobi (Mongolia)	5,2	3,6	146	5,7
Gobi (China) a-Si	4,6	4,6	161	5,4
Gobi (China) CdTe	3,9	4,4	154	5,1

Notes: * Corrector loss includes cable loss, transformer and static var compensator (SVC) loss.
** Excluding all losses from PV power output.
*** Including efficiency of transformer and SVC.

Table A.18 Requirements of 100 MW PV system components

	Sahara Nema	Ouarzazate	Negev	Thar	Sonoran Sandy	Great Huhhot	Gobi Sainshand	Gobi Huhhot	Gobi Huhhot	Gobi Huhhot	Gobi (tracking)
Land requirement (km²)											
Tilt = 10°	1,24	1,33	1,34	1,29	1,31	1,16	1,43	1,48	2,72	2,05	3,66
Tilt = 20°	1,56	1,73	1,74	1,66	1,69	1,47	1,91	2,00	3,62	2,76	(Tracking)
Tilt = 30°	1,83	2,07	2,10	1,98	2,03	1,76	2,33	2,46	4,40	3,38	
Tilt = 40°	2,05	2,36	2,39	2,24	2,30	2,00	2,69	2,85	5,06	3,91	
PV module	m-Si	m-Si	m-Si	m-Si	m-Si	m-Si	m-Si	m-Si	a-Si	CdTe	m-Si
piece	848 500	848 500	848 500	848 500	848 500	848 500	848 500	848 500	1 909 100	1 616 200	785 500
Capacity (MW)	100,8	100,8	100,8	100,8	100,8	100,8	100,8	100,8	109,6	104,0	93,3
Array support structure (tonne)											
Tilt = 10°	8 375	8 375	8 375	8 375	8 375	8 375	8 375	8 375	17 936	16 302	10 819
Tilt = 20°	8 693	8 693	8 693	8 693	8 693	8 693	8 693	8 693	18 618	17 667	(Tracking)
Tilt = 30°	9 755	9 755	9 755	9 755	9 755	9 755	9 755	9 755	19 255	18 690	
Tilt = 40°	10 871	10 871	10 871	10 871	10 871	10 871	10 871	10 871	22 591	19 755	
Foundation (m³)											
Tilt = 10°	39 913	39 913	39 913	39 913	39 913	39 913	39 913	39 913	71 902	45 733	46 061
Tilt = 20°	39 913	39 913	39 913	39 913	39 913	39 913	39 913	39 913	89 804	45 733	(Tracking)
Tilt = 30°	59 578	59 578	59 578	59 578	59 578	59 578	59 578	59 578	110 455	68 267	
Tilt = 40°	84 829	84 829	84 829	84 829	84 829	84 829	84 829	84 829	160 789	97 200	
Cable (30°) (km)											
600 V CV 2 sq	1 434	1 434	1 434	1 434	1 434	1 434	1 434	1 434	16 521	10 284	621
600 V CV 8 sq double core	173	173	173	173	173	173	173	173	584	487	219
600 V CV 60 sq	87	98	99	94	96	90	109	114	497	203	229
6,6 kV CVT 22 sq	27	30	30	29	29	28	33	35	44	33	46
6,6 kV CV 200 sq	38	36	36	42							
110 kV CV 150 sq	26	27	27	27	27	26	29	29	43	35	18
Trough (30°) (m³)	32 000	33013	33108	32645	32823	32215	34 052	34492	80511	50457	45 288
Common apparatus											
Inverter with transformer (set)	202										
6,6 kV circuit breaker (set)	208										
110 kV/6,6 kV transformer (set)	5										
110 kV disconnecting SW (set)	18										
110 kV GIS (set)	10										
SVC (set)	2										
Common power board (set)	1										
Power transmission corresponding to 100 MW PV system Transmission line											
110 kV TACSR 410 mm² (km)	401										
AC 70 mm² (km)	33										
Pylon (steel) (km)*	2 229										
Foundation (km)*	5 150										

Table A.19 Annual cost of a 100 MW VLS-PV system in the Sahara Desert (Nema)

Tilt angle = (degrees)	10	20	30	40
Annual power generation (GWh/year)	175	176	173	165
Module price = 4,0 USD/W				
Annual cost (MUSD/year)	28,6	28,7	28,8	29,0
PV module (MUSD/year)	19,9	19,9	19,9	19,9
BOS (MUSD/year)	3,5	3,5	3,6	3,8
Transportation (MUSD/year)	0,6	0,6	0,6	0,6
Transmission (MUSD/year)	0,2	0,2	0,2	0,2
Construction (MUSD/year)	0,3	0,3	0,3	0,3
O&M plus overhead (MUSD/year)	0,5	0,5	0,5	0,5
Property tax (MUSD/year)	3,7	3,7	3,7	3,8
Module price = 3,0 USD/W				
Annual cost (MUSD/year)	22,5	22,5	22,7	22,9
PV module (MUSD/year)	14,9	14,9	14,9	14,9
BOS (MUSD/year)	3,2	3,3	3,4	3,5
Transportation (MUSD/year)	0,6	0,6	0,6	0,6
Transmission (MUSD/year)	0,2	0,2	0,2	0,2
Construction (MUSD/year)	0,3	0,3	0,3	0,3
O&M plus overhead (MUSD/year)	0,4	0,4	0,4	0,4
Property tax (MUSD/year)	2,9	2,9	2,9	2,9
Module price = 2,0 USD/W				
Annual cost (MUSD/year)	16,4	16,4	16,6	16,8
PV module (MUSD/year)	10,0	10,0	10,0	10,0
BOS (MUSD/year)	3,0	3,1	3,2	3,3
Transportation (MUSD/year)	0,6	0,6	0,6	0,6
Transmission (MUSD/year)	0,2	0,2	0,2	0,2
Construction (MUSD/year)	0,3	0,3	0,3	0,3
O&M plus overhead (MUSD/year)	0,3	0,3	0,3	0,3
Property tax (MUSD/year)	2,1	2,1	2,1	2,1
Module price = 1,0 USD/W				
Annual cost (MUSD/year)	10,2	10,3	10,4	10,6
PV module (MUSD/year)	5,0	5,0	5,0	5,0
BOS (MUSD/year)	2,8	2,8	2,9	3,1
Transportation (MUSD/year)	0,6	0,6	0,6	0,6
Transmission (MUSD/year)	0,2	0,2	0,2	0,2
Construction (MUSD/year)	0,3	0,3	0,3	0,3
O&M plus overhead (MUSD/year)	0,2	0,2	0,2	0,2
Property tax (MUSD/year)	1,3	1,3	1,3	1,3

Table A.20 Annual cost of a 100 MW VLS-PV system in the Sahara Desert (Ouarzazate)

Tilt angle = (degrees)	10	20	30	40
Annual power generation (GWh/year)	148	153	155	152
Module price = 4,0 USD/W				
Annual cost (MUSD/year)	29,9	29,9	30,1	30,3
PV module (MUSD/year)	19,9	19,9	19,9	19,9
BOS (MUSD/year)	3,5	3,5	3,6	3,8
Transportation (MUSD/year)	0,6	0,6	0,6	0,6
Transmission (MUSD/year)	0,2	0,2	0,2	0,2
Construction (MUSD/year)	0,3	0,3	0,3	0,3
O&M plus overhead (MUSD/year)	1,7	1,7	1,7	1,7
Property tax (MUSD/year)	3,7	3,7	3,7	3,8
Module price = 3,0 USD/W				
Annual cost (MUSD/year)	23,4	23,5	23,7	23,9
PV module (MUSD/year)	14,9	14,9	14,9	14,9
BOS (MUSD/year)	3,2	3,3	3,4	3,6
Transportation (MUSD/year)	0,6	0,6	0,6	0,6
Transmission (MUSD/year)	0,2	0,2	0,2	0,2
Construction (MUSD/year)	0,3	0,3	0,3	0,3
O&M plus overhead (MUSD/year)	1,3	1,3	1,3	1,4
Property tax (MUSD/year)	2,9	2,9	2,9	2,9
Module price = 2,0 USD/W				
Annual cost (MUSD/year)	17,1	17,1	17,3	17,5
PV module (MUSD/year)	10,0	10,0	10,0	10,0
BOS (MUSD/year)	3,0	3,1	3,2	3,4
Transportation (MUSD/year)	0,6	0,6	0,6	0,6
Transmission (MUSD/year)	0,2	0,2	0,2	0,2
Construction (MUSD/year)	1,9	1,9	1,9	2,0
O&M plus overhead (MUSD/year)	1,0	1,0	1,0	1,0
Property tax (MUSD/year)	2,1	2,1	2,1	2,1
Module price = 1,0 USD/W				
Annual cost (MUSD/year)	10,7	10,7	10,9	11,1
PV module (MUSD/year)	5,0	5,0	5,0	5,0
BOS (MUSD/year)	2,8	2,8	2,9	3,1
Transportation (MUSD/year)	0,6	0,6	0,6	0,6
Transmission (MUSD/year)	0,2	0,2	0,2	0,2
Construction (MUSD/year)	0,3	0,3	0,3	0,3
O&M plus overhead (MUSD/year)	0,6	0,6	0,6	0,6
Property tax (MUSD/year)	1,3	1,3	1,3	1,3

Table A.21 Annual cost of a 100 MW VLS-PV system in the Negev Desert

Tilt angle = (degrees)	10	20	30	40
Annual power generation (GWh/year)	141	145	146	143
Module price = 4,0 USD/W				
Annual cost (MUSD/year)	31,9	32,0	32,6	33,3
PV module (MUSD/year)	19,9	19,9	19,9	19,9
BOS (MUSD/year)	4,6	4,7	5,1	5,7
Transportation (MUSD/year)	0,6	0,6	0,6	0,6
Transmission (MUSD/year)	0,2	0,2	0,2	0,2
Construction (MUSD/year)	1,9	1,9	1,9	2,0
O&M plus overhead (MUSD/year)	0,8	0,8	0,8	0,8
Property tax (MUSD/year)	3,9	3,9	4,0	4,1
Module price = 3,0 USD/W				
Annual cost (MUSD/year)	25,7	25,8	26,4	27,2
PV module (MUSD/year)	14,9	14,9	14,9	14,9
BOS (MUSD/year)	4,3	4,4	4,9	5,5
Transportation (MUSD/year)	0,6	0,6	0,6	0,6
Transmission (MUSD/year)	0,2	0,2	0,2	0,2
Construction (MUSD/year)	1,9	1,9	1,9	2,0
O&M plus overhead (MUSD/year)	0,7	0,7	0,7	0,7
Property tax (MUSD/year)	3,1	3,1	3,2	3,2
Module price = 2,0 USD/W				
Annual cost (MUSD/year)	19,6	19,7	20,3	21,1
PV module (MUSD/year)	10,0	10,0	10,0	10,0
BOS (MUSD/year)	4,1	4,2	4,7	5,3
Transportation (MUSD/year)	0,6	0,6	0,6	0,6
Transmission (MUSD/year)	0,2	0,2	0,2	0,2
Construction (MUSD/year)	1,9	1,9	1,9	2,0
O&M plus overhead (MUSD/year)	0,6	0,6	0,6	0,6
Property tax (MUSD/year)	2,3	2,3	2,3	2,4
Module price = 1,0 USD/W				
Annual cost (MUSD/year)	13,5	13,6	14,2	14,9
PV module (MUSD/year)	5,0	5,0	5,0	5,0
BOS (MUSD/year)	3,9	4,0	4,4	5,0
Transportation (MUSD/year)	0,6	0,6	0,6	0,6
Transmission (MUSD/year)	0,2	0,2	0,2	0,2
Construction (MUSD/year)	1,9	1,9	1,9	2,0
O&M plus overhead (MUSD/year)	0,5	0,5	0,5	0,5
Property tax (MUSD/year)	1,4	1,4	1,5	1,6

Table A.22 Annual cost of a 100 MW VLS-PV system in the Thar Desert (Jodhpur)

Tilt angle = (degrees)	10	20	30	40
Annual power generation (GWh/year)	152	156	157	155
Module price = 4,0 USD/W				
Annual cost (MUSD/year)	29,3	29,4	29,8	30,4
PV module (MUSD/year)	19,9	19,9	19,9	19,9
BOS (MUSD/year)	4,2	4,3	4,7	5,1
Transportation (MUSD/year)	0,6	0,6	0,6	0,6
Transmission (MUSD/year)	0,2	0,2	0,2	0,2
Construction (MUSD/year)	0,1	0,1	0,1	0,1
O&M plus overhead (MUSD/year)	0,5	0,5	0,5	0,5
Property tax (MUSD/year)	3,8	3,8	3,9	4,0
Module price = 3,0 USD/W				
Annual cost (MUSD/year)	23,2	23,2	23,7	24,2
PV module (MUSD/year)	14,9	14,9	14,9	14,9
BOS (MUSD/year)	4,,0	4,0	4,4	4,9
Transportation (MUSD/year)	0,6	0,6	0,6	0,6
Transmission (MUSD/year)	0,2	0,2	0,2	0,2
Construction (MUSD/year)	0,1	0,1	0,1	0,1
O&M plus overhead (MUSD/year)	0,4	0,4	0,4	0,4
Property tax (MUSD/year)	3,0	3,0	3,1	3,1
Module price = 2,0 USD/W				
Annual cost (MUSD/year)	17,1	17,1	17,6	18,1
PV module (MUSD/year)	10,0	10,0	10,0	10,0
BOS (MUSD/year)	3,8	3,8	4,2	4,7
Transportation (MUSD/year)	0,6	0,6	0,6	0,6
Transmission (MUSD/year)	0,2	0,2	0,2	0,2
Construction (MUSD/year)	0,1	0,1	0,1	0,1
O&M plus overhead (MUSD/year)	0,3	0,3	0,3	0,3
Property tax (MUSD/year)	2,2	2,2	2,3	2,3
Module price = 1,0 USD/W				
Annual cost (MUSD/year)	10,9	11,0	11,4	12,0
PV module (MUSD/year)	5,0	5,0	5,0	5,0
BOS (MUSD/year)	3,5	3,6	4,0	4,4
Transportation (MUSD/year)	0,6	0,6	0,6	0,6
Transmission (MUSD/year)	0,2	0,2	0,2	0,2
Construction (MUSD/year)	0,1	0,1	0,1	0,1
O&M plus overhead (MUSD/year)	0,2	0,2	0,2	0,2
Property tax (MUSD/year)	1,4	1,4	1,5	1,6

Table A.23 Annual cost of a 100 MW VLS-PV system in the Sonoran Desert (Chihuahua)

Tilt angle = (degrees)	10	20	30	40
Annual power generation (GWh/year)	145	150	151	148
Module price = US4,0/W				
Annual cost (MUSD/year)	29,1	29,1	29,5	29,9
PV module (MUSD/year)	19,9	19,9	19,9	19,9
BOS (MUSD/year)	4,0	4,1	4,4	4,7
Transportation (MUSD/year)	0,0	0,0	0,0	0,0
Transmission (MUSD/year)	0,2	0,2	0,2	0,2
Construction (MUSD/year)	0,6	0,6	0,6	0,6
O&M plus overhead (MUSD/year)	0,5	0,5	0,5	0,5
Property tax (MUSD/year)	3,8	3,8	3,9	3,9
Module price = 3,0 USD/W				
Annual cost (MUSD/year)	22,9	23,0	23,3	23,8
PV module (MUSD/year)	14,9	14,9	14,9	14,9
BOS (MUSD/year)	3,8	3,8	4,1	4,5
Transportation (MUSD/year)	0,0	0,0	0,0	0,0
Transmission (MUSD/year)	0,2	0,2	0,2	0,2
Construction (MUSD/year)	0,6	0,6	0,6	0,6
O&M plus overhead (MUSD/year)	0,4	0,4	0,4	0,4
Property tax (MUSD/year)	3,0	3,0	3,0	3,1
Module price = 2,0 USD/W				
Annual cost (MUSD/year)	16,8	16,9	17,3	17,7
PV module (MUSD/year)	10,0	10,0	10,0	10,0
BOS (MUSD/year)	3,6	3,6	3,9	4,3
Transportation (MUSD/year)	0,0	0,0	0,0	0,0
Transmission (MUSD/year)	0,2	0,2	0,2	0,2
Construction (MUSD/year)	0,6	0,6	0,6	0,6
O&M plus overhead (MUSD/year)	0,3	0,3	0,3	0,3
Property tax (MUSD/year)	2,2	2,2	2,2	2,3
Module price = 1,0 USD/W				
Annual cost (MUSD/year)	10,7	10,7	11,1	11,5
PV module (MUSD/year)	5,0	5,0	5,0	5,0
BOS (MUSD/year)	3,3	3,4	3,7	4,0
Transportation (MUSD/year)	0,0	0,0	0,0	0,0
Transmission (MUSD/year)	0,2	0,2	0,2	0,2
Construction (MUSD/year)	0,6	0,6	0,6	0,6
O&M plus overhead (MUSD/year)	0,2	0,2	0,2	0,2
Property tax (MUSD/year)	1,3	1,3	1,4	1,5

Table A.24 Annual cost of a 100 MW VLS-PV system in the Great Sandy Desert (Port Headland)

Tilt angle = (degrees)	10	20	30	40
Annual power generation (GWh/year)	160	162	160	154
Module price = 4,0 USD/W				
Annual cost (MUSD/year)	34,0	34,1	34,6	35,2
PV module (MUSD/year)	19,9	19,9	19,9	19,9
BOS (MUSD/year)	4,2	4,2	4,6	5,0
Transportation (MUSD/year)	0,0	0,0	0,0	0,0
Transmission (MUSD/year)	0,2	0,2	0,2	0,2
Construction (MUSD/year)	4,7	4,8	4,9	4,9
O&M plus overhead (MUSD/year)	1,1	1,1	1,1	1,1
Property tax (MUSD/year)	3,8	3,8	3,9	4,0
Module price = 3,0 USD/W				
Annual cost (MUSD/year)	27,8	27,9	28,4	29,0
PV module (MUSD/year)	14,9	14,9	14,9	14,9
BOS (MUSD/year)	3,9	4,0	4,3	4,8
Transportation (MUSD/year)	0,0	0,0	0,0	0,0
Transmission (MUSD/year)	0,2	0,2	0,2	0,2
Construction (MUSD/year)	4,7	4,8	4,9	4,9
O&M plus overhead (MUSD/year)	1,0	1,0	1,0	1,0
Property tax (MUSD/year)	3,0	3,0	3,1	3,1
Module price = 2,0 USD/W				
Annual cost (MUSD/year)	21,8	21,9	22,4	22,9
PV module (MUSD/year)	10,0	10,0	10,0	10,0
BOS (MUSD/year)	3,7	3,8	4,1	4,6
Transportation (MUSD/year)	0,0	0,0	0,0	0,0
Transmission (MUSD/year)	0,2	0,2	0,2	0,2
Construction (MUSD/year)	4,7	4,8	4,9	4,9
O&M plus overhead (MUSD/year)	0,9	0,9	0,9	1,0
Property tax (MUSD/year)	2,2	2,2	2,2	2,3
Module price = 1,0 USD/W				
Annual cost (MUSD/year)	15,6	15,7	16,2	16,8
PV module (MUSD/year)	5,0	5,0	5,0	5,0
BOS (MUSD/year)	3,5	3,5	3,9	4,3
Transportation (MUSD/year)	0,0	0,0	0,0	0,0
Transmission (MUSD/year)	0,2	0,2	0,2	0,2
Construction (MUSD/year)	4,7	4,8	4,9	4,9
O&M plus overhead (MUSD/year)	0,8	0,8	0,8	0,9
Property tax (MUSD/year)	1,4	1,4	1,4	1,5

Table A.25 Annual cost of a 100 MW VLS-PV system in the Gobi Desert (Huhhot)

Tilt angle = (degrees)	10	20	30	40
Annual power generation (GWh/year)	138	145	149	150
Module price = 4,0 USD/W				
Annual cost (MUSD/year)	28,7	28,8	28,9	29,0
PV module (MUSD/year)	19,9	19,9	19,9	19,9
BOS (MUSD/year)	3,6	3,6	3,7	3,8
Transportation (MUSD/year)	0,6	0,6	0,6	0,6
Transmission (MUSD/year)	0,2	0,2	0,2	0,2
Construction (MUSD/year)	0,2	0,2	0,2	0,2
O&M plus overhead (MUSD/year)	0,5	0,5	0,5	0,5
Property tax (MUSD/year)	3,7	3,7	3,8	3,8
Module price = 3,0 USD/W				
Annual cost (MUSD/year)	22,6	22,6	22,7	22,8
PV module (MUSD/year)	14,9	14,9	14,9	14,9
BOS (MUSD/year)	3,4	3,4	3,5	3,6
Transportation (MUSD/year)	0,6	0,6	0,6	0,6
Transmission (MUSD/year)	0,2	0,2	0,2	0,2
Construction (MUSD/year)	0,2	0,2	0,2	0,2
O&M plus overhead (MUSD/year)	0,4	0,4	0,4	0,4
Property tax (MUSD/year)	2,9	2,9	2,9	2,9
Module price = 2,0 USD/W				
Annual cost (MUSD/year)	16,5	16,5	16,6	16,8
PV module (MUSD/year)	10,0	10,0	10,0	10,0
BOS (MUSD/year)	3,2	3,2	3,3	3,4
Transportation (MUSD/year)	0,6	0,6	0,6	0,6
Transmission (MUSD/year)	0,2	0,2	0,2	0,2
Construction (MUSD/year)	0,2	0,2	0,2	0,2
O&M plus overhead (MUSD/year)	0,3	0,3	0,3	0,3
Property tax (MUSD/year)	2,1	2,1	2,1	2,1
Module price = 1,0 USD/W				
Annual cost (MUSD/year)	10,3	10,4	10,5	10,6
PV module (MUSD/year)	5,0	5,0	5,0	5,0
BOS (MUSD/year)	2,9	2,9	3,0	3,1
Transportation (MUSD/year)	0,6	0,6	0,6	0,6
Transmission (MUSD/year)	0,2	0,2	0,2	0,2
Construction (MUSD/year)	0,2	0,2	0,2	0,2
O&M plus overhead (MUSD/year)	0,2	0,2	0,2	0,2
Property tax (MUSD/year)	1,3	1,3	1,3	1,3

Table A.26 Annual cost of a 100 MW VLS-PV system in the Gobi Desert (Sainshand)

Tilt angle = (degrees)	10	20	30	40
Annual power generation (GWh/year)	131	140	146	149
Module price = 4,0 USD/W				
Annual cost (MUSD/year)	28,7	28,8	28,9	29,2
PV module (MUSD/year)	19,9	19,9	19,9	19,9
BOS (MUSD/year)	3,1	3,1	3,3	3,4
Transportation (MUSD/year)	0,6	0,6	0,6	0,6
Transmission (MUSD/year)	0,2	0,2	0,2	0,2
Construction (MUSD/year)	0,8	0,8	0,8	0,9
O&M plus overhead (MUSD/year)	0,5	0,5	0,5	0,5
Property tax (MUSD/year)	3,6	3,7	3,7	3,7
Module price = 3,0 USD/W				
Annual cost (MUSD/year)	22,5	22,6	22,8	23,0
PV module (MUSD/year)	14,9	14,9	14,9	14,9
BOS (MUSD/year)	2,8	2,9	3,0	3,2
Transportation (MUSD/year)	0,6	0,6	0,6	0,6
Transmission (MUSD/year)	0,2	0,2	0,2	0,2
Construction (MUSD/year)	0,8	0,8	0,8	0,9
O&M plus overhead (MUSD/year)	0,4	0,4	0,4	0,4
Property tax (MUSD/year)	2,8	2,8	2,9	2,9
Module price = 2,0 USD/W				
Annual cost (MUSD/year)	16,5	16,5	16,7	16,9
PV module (MUSD/year)	10,0	10,0	10,0	10,0
BOS (MUSD/year)	2,6	2,7	2,8	3,0
Transportation (MUSD/year)	0,6	0,6	0,6	0,6
Transmission (MUSD/year)	0,2	0,2	0,2	0,2
Construction (MUSD/year)	0,8	0,8	0,8	0,9
O&M plus overhead (MUSD/year)	0,3	0,3	0,3	0,3
Property tax (MUSD/year)	2,0	2,0	2,0	2,1
Module price = 1,0 USD/W				
Annual cost (MUSD/year)	10,3	10,4	10,6	10,8
PV module (MUSD/year)	5,0	5,0	5,0	5,0
BOS (MUSD/year)	2,4	2,4	2,5	2,7
Transportation (MUSD/year)	0,6	0,6	0,6	0,6
Transmission (MUSD/year)	0,2	0,2	0,2	0,2
Construction (MUSD/year)	0,8	0,8	0,8	0,9
O&M plus overhead (MUSD/year)	0,2	0,2	0,2	0,2
Property tax (MUSD/year)	1,2	1,2	1,2	1,2

Table A.27 Generation cost of a 100 MW VLS-PV system (US cents/kWh)

	Sahara		Negev	Thar	Sonoran	Great Sandy	Gobi	
	Mauritania Nema	Morocco Ouarzazate	Israel	India Jodhpur	Mexco Chihuahua	Australia Port Headland	China Hohhot	Mongolia Sainshand
Module price = 1 USD/W								
Tilt angle = 10°	5,8	7,2	9,6	7,2	7,3	9,8	7,5	7,9
Tilt angle = 20°	**5,8**	**7,0**	**9,4**	**7,0**	**7,2**	**9,7**	7,2	7,4
Tilt angle = 30°	6,0	7,0	9,8	7,3	7,4	10,1	**7,0**	**7,2**
Tilt angle = 40°	6,4	7,3	10,5	7,7	7,8	10,9	7,1	7,2
Module price = 2 USD/W								
Tilt angle = 10°	9,3	11,5	13,9	11,3	11,6	13,6	12,0	12,6
Tilt angle = 20°	**9,3**	**11,2**	**13,6**	**11,0**	**11,3**	**13,5**	11,4	11,8
Tilt angle = 30°	9,6	11,2	14,0	11,2	11,5	14,0	**11,1**	**11,4**
Tilt angle = 40°	10,2	11,5	14,8	11,7	11,9	14,9	11,2	11,4
Module price = 3 USD/W								
Tilt angle = 10°	12,8	15,8	18,3	15,3	15,8	17,4	16,4	17,2
Tilt angle = 20°	**12,8**	**15,3**	**17,8**	**14,9**	**15,4**	**17,3**	15,6	16,1
Tilt angle = 30°	13,1	15,3	18,2	15,0	15,5	17,8	**15,2**	**15,6**
Tilt angle = 40°	13,8	15,7	19,0	15,6	16,1	18,8	15,2	15,5
Module price = 4 USD/W								
Tilt angle = 10°	16,3	20,1	22,6	19,3	20,0	21,3	20,9	21,9
Tilt angle = 20°	**16,3**	**19,5**	**22,1**	**18,8**	**19,5**	**21,1**	19,8	20,5
Tilt angle = 30°	16,7	19,4	22,4	18,9	19,6	21,6	**19,3**	**19,8**
Tilt angle = 40°	17,6	19,9	23,3	19,5	20,2	22,8	19,3	19,6

Table A.28 System components (for 30°)

PV module	m-Si	a-Si	CdTe	m-Si
Number of modules (10^3)	849	1 909	1 616	786
System capacity (MW)	100,8	109,6	104,0	93,3
Area (km^2)	2,3	4,4	3,4	3,7
Array support structure (tonne)	9 800	Fixed flat plate 19 300	18 700	Tracking 10 800
Foundation (k tonne)	60	110	68	46

Table A.29 Annual cost for a 100 MW VLS-PV system with m-Si modules

Tilt angle = (degrees)	10	20	30	40
Annual power generation (GWh/year)	138	145	149	150
Module price = 4,0 USD/W				
Annual cost (MUSD/year)	28,9	29,0	29,1	29,3
PV module (MUSD/year)	19,9	19,9	19,9	19,9
BOS (MUSD/year)	3,6	3,6	3,7	3,8
Transportation (MUSD/year)	0,8	0,8	0,8	0,9
Transmission (MUSD/year)	0,2	0,2	0,2	0,2
Construction (MUSD/year)	0,2	0,2	0,2	0,2
O&M plus overhead (MUSD/year)	0,5	0,5	0,5	0,5
Property tax (MUSD/year)	3,7	3,7	3,8	3,8
Module price = 3,0 USD/W				
Annual cost (MUSD/year)	22,8	22,8	23,0	23,2
PV module (MUSD/year)	14,9	14,9	14,9	14,9
BOS (MUSD/year)	3,4	3,4	3,5	3,6
Transportation (MUSD/year)	0,8	0,8	0,8	0,9
Transmission (MUSD/year)	0,2	0,2	0,2	0,2
Construction (MUSD/year)	0,2	0,2	0,2	0,2
O&M plus overhead (MUSD/year)	0,4	0,4	0,4	0,4
Property tax (MUSD/year)	2,9	2,9	2,9	2,9
Module price = 2,0 USD/W				
Annual cost (MUSD/year)	16,7	16,7	16,9	17,1
PV module (MUSD/year)	10,0	10,0	10,0	10,0
BOS (MUSD/year)	3,2	3,2	3,3	3,4
Transportation (MUSD/year)	0,8	0,8	0,8	0,9
Transmission (MUSD/year)	0,2	0,2	0,2	0,2
Construction (MUSD/year)	0,2	0,2	0,2	0,2
O&M plus overhead (MUSD/year)	0,3	0,3	0,3	0,3
Property tax (MUSD/year)	2,1	2,1	2,1	2,1
Module price = 1,0 USD/W				
Annual cost (MUSD/year)	10,5	10,6	10,7	10,9
PV module (MUSD/year)	5,0	5,0	5,0	5,0
BOS (MUSD/year)	2,9	2,9	3,0	3,1
Transportation (MUSD/year)	0,8	0,8	0,8	0,9
Transmission (MUSD/year)	0,2	0,2	0,2	0,2
Construction (MUSD/year)	0,2	0,2	0,2	0,2
O&M plus overhead (MUSD/year)	0,2	0,2	0,2	0,2
Property tax (MUSD/year)	1,3	1,3	1,3	1,3

Table A.30 Annual cost for a 100 MW VLS-PV system with a-Si modules

Tilt angle = (degrees)	10	20	30	40
Annual power generation (GWh/year)	149	157	161	162
Module price = 4,0 USD/W				
Annual cost (MUSD/year)	32,8	33,0	33,2	33,6
PV module (MUSD/year)	21,7	21,7	21,7	21,7
BOS (MUSD/year)	4,5	4,6	4,7	5,0
Transportation (MUSD/year)	1,4	1,4	1,5	1,6
Transmission (MUSD/year)	0,2	0,2	0,2	0,2
Construction (MUSD/year)	0,4	0,4	0,4	0,5
O&M plus overhead (MUSD/year)	0,5	0,5	0,5	0,5
Property tax (MUSD/year)	4,1	4,2	4,2	4,2
Module price = 3,0 USD/W				
Annual cost (MUSD/year)	26,1	26,3	26,5	26,9
PV module (MUSD/year)	16,2	16,2	16,2	16,2
BOS (MUSD/year)	4,2	4,3	4,4	4,7
Transportation (MUSD/year)	1,4	1,4	1,5	1,6
Transmission (MUSD/year)	0,2	0,2	0,2	0,2
Construction (MUSD/year)	0,4	0,4	0,4	0,5
O&M plus overhead (MUSD/year)	0,4	0,4	0,4	0,4
Property tax (MUSD/year)	3,3	3,3	3,3	3,3
Module price = 2,0 USD/W				
Annual cost (MUSD/year)	19,5	19,7	19,9	20,3
PV module (MUSD/year)	10,8	10,8	10,8	10,8
BOS (MUSD/year)	4,0	4,1	4,2	4,5
Transportation (MUSD/year)	1,4	1,4	1,5	1,6
Transmission (MUSD/year)	0,2	0,2	0,2	0,2
Construction (MUSD/year)	0,4	0,4	0,4	0,5
O&M plus overhead (MUSD/year)	0,3	0,3	0,3	0,3
Property tax (MUSD/year)	2,4	2,4	2,4	2,4
Module price = 1,0 USD/W				
Annual cost (MUSD/year)	12,8	13,0	13,2	13,6
PV module (MUSD/year)	5,4	5,4	5,4	5,4
BOS (MUSD/year)	3,7	3,8	3,9	4,2
Transportation (MUSD/year)	1,4	1,4	1,5	1,6
Transmission (MUSD/year)	0,2	0,2	0,2	0,2
Construction (MUSD/year)	0,4	0,4	0,4	0,5
O&M plus overhead (MUSD/year)	0,2	0,2	0,2	0,2
Property tax (MUSD/year)	1,5	1,5	1,5	1,5

Table A.31 Annual cost for a 100 MW VLS-PV system with CdTe modules

Tilt angle = (degrees)	10	20	30	40
Annual power generation (GWh/year)	141	149	154	155
Module price = 4,0 USD/W				
Annual cost (MUSD/year)	30,8	31,0	31,1	31,4
PV module (MUSD/year)	20,6	20,6	20,6	20,6
BOS (MUSD/year)	4,1	4,2	4,3	4,5
Transportation (MUSD/year)	1,1	1,1	1,2	1,3
Transmission (MUSD/year)	0,2	0,2	0,2	0,2
Construction (MUSD/year)	0,4	0,4	0,4	0,4
O&M plus overhead (MUSD/year)	0,5	0,5	0,5	0,5
Property tax (MUSD/year)	3,9	3,9	3,9	4,0
Module price = 3,0 USD/W				
Annual cost (MUSD/year)	24,5	24,6	24,8	25,0
PV module (MUSD/year)	15,4	15,4	15,4	15,4
BOS (MUSD/year)	3,9	4,0	4,1	4,2
Transportation (MUSD/year)	1,1	1,1	1,2	1,3
Transmission (MUSD/year)	0,2	0,2	0,2	0,2
Construction (MUSD/year)	0,4	0,4	0,4	0,4
O&M plus overhead (MUSD/year)	0,4	0,4	0,4	0,4
Property tax (MUSD/year)	3,1	3,1	3,1	3,1
Module price = 2,0 USD/W				
Annual cost (MUSD/year)	18,2	18,3	18,5	18,7
PV module (MUSD/year)	10,3	10,3	10,3	10,3
BOS (MUSD/year)	3,7	3,8	3,9	4,0
Transportation (MUSD/year)	1,1	1,1	1,2	1,3
Transmission (MUSD/year)	0,2	0,2	0,2	0,2
Construction (MUSD/year)	0,4	0,4	0,4	0,4
O&M plus overhead (MUSD/year)	0,3	0,3	0,3	0,3
Property tax (MUSD/year)	2,2	2,2	2,3	2,3
Module price = 1,0 USD/W				
Annual cost (MUSD/year)	11,9	12,0	12,2	12,4
PV module (MUSD/year)	5,1	5,1	5,1	5,1
BOS (MUSD/year)	3,4	3,5	3,6	3,7
Transportation (MUSD/year)	1,1	1,1	1,2	1,3
Transmission (MUSD/year)	0,2	0,2	0,2	0,2
Construction (MUSD/year)	0,4	0,4	0,4	0,4
O&M plus overhead (MUSD/year)	0,2	0,2	0,2	0,2
Property tax (MUSD/year)	1,4	1,4	1,4	1,4

Table A.32 Annual cost for a 100 MW sun-tracking VLS-PV system

Module price (USD/W)	4,0	3,0	2,0	1,0
Annual power generation (GWh/year)	163	163	163	163
Annual cost (MUSD/year)	27,3	21,5	15,9	10,2
PV module (MUSD/year)	18,4	13,8	9,2	4,6
BOS (MUSD/year)	3,6	3,3	3,2	3,0
Transportation (MUSD/year)	0,8	0,8	0,8	0,8
Transmission (MUSD/year)	0,2	0,2	0,2	0,2
Construction (MUSD/year)	0,2	0,2	0,2	0,2
O&M plus overhead (MUSD/year)	0,5	0,4	0,3	0,2
Property tax (MUSD/year)	3,5	2,7	2,0	1,2

Table A.33 Generation cost of 100 MW VLS-PV systems at optimal tilt angle (30º)

PV module	m-Si	a-Si	CdTe	m-Si (Tracking)
Module price				
1 USD/W	6,8	8,2	7,9	6,3
2 USD/W	11,3	12,3	12,1	9,8
3 USD/W	15,4	16,4	16,2	13,2
4 USD/W	19,5	20,6	20,4	16,7

Table A.34 Total energy requirement and energy payback time of a 100 MW sun-tracking VLS-PV system using m-Si PV modules

Total requirement energy (TJ)	3 736
PV modules	1 538
Array support	1 130
Foundation and trough	160
Cable	39
Transportation	281
Transmission	547
Other components	28
Energy payback time (year)	1,9

Table A.35 Life-cycle CO_2 emissions and life-cycle CO_2 emission rate of a sun-tracking 100 MW PV system with m-Si PV modules

Total requirement energy (t-C)	69 741
PV modules	24 185
Array support	27 963
Foundation and trough	3 966
Cable	617
Transportation	5 257
Transmission	6 906
Other components	571
CO_2 emission rate (g-C/kWh)	13,3

Figure A.37 Examples of the surface of the Gobi Desert

A.2 RESOURCE ANALYSIS OF SOLAR ENERGY BY USING SATELLITE IMAGES

A.2.1 Introduction

There is high irradiation and vast open land in desert areas, but these regions are frequently unsuitable for constructing buildings. Figure A.37 shows examples of different types of desert surfaces. For example, sand dunes, mountains and valleys provide different conditions that must be taken into account when installing VLS-PV systems. However, gravel desert (for example, in the Gobi Desert) also exists in desert areas and is suitable land for VLS-PV systems because of its hardness and flatness.

This section will discuss how to find land where VLS-PV systems can be installed in arid regions, and will provide a detailed classification algorithm to find suitable areas by using a remote sensing approach with satellite images. Two main purposes are outlined:

1 Find suitable places for very large scale PV systems.
2 Investigate how much PV potential desert regions have.

Table A.36 Main characteristics of satellite images

	JERS-1	Landsat-7
Analysed image	Band 1, 2, 3	Band 2 ,3, 4
Resolution	18.3 m × 24.2 m	30.0 m × 30.0 m
Swathe width	75 km	180 km
Subject desert	Gobi Desert	Gobi, Sahara, Negev, Sonora, Thar, Great Sandy deserts

In addition, the potential of a resource for developing a VLS-PV system will be presented.

A.2.2 Subject area

Case study deserts are located in the Gobi, Sahara, Negev, Sonora, Thar and Great Sandy deserts, as shown in Figure A.38. Since these deserts have many types of land surface (for example, stone, sand dunes, water basins and mountain areas), detailed investigations are necessary. In addition, land surface features differ in every desert. The categories of soil especially differ according to their location. Forest and moving areas such as sand dune zones are not the most suitable regions in which to site VLS-PV systems. Finding the right locations is the priority of this study.

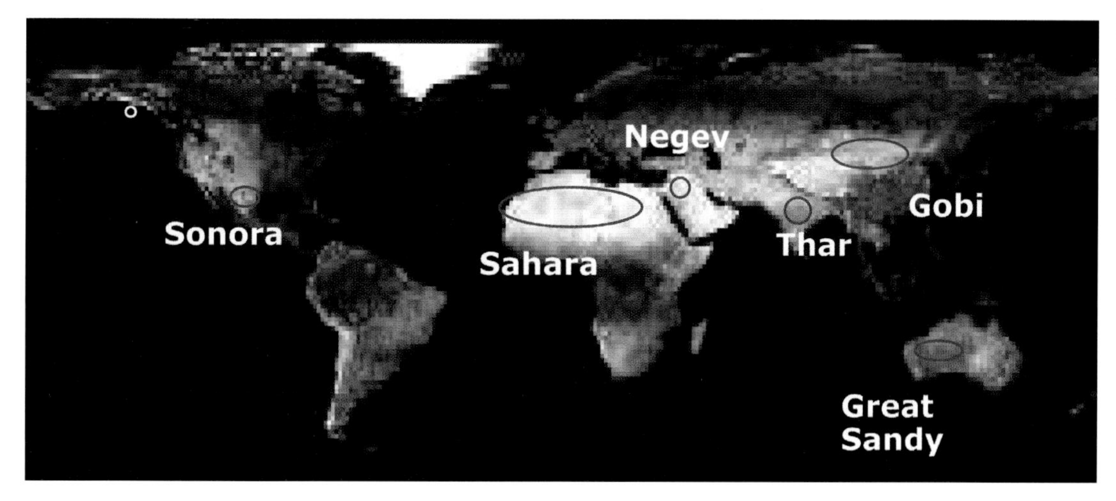

Figure A.38 Six major world deserts as subject area

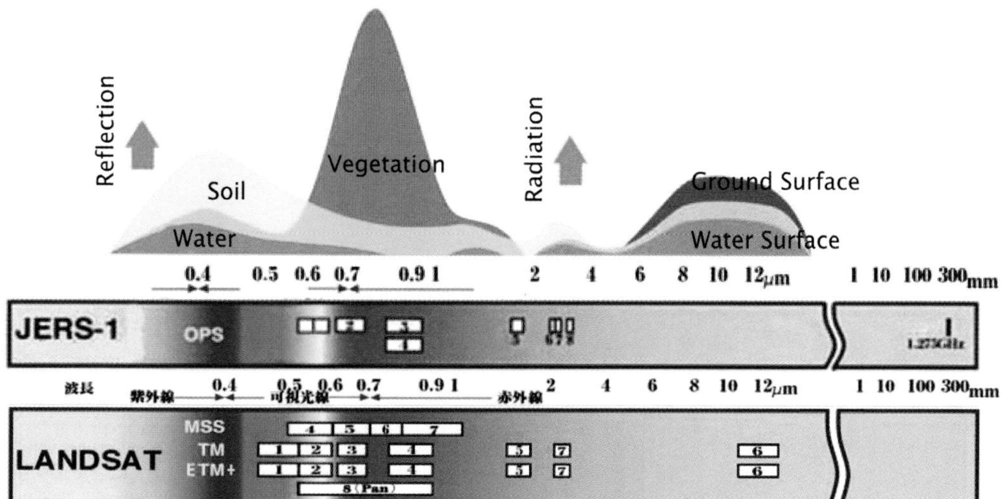

Figure A.39 Examples of reflecting features and radiation characteristics

A.2.3 Method and results

A.2.3.1 Satellite images used

JERS-1 and Landsat-7 satellite images were chosen for analysis.[1] The characteristics of these images are shown in Table A.36 and Figure A.39. Although the JERS-1 images are known as high-resolution satellite images, it is difficult to obtain all images of these deserts, and it was possible to get a JERS-1 image of only the Gobi Desert. Therefore, the Gobi Desert has been investigated with both JERS-1 and Landsat-7 images. All of the other deserts have been investigated with images from Landsat-7. Incidentally, the great quantity of images produced gives Landsat-7 subjects an advantage.

A.2.3.2 Scheme of evaluation

The following five steps were adapted to evaluate arid land areas and to remove the areas unsuitable for PV systems:

1 Obtain satellite images.
2 Divide images by vegetation index. Vegetation index is evaluated by near infrared ray (NIR) and red light taken by satellite. It is possible to divide water, steppe and meadow.
3 Obtain a ground cover classification. A statistical method, the maximum likelihood classifier (MLC), is employed to classify ground. Training data are set by authors, and object images are classified into seven patterns: cloud; desert steppe or steppe; forest; water; dune or desert steppe; shadow; and rock desert.
4 Hill classification is calculated by extracting the edge of hills. Edge is calculated by three processes: Laplacian filter, binarization processing and morphology filter.
5 Ground truth: to find the accuracy of this analysis, evaluation results and site conditions are compared by checking against the real site.

Finally, suitable areas in six world deserts for PV systems are evaluated and mapped.

A.2.3.3 Vegetation index

A Modified Soil Adjusted Vegetation Index (MSAVI) is calculated in order to investigate the vegetation level. MSAVI was analysed by using reflectance values from two channels (NIR:860nm and RED:672nm). MSAVI is calculated by Equation A.13:

$$MSAVI = \frac{2NIR + 1 - \sqrt{(2NIR + 1)^2 - 8(NIR - RED)}}{2} \quad [A.13]$$

Figure A.40 Example of vegetation classification by vegetation index

MSAVI evaluates the density and quantity of vegetation. When vegetation cover is minimal, soil reflectance usually increases in both the red and infrared channels. Other indices are available to describe this soil–vegetation system more adequately. To minimize the effect of bare soil, MSAVI was developed by Qi et al (1994),[2] who extracted various images as training data based on results that were previously investigated in detail by Michael et al (2000).[3] According to these training data sets, three types of suitable vegetation level were highlighted, shown in Figure A.41. The vegetation levels are divided into six patterns, and the three types of suitable vegetation levels comprise superior steppe line, inferior steppe line and desert line. The six major world deserts were classified according to these threshold values.

However, care must be taken (and problems may be encountered) when applying MSAVI. The problems lie in seasonal changes in MSAVI as a result of the difference in satellite image dates. In order to correct the defect, it is estimated that the seasonal changes in MSAVI are based on the global four-minute Advanced

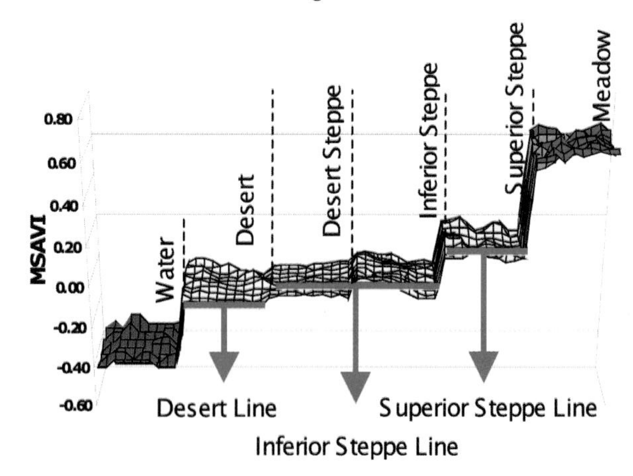

Figure A.41 Suitable vegetation levels for VLS-PV systems

Very High Resolution Radiometer – Normalized Difference Vegetation Index (AVHRR-NDVI) data set. This data set was already developed by the Centre for Environmental Remote Sensing (CEReS). Based on this data set, the seasonal changes in vegetation level are analysed. This method is adapted to the Landsat-7 image. The method that was adapted to JERS-1 was already developed in a previous paper.[4] Each seasonal change in vegetation level is shown in Figure A.42. Although these vegetation lines are different, these two threshold lines do not fluctuate if we focus on desert and inferior steppe lines. Therefore, the superior steppe threshold line is only adapted to seasonal change in vegetation levels.

A.2.3.4 Classification of the Gobi Desert

The surface of the Gobi Desert is classified into five patterns according to the maximum likelihood classifier method:

$$L(x,c) = \frac{1}{(2\pi)^{\frac{k}{2}}\,{}_cV^{1/2}} \exp\left\{-\frac{1}{2} d^2_M(x,c)\right\} \quad [A.14]$$

c: classification class
$L(x,c)$: likelihood
$d^2_M(x,c)$: Maharanobis distance
${}_cV$: covariance matrix.

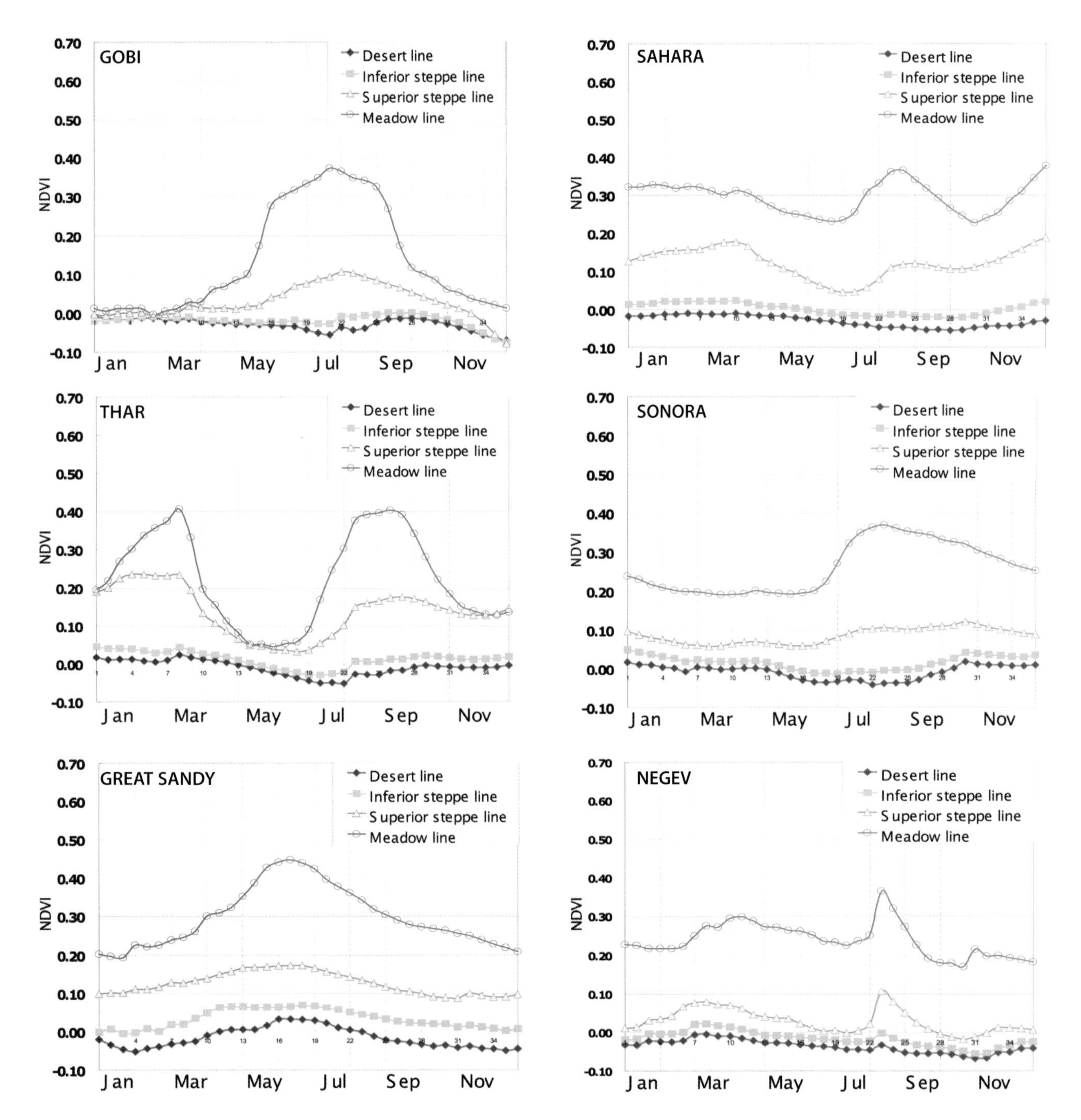

Figure A.42 Seasonal changes in vegetation levels

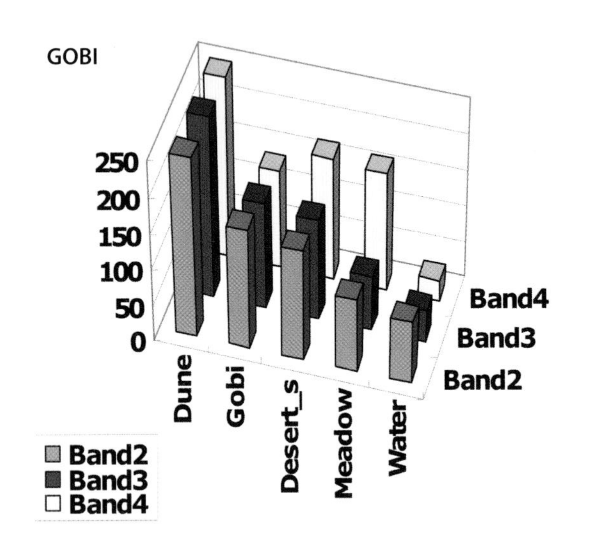

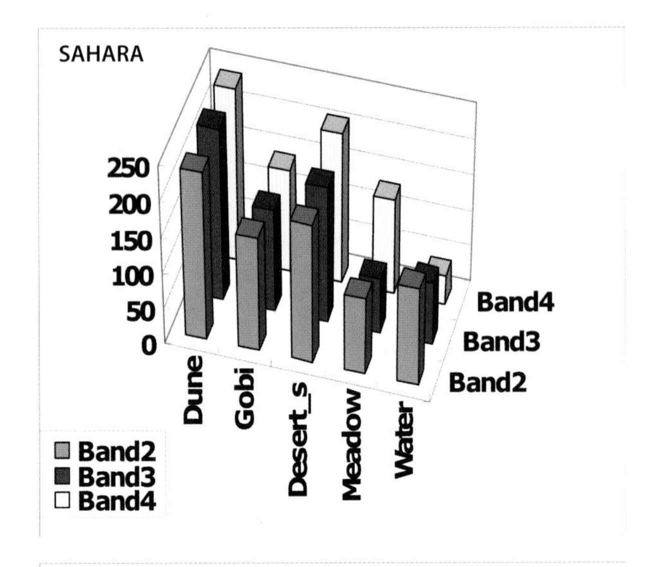

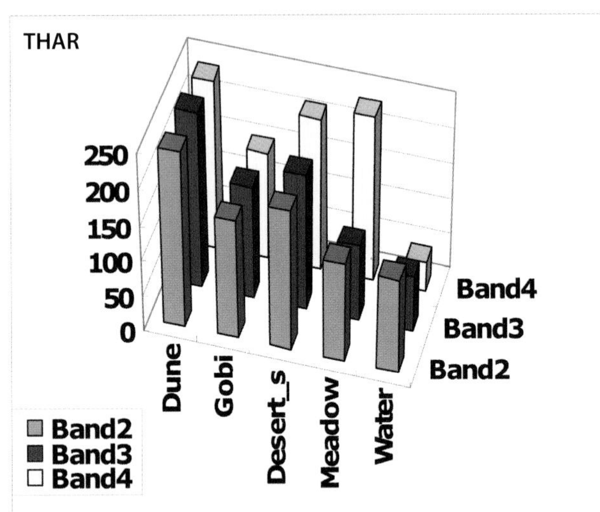

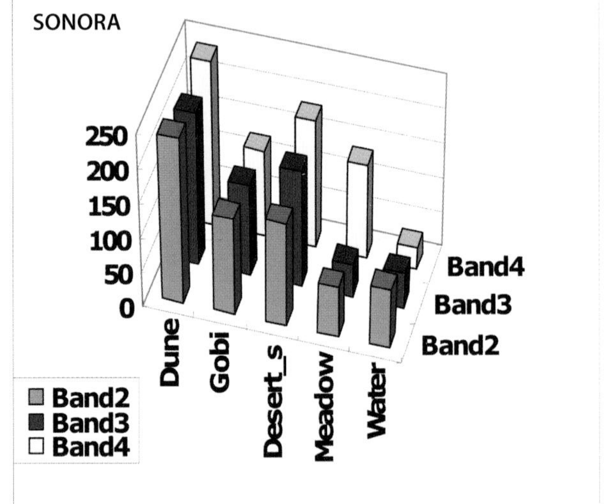

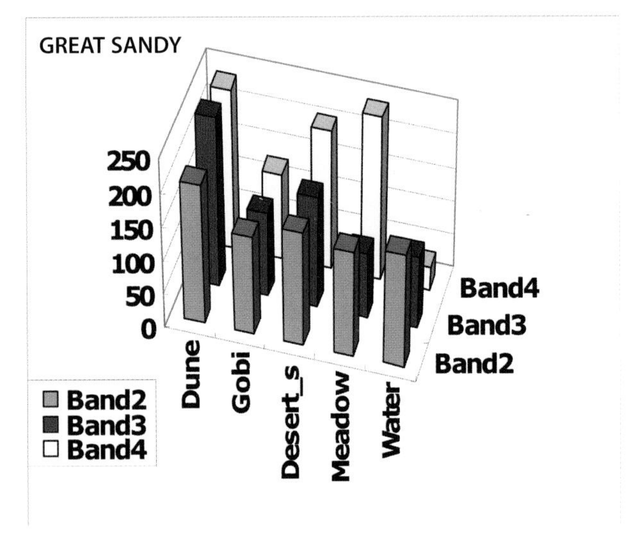

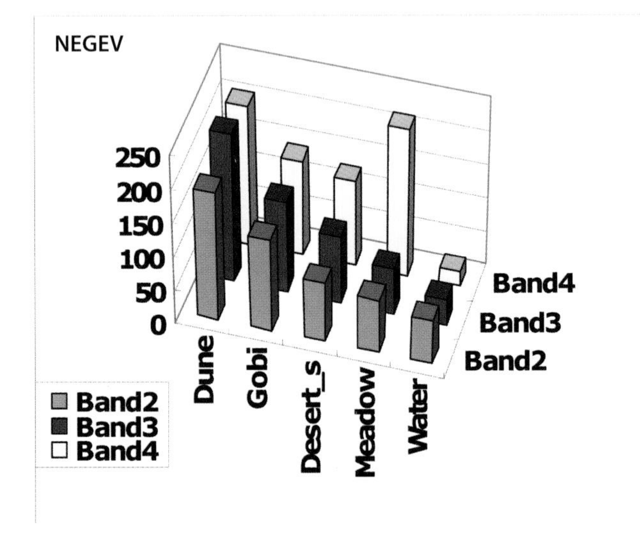

Figure A.43 Parameters of the maximum likelihood classifier

A.2.3.5 Filtering of satellite images

The most suitable areas for PV system installation are flat surfaces. Hills were extracted using Gaussian and Laplacian filters for band 3. Hills include valleys and flood traces. In addition, the processed images were divided into black and white according to threshold level and morphology filtering. Morphology filtering includes dilation and erosion. Secondly, the circumference of the neighbourhood's edge line was painted white. These methods were useful in solving a misclassification problem that was caused by relief shading.

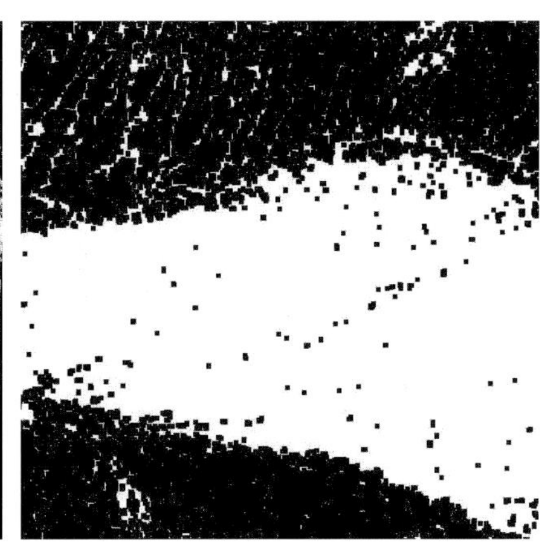

Figure A.44 Result of extracted edge line by satellite image

A.2.4 A resource evaluation of a PV system

According to these methods, estimations for suitable land for PV system installation were developed by integrating three processed images, which were calibrated by using MSAVI, MLC and edge lines extraction. In this study, only the estimation results of Landsat-7 were presented because the overall results were extensive, making it necessary to present only the latest results of Landsat-7. The area percentage of suitable land for the JERS-1 image has been already estimated as 40 per cent in a previous paper.[4]

Figure A.45 shows the results for part of the Sahara Desert. The authors of the study qualified results as S rank or A rank. S rank means that grass is shorter than 30 cm, land is hard and there are no hills nearby. A rank is almost same as S rank, but unsuitable area is also included, such as hills, dunes or water.

The authors have indicated three results from the parameters of the three vegetation levels. The results of suitable land for each desert are shown in Table A.37. These were characteristic results. The Gobi Desert has a large suitable area in the desert line, and the Great Sandy Desert has huge potential in the superior steppe line. We can forecast that PV system installation for suitable land has significant potential for electric power generation.

A.2.5 Ground truth

Ground truth was examined to evaluate the accuracy of the estimated results. Ground truth operation compares estimated results with actual field examination by using a global positioning system (GPS) receiver. Figure A.47 shows the results of the actual ground truth. The site is located in Tunisia. The line shows the track of the ground truth, and white points locate the actual field photograph in the figure. The precise latitude and longitude were measured every second with a GPS receiver, and the records and photographs were synchronized by time.

A GPS-photograph data set was developed with measured data. In the data set, each photograph was evaluated with S and A ranks:

- S rank: most suitable ground condition due to stable surface geometry.
- A rank: although surface geometry is S rank, the hills are visible to the eye well ahead.

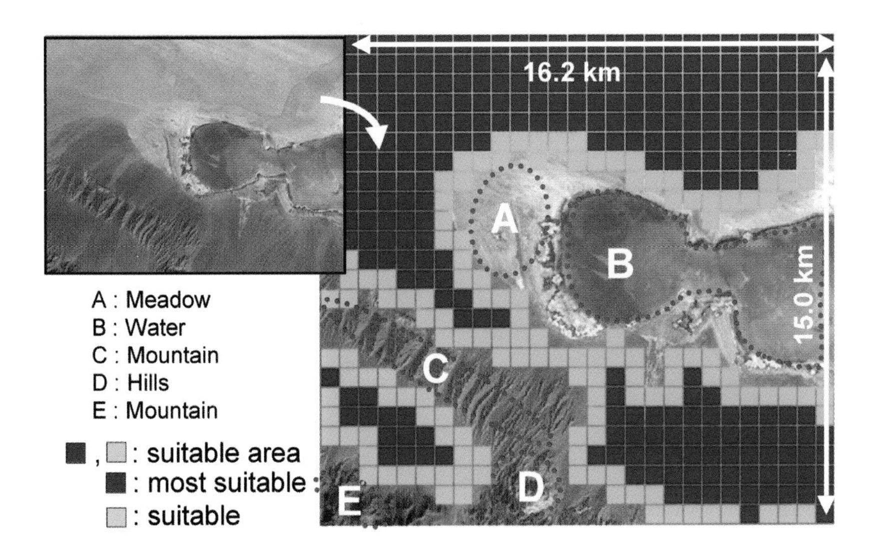

Figure A.45 Estimated results of part of the Sahara Desert

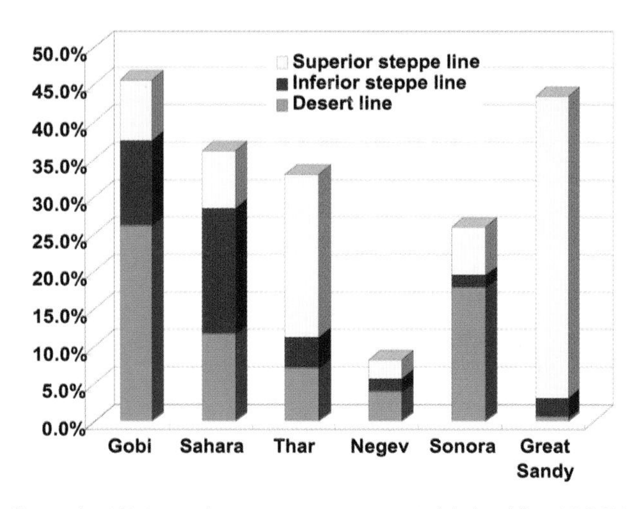

Figure A.46 Estimated area percentage as suitable land for a VLS-PV system using Landsat-7

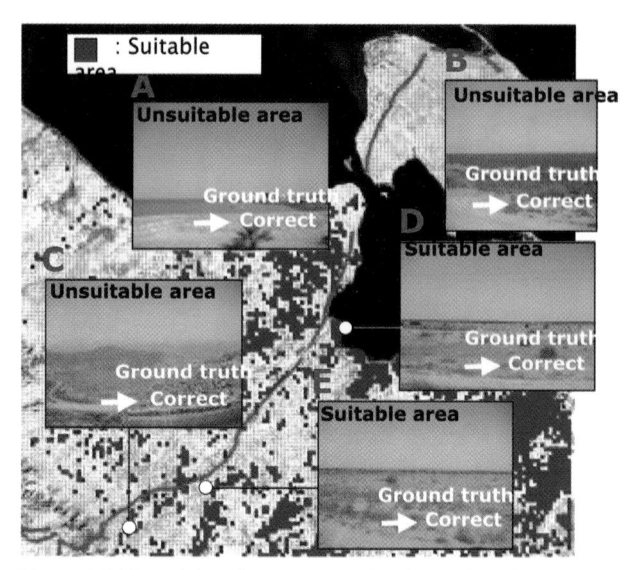

Figure A.47 Examining the accuracy of estimated results

Photograph D was evaluated as an S rank, indicating a suitable area, which was estimated by classification algorithm. Photograph C, evaluated as an A rank, indicated the presence of an unsuitable area. Therefore, sufficient results were obtained. However, although the ground condition is evaluated as S in the actual field photograph, a point evaluated as unsuitable ground surface also existed. This is attributed to the fact that the threshold level for extracting suitable ground conditions is tightly configured. If the threshold level for extracting suitable ground condition is configured loosely, unsuitable ground surface is extracted as suitable ground condition. Moreover, the seasonal changes of NDVI make analysis difficult. Under such circumstances, it would appear that presumed accuracy drops to a lower value. From these points, it also appears that tightening the threshold level is better than loosening it; but this needs to be investigated in detail. Additional ground truth is expected to develop accuracy of estimation by increasing the number of GPS photograph data sets.

A.2.6 Conclusions

This case study has described the process of finding a suitable area for large PV system installations through the use of remote sensing.

An algorithm using a vegetation index, the maximum likelihood classifier method and the edge line extraction method was developed. These methods were adapted to the six world major deserts, and the six deserts were investigated and evaluated in detail. Each parameter of the algorithm was established for each desert. The actual ground truth was also examined in order to determine the accuracy of the estimated results.

It is concluded that the proposed method provides sufficient information for planning PV system installations. In addition, the resources for solar photovoltaic generation systems in these deserts has been evaluated as having huge potential (see Figure A.48). The total potential of the six deserts is much larger than the total primary energy supply in 2000.

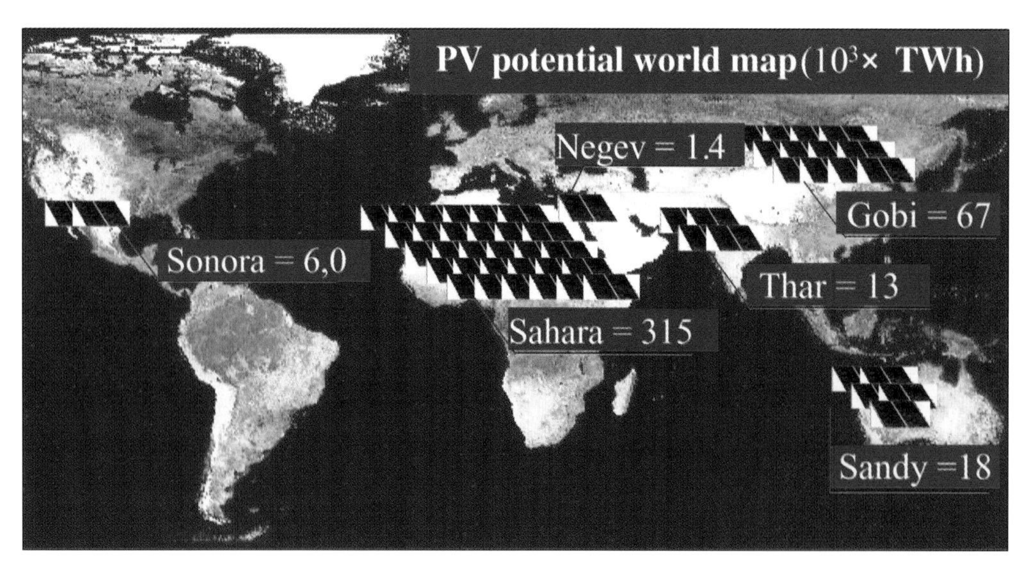

Figure A.48 PV potential world map

Table A.37 The resource potential of six world deserts

Desert	Analysed area (× 10^4 km^2)	Average solar irradiation ($MJ/m^2/day$)	Annual yield (hours)	Suitable condition zone	Percentage of suitable area	PV capacity (TW)	Annual power generation (10^3 TWh)
Gobi	162,8	17,6	1 789,1	Only desert	26,3	30,0	38,6
				Including inferior	37,6	42,9	55,3
				Including inferior, superior	45,8	52,2	66,9
Sahara	743,0	26,5	2 460,2	Only desert	11,7	59,8	100,0
				Including inferior	28,5	146,4	248,1
				Including inferior, superior	36,3	186,4	315,0
Thar	44,3	21,6	2 188,6	Only desert	7,2	1,6	2,4
				Including inferior	11,3	2,7	4,2
				Including inferior, superior	33,1	8,6	13,1
Negev	17,8	22,4	2 275,7	Only desert	3,9	0,3	0,6
				Including inferior	5,7	0,5	0,8
				Including inferior, superior	8,2	0,9	1,4
Sonora	18,7	20,7	2 100,0	Only desert	17,8	2,4	3,6
				Including inferior	19,4	3,2	4,7
				Including inferior, superior	26,1	4,1	6,0
Great Sandy	35,4	23,5	2 383,0	Only desert	0,6	0,2	0,3
				Including inferior	3,1	0,9	1,4
				Including inferior, superior	43,5	10,9	18,1

However, the following points are left as future problems. One difficulty with this method is deciding upon the parameters for the deserts. Correcting this defect requires more actual ground truth work in order to improve the accuracy of the estimation algorithm. Further work using different approaches is therefore necessary.

REFERENCES

1. Global Land Cover Factory retains ownership of Landsat-7 data; see http://glcfapp.umiacs.umd.edu:8080/esdi/index.jsp
2. Qi, J., Chehbouni, A., Huete, A. R. and Kerr, Y. H. (1994) 'Modified Soil Adjusted Vegetation Index (MSAVI)', *Remote Sensing of Environment*, vol 48, pp119–126
3. Burkart, M., Itzerott, S. and Zebisch, M. (2000) 'Classification of vegetation by chronosequences of NDVI from remote sensing and field data: The example of Uvs Nuur Basin', *Berliner geowiss*. Abh., A 205, pp39–50
4. Sakakibara, K., Ito, M. and Kurokawa, K. (2004) 'A modified resource analysis of very large scale PV (VLS-PV) system on the Gobi Desert by a remote sensing approach', *Proceedings of PVSEC14*, pp671–672

A.3 THE HIGHEST EFFICIENCY PV MODULE BY PRACTICAL CONCENTRATOR TECHNOLOGIES, PERFORMANCE, RELIABILITY AND APPLICATIONS TO DESERTS

A.3.1 Introduction

The efficiency record for PV modules published in 1996 by the University of South Wales in Australia, achieved by the PERL mono-crystalline silicon solar cells, has been updated after nine years.[1,2] This has been achieved with 550X and 400X (geometrical concentration ratio) concentrator PV modules with III-V multi-junction solar cells, innovative packaging technologies and the world's first dome-shaped non-imaging Fresnel lens made by practical injection moulding.

The motivation for developing high efficiency photovoltaic concentrator systems is to generate the maximum electrical power with the minimum module area and to lower the cost of photovoltaic power generation,[3] including material cost, land preparation cost and construction cost. The concept of concentrator photovoltaics is to use a lens to focus the sunlight onto a small but highly efficient solar cell. In this way, sunlight can be collected from a large area using cheap materials, such as plastic; but the power conversion is performed by a specialized high performance solar cell. The concentrator system described here has a concentration ratio of 400X or 550X (geometrical concentration ratio), meaning that the area over which sunlight is collected is 400 or 550 times larger than the area covered by the solar cells. This concentrator system uses triple-junction InGaP/InGaAs/Ge cells that are usually used on a spacecraft and would be prohibitively expensive for terrestrial application if they were employed alone. However, the use of lenses made from readily manufactured materials greatly reduces the area of expensive solar cell that is required, lowering the cost sufficiently that high efficiency photovoltaics can be used for terrestrial applications. The use of high efficiency PV is particularly important since it is desirable to generate as much power as possible from a module area so that it will reduce the area-dependent cost.

A.3.2 Technology overview

The module described in this section is shown in Figure A.49 and was developed using the following new technologies. The lower right-hand module with two lines of lenses is a 550X 150 W module, and the rest are 400X 200 W modules. The total rating is 1 550 W.

A.3.2.1 Module concept

The concentrator module is designed with ease of assembly in mind. All of the technologically complex components are packaged into a receiver so that a series of receivers and lenses can be assembled with standard tools, using local materials and a local workforce. The concept is similar to the computer and automobile assembly industries, where key components are imported, but the product is assembled locally. It is anticipated that this approach will reduce the manufacturing cost of the module.[4]

Figure A.49 550X and 400X on an open-loop tracker

A.3.2.2 Solar cell

We have successfully fabricated high efficiency concentrator InGaP/InGaAs/Ge three-junction solar cells designed for a 500 sun irradiation application. The efficiency by in-house measurement peaked at 489 sun irradiation and 38,9 %, as shown in Figures A.51 and A.52. The low loss tunnel junction interconnections and precise lattice matching are some of the features that have enabled such a high efficiency to be attained.

The solar simulator was equipped with both a Xe lamp and a halogen lamp and adjusted AM 1.5 G

Figure A.50 The receiver, a key component of concentrator technology

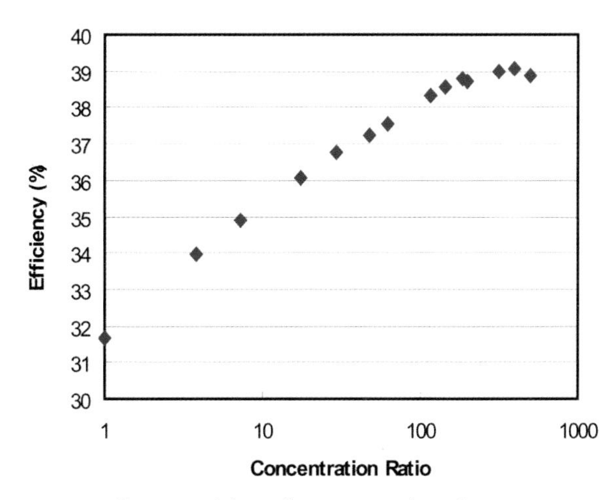

Figure A.51 Efficiency of the cell versus number of suns

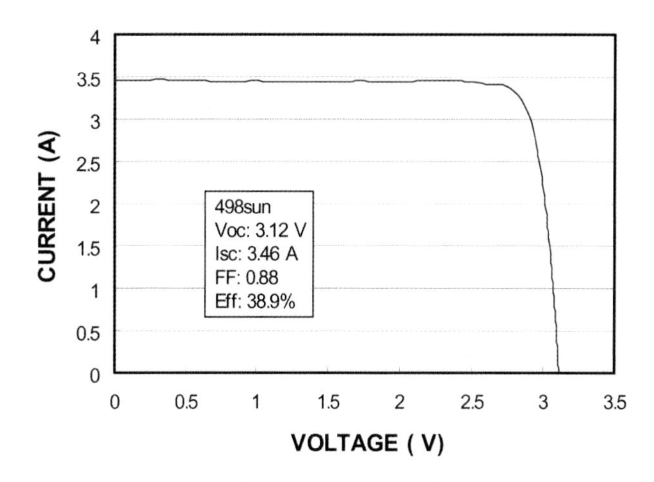

Figure A.52 I-V curve of the three-junction cell at 498 sun irradiation

spectrum. The chromatic aberrations of the simulator were evaluated by single junction Ge, InGaAs and InGaP cells with pin holes.[5] The error due to the chromatic aberration was analysed by a multi-unit SPICE model.[6] The actual efficiency under uniform illumination is thought to be approximately 1 % lower.

The cell shown in Figure A.53 is manufactured at Sharp Corporation using metal organic vapour-phase epitaxy (MOVPE). The new concentrator solar cell had a wide-gap tunnel junction with a double-hetero structure that enables only 1 m Ω resistance at 500 sun irradiation of current density.[7] The overall size was 7 mm × 9 mm, with a 7 mm square aperture area (see Figure A.54).

A.3.2.3 Packaging technology

A new packaging structure for III-V concentrator solar cells was developed, applicable mainly to Fresnel lens concentrator modules, but which may also be used in dish concentrator systems.[8] With this new structure, the concentrator module can generate the power without the help of heat sinks or other external cooling devices.[9]

Figure A.55 indicates how the heat spreading structure works efficiently without the help of heat sinks. It is true that the cell is exposed to a high energy flow and is potentially heated up to 1 400 °C under 500X concentration if it is insulated. However, the total energy that the module itself receives is the same as that of a flat-plate system and the cell temperature could be lowered almost equivalent to that of a flat-plate system if the heat received by the cell is efficiently transferred and spread in the base plate.

The first step is to solder tab sheets to the cell chip. The soldering technologies are chosen to connect the concentrator solar cells to the back electrodes. The key technology is soldering the front tab directly to the fragile cell chip. The commonly used technology is wire bonding or silver-epoxy bonding. These are convenient to avoid damage to the fragile concentrator cells. However, both methods have higher contact resistance and may not be scaled to a higher concentration ratio. The wire-bonding technology presents the problem of re-crystallization of

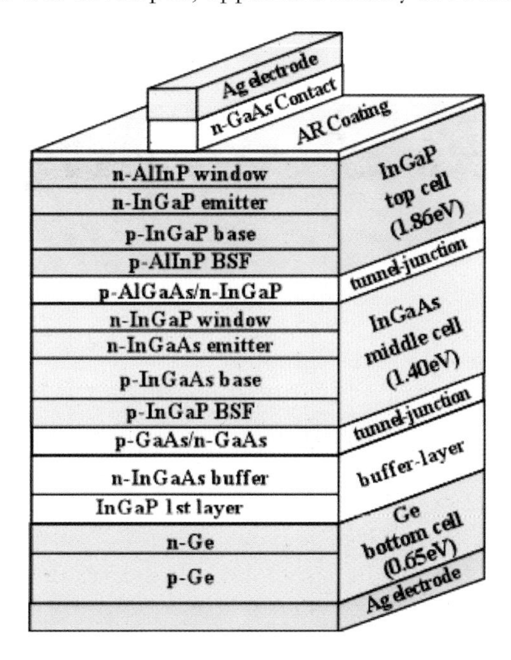

Figure A.53 Structure of the three-junction concentrator cell

Figure A.54 A bare III-V three-junction concentrator chip

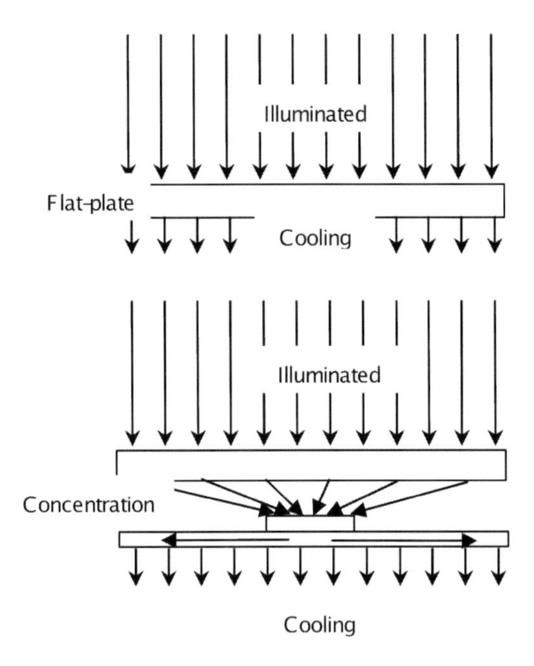

Figure A.55 Heat spreading concept for a concentrator module

Note: Notice that the total energy will be the same (as low as 1 kW/m^2) if the concentrated heat is spread efficiently.

the wires at a high current density. The silver-epoxy bonding showed a decrease in the contact points and an increase in the series resistance after heat cycling.

The second step is the lamination of the solar cell to the aluminium base plate. The material for laminating the solar cell to the base plate was heat conductive epoxy. It contained a heat-conductive medium uniformly dispersed in the epoxy. One of the difficulties of this process was that heat conductive epoxy required an extremely high pressure to bond the solar cell. Since the solar cell was fragile – 150 µm thin germanium – it cracked or became damaged only by handling with stainless steel tweezers. Precise control of pressure and temperature was essential.

Thanks to the well-controlled process, the temperature rise on the cell surface remained 15 °K with 550X (geometrical concentration ratio) application, while maintaining good electrical insulation. The voltage drop due to the temperature rise can be ignored. The insulation capability is over 2,0 kV. A 4 to 5 kV product is being developed for high system voltage application, maintaining cooling capability. Various environmental tests, including hot-wet, heat cycling and freezing cycling tests described in Institute of Electrical and Electronics Engineers (IEEE) Standard 1513-2001 did not decrease endurance voltage below 2,0 kV.

Another important component is the sealing polymer for the solar cell. A special material with examined treatment was chosen to completely isolate the environmentally sensitive III-V solar cells from moisture. The endurance against concentrated flux or ultraviolet (UV) light, as well as compound stress with water condensation, was examined and showed more than 20 years of lifetime in the desert area.[10]

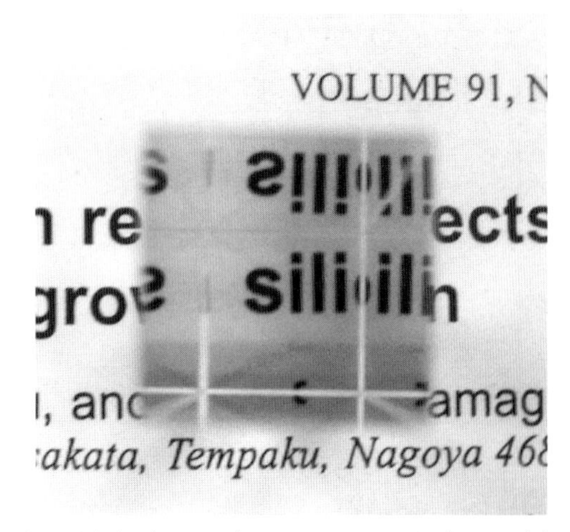

Figure A.56 A kaleidoscope homogenizer: an indispensable optical device for multi-junction concentrator solar cells

A.3.2.4 Homogenizer

The homogenizer is an optical kaleidoscope optic that mixes the concentrated rays into a well-defined square and uniform concentrated flux (see Figure A.56).[11,12] This is an essential optical device for III-V multi-junction solar cells, whose efficiency drops sharply by chromatic aberrations of the lens or excessively concentrated flux.

Another advantage of the homogenizer is to enlarge mechanical assembly tolerance. Common technologies use optical alignment procedures in the fabrication of concentrator modules so that the concentrated spot is well aligned to the cell centre and will collect all concentrated flux. The homogenizer, on the other hand, allowed normal assembly technology common to various machines and did not require the use of the optical alignment technique.

A.3.2.5 Primary optics

Another new concentrator optic is a non-imaging dome-shaped Fresnel lens.[13] The non-imaging Fresnel

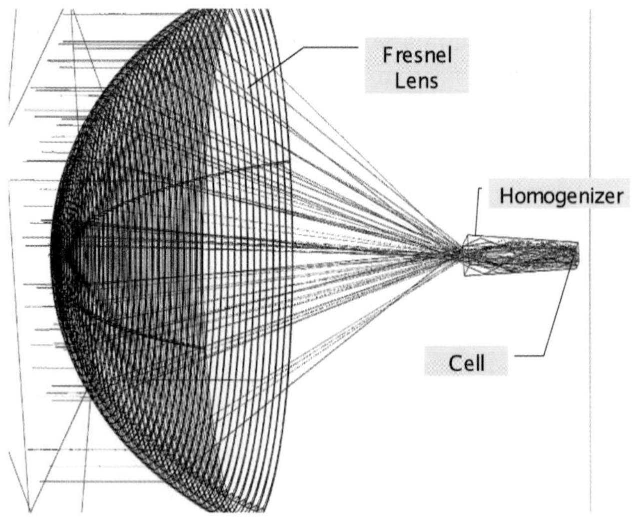

Figure A.57 Ray-tracing for concentrator optics

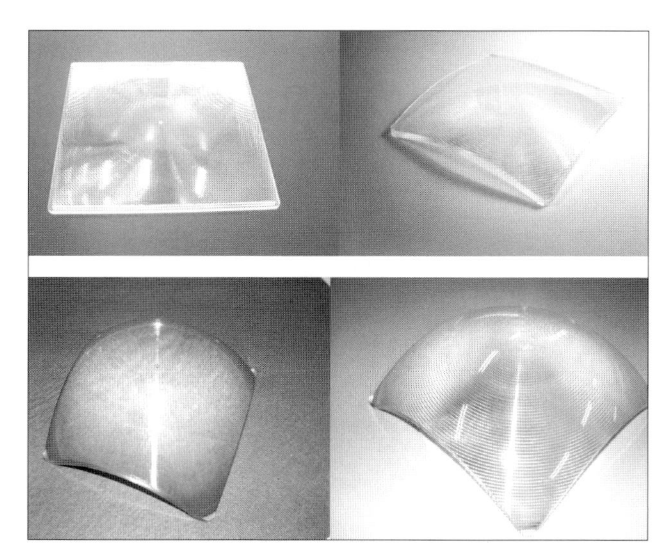

Figure A.58 Generations of injection-moulded Fresnel lenses: (top left) first-generation flat design (400X, 77,3 % of peak efficiency, 2001); (top right): second-generation half-dome design (400X, 81,5 % of peak efficiency, 2002); (bottom left) third-generation full-dome design made by collapsible moulding die (400X, 85,4 % of peak efficiency); (bottom right) fourth-generation full-dome design made by collapsible moulding die (556X, 91 % of theoretical efficiency)

lens allows a wide acceptance half angle, while keeping the same optical efficiency with minimum chromatic aberration (see Figures A.57 and A.58).

The conventional approach of curving or thermal pressing is precise but not productive. Injection moulding, however, is capable of manufacturing thousands of lenses in a single day by a single machine. The drawback of this method is the difficulty of creating a precise prism angle and flat facets. The maximum efficiency was a little above 80 % and overall efficiency was 73 % by conventional injection moulding technology. After improving the conditions, the efficiency was raised to 85,4 %.

A.3.2.6 Module body

Since the receiver allows a designed assembly tolerance of up to 1,75 mm, there is no need for special optical alignment. The actual assembly tolerance is 3,75 mm, with the combination of the 550X dome-shaped Fresnel lens. Even local mechanical industries can assemble the main body, such as shown in Figure A.59.

A.3.2.7 Tracker

Any two-axis tracker that has a sub-degree level of tracking accuracy and appropriate robustness against the environment is applicable to the technology. An open-loop clock-driven tracker was fabricated and tested.[13] There were no troubles except for stowing action by strong winds for one year of operation in both the Inuyama and Toyohashi sites.

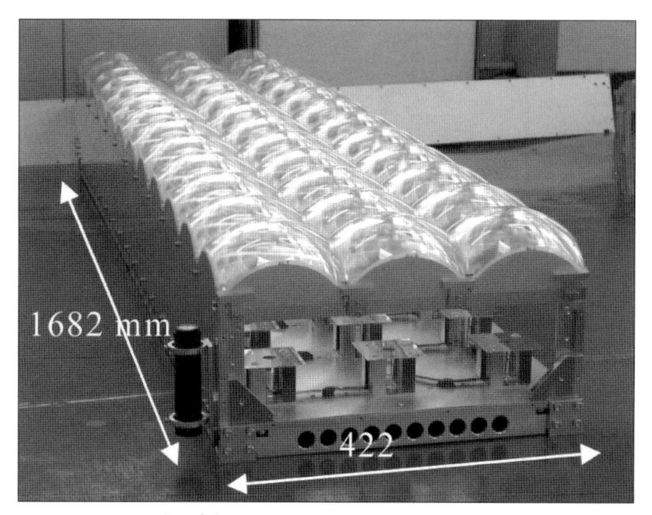

Figure A.59 Inside of the 400X concentrator module with 36 receivers connected in series

A.3.2.8 Proposed system

The new system that promises high efficiency PV generation consists of:

- use of III-V multi-junction solar concentrator cells that are available to several companies in the world;
- high concentration at around 500X;
- use of dome-shaped Fresnel lenses (they made by injection moulding of PMMA);
- use of kaleidoscope glass homogenizers (these are not concentrators like CPCs, but give a uniform flux and prevent the conversion losses that stem from chromatic aberration and surface voltage variation);
- solar cells that are directly laminated to the chassis by epoxy glue;
- robust sealing that protects environmentally sensitive III-V solar cells against humidity and concentration of UV flux.

The receiver concept covers the last three points above. With this receiver, there is no need of special optical alignment. Even local mechanical industries can assemble the main body. The new receiver consists of the following advanced technologies:

- super-high pressure and vacuum-free lamination of the solar cell that suppresses the temperature rise to only 15° (including temperature rise in solder and tab metals) under 550X geometrical concentration illumination of sun beam;[14,15]
- direct and voids-free soldering technologies of the fat metal ribbon to the solar cell, suppressing hot spots and reducing the resistance, thereby allowing a current 400 times higher than normal non-concentration operation to be passed;[8]
- a new encapsulating polymer that survives exposure to high concentration UV and heat cycles;[16,11]
- a beam-shaping technology that illuminates the

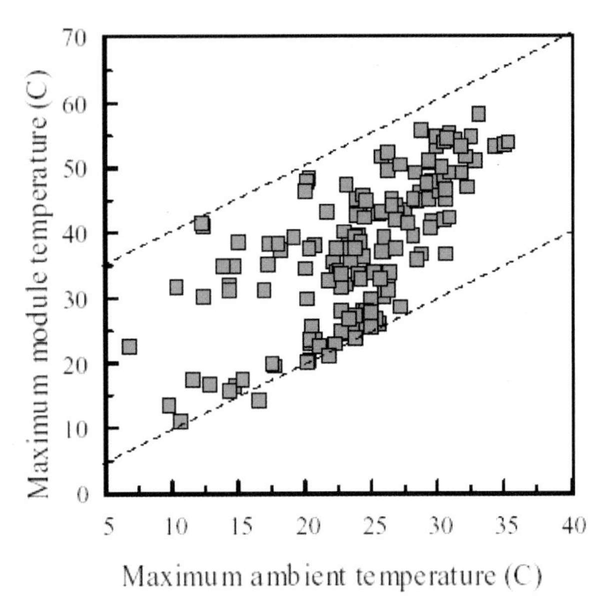

Figure A.60 Module temperature versus ambient temperature

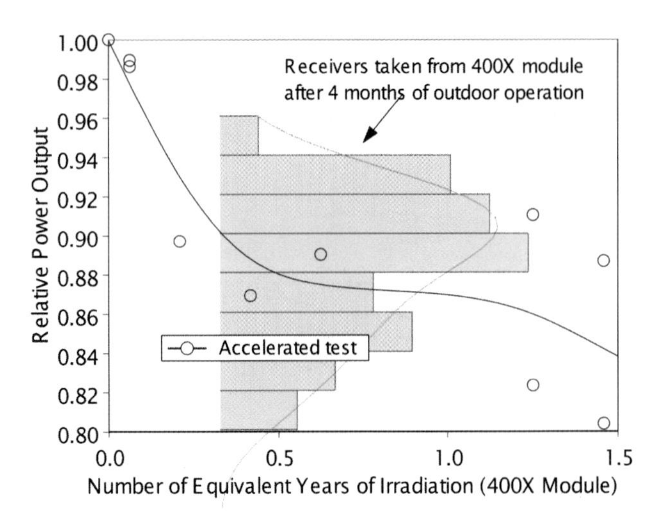

Figure A.61 Comparison between the accelerated test and the outside exposure test

square aperture of the solar cell from a round concentration spot;[17,11]

- homogenizer technologies that give a uniform flux and prevent the conversion losses that stem from chromatic aberration and surface voltage variation;[12,18]
- an assembly tolerance of up to 1,75 mm (there is no need for special optical alignment so even local mechanical industries can assemble the main body).

A.3.3 Reliability in the desert

In order to determine the environmental testing conditions, the temperature of the module body has been monitored over a period of one year. Figure A.60 shows the relationship between the maximum ambient temperature and the maximum 400X module temperature of that day. The ambient temperature was measured by an aspirating thermometer placed in the shaded area. The module was disconnected from the inverter, meaning that the generated power was disposed in the module. There are very few days when the maximum module temperature exceeded the maximum ambient temperature by more than 30°. Considering that the maximum module temperature of the flat-plate system often rises up to 30° higher than the maximum ambient temperature at the same site, the general requirement for the durability of the concentrator module is not higher than that of the flat-plate system. In most cases, it will be safe to say that the concentrator module is as reliable as the flat-plate module if they both fail at the same point in environmental chamber tests.

The concentrator receivers were examined using the common environmental chamber tests (85 °C, 85 % RH, 1 000 Hr and 90 °C to –40 °C, 500 cycles) and were passed two or more times. However, there are some exceptions. One is the Fresnel lens. The commonly accepted environmental chamber test of 85 °C and 85 % RH is too extreme for lens materials. Instead, the environmental chamber test is conducted at a lower temperature – for example, 60 °C,[18] but with increased test duration. Recently, it was found that the residual stress in a moulded PMMA lens induced fine cracks under a freezing cycle environment.[11] This problem was solved by removing stress during injection moulding.

Apart from the environmental chamber test, durability against strong winds and hail were examined. The module showed no damage against repeated static pressure equivalent to 60 m/s wind velocity and 90 km/hr hail impacts.

Another reliability issue inherent to the high concentration III-V solar cell application is the complete protection of the cell against humidity,[19] and a III-V solar cell is more chemically reactive than the well-established back-contact concentrator silicon solar cell. In order to passivate the cell, it should be protected by using a transparent sealing polymer. However, most transparent polymers do not survive strong UV irradiation. Degradation of the sealing polymer was often found to be accelerated by the presence of moisture.

For the purpose of the accelerated degradation test, a new concentrated UV test chamber was developed.[16] The accelerated life defined by the total amount of UV-A irradiation was confirmed by a field exposure test. The module containing 36 concentrator receivers and operated in the field for four months was disassembled afterwards and each receiver was evaluated by a 50X solar simulator. The result is shown as a horizontal histogram in Figure A.61. At the same time, other receivers were examined using the UV accelerated test. The result is shown as a line plot with circle symbols and meets the result of the outside exposure test.

We are developing and testing a robust sealing structure against concentrated sunlight. We found that the

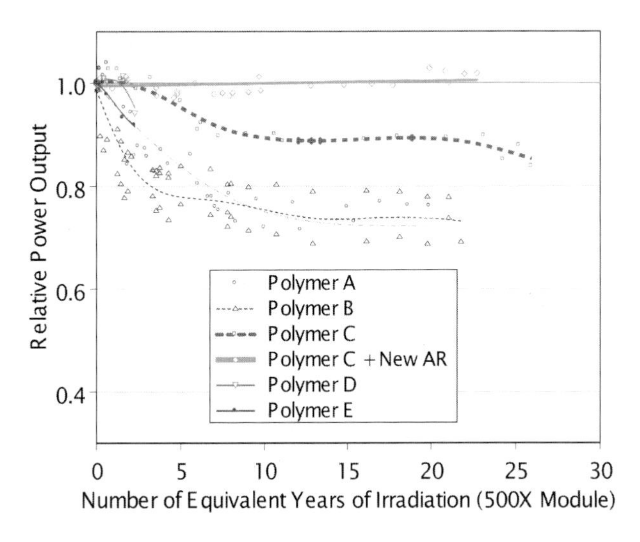

Figure A.62 Accelerated test (equivalent to 20 years of installation in the desert area) by compound stress of simultaneous concentrated UV irradiation and water condensation

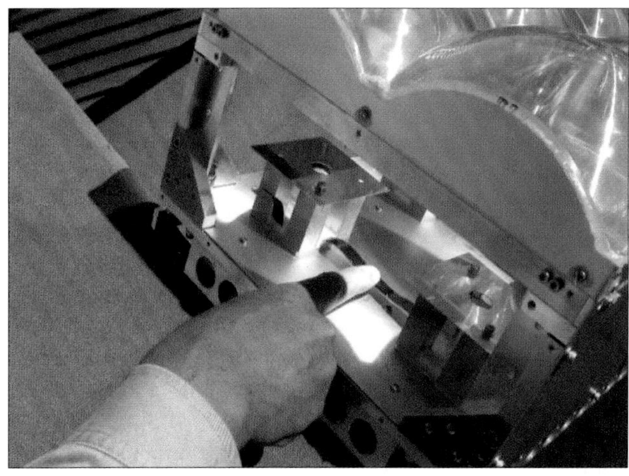

Figure A.63 Off-axis beam test (thanks to the optical characteristics of the non-imaging Fresnel lens, etc., fingers will not be burned by an off-axis beam)

sealing structure that passed the environmental chamber test (hot-wet, thermal cycle and freezing cycle) did not always survive the concentrated sunlight test. Some polymers (weather-tough silicone resin) degraded after several months of equivalent accumulated irradiation, as shown in Figure A.62. The Polymer B present in the 400X module shown in Figure A.49 has recently shown rapid degradation due to the compound stress of moisture and concentrated UV in a cold environment, although it survived various environmental chamber tests, a water-dipping test and the UV irradiation test in a hot dry environment. We are now testing new materials, including Polymer A (400X module) that showed good performance in a 'dry' environment but rapid degradation in a wet environment, and Polymer C (550X module) that showed good endurance even in a wet environment. The new polymer material, combined with a new sealing structure and the robust cell design, is expected to result in a concentrator receiver that may survive more than 20 years of concentrated irradiation.

One of the concerns of the reliability of concentrator modules is safety against accidents. For example, the concentrated solar beam may burn the interior of the module components, including internal cables, if the solar tracker encounters problems and an off-axis beam strays off the solar cell. This was true in the early history of the concentrators. The current 550X non-imaging concentrator Fresnel lens maintains beam intensity within ±0,9° of tracking error; but the beam intensity substantially drops out of that acceptance angle. These characteristics help to reduce the thermal flux from off-axis beam to unwanted components. The homogenizer glass rod shifts the focal surface from the cell area and thus reduces intensity of the off-axis beam. The temperature rise by an off-axis beam under 850 W/m^2 direct normal irradiance was typically 15° K (see Figure A.63).

A.3.4 Outdoor characteristics

A.3.4.1 Uncorrected efficiency

Both 400X and 550X modules were evaluated in four test sites. Three sites were operated by an independent organization. Table A.38 summarizes measured efficiency in three different sites. All of the results were 'uncorrected'. The peak uncorrected efficiency for the 7 056 cm^2 400X module with 36 solar cells connected in series was 26,6 %, measured in-house. The peak uncorrected efficiencies for the same type of module with six solar cells connected in series and a 1 176 cm^2 area measured by Fraunhofer ISE and National Renewable Energy Laboratory (NREL) were 27,4 % and 24,9 %, respectively. The peak uncorrected efficiency for the 550X and 5 445cm^2 module with 20 solar cells connected in series was 28,4 %. Different from flat-plate modules, the correction procedure is not agreed internationally.

Figure A.64 shows a typical outdoor I–V curve of the 550X module. In spite of the fact that all 20 receivers were connected in series with a standard assembly tolerance and without optical alignment procedures, the current mismatching is hardly seen.

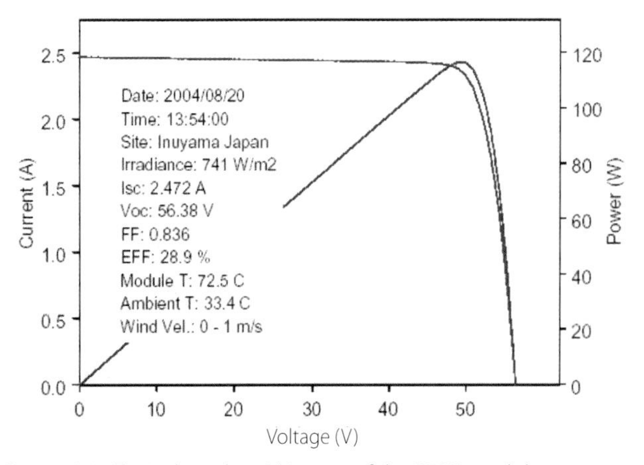

Figure A.64 Typical outdoor I-V curve of the 550X module

Table A.38 Uncorrected peak efficiency measurement

Concentration	Area (cm^2)	Site	Ambient	Uncorrected efficiency	Direct normal irradiance (DNI) (W/m^2)
400 X	7,056	Inuyama, Japan manufacturer	29 °C	27,6 %	810
400 X	7,056	Toyohashi, Japan independent	7 °C	25,9 %	645
400 X	1,176	Fraunhofer ISE, Germany independent	19 °C	27,4 %	839
400 X	1,176	NREL, US independent	29 °C	24,9 %	940
550 X	5,445	Inuyama, Japan manufacturer	33 °C	28,9 %	741
550 X	5,445	Toyohashi, Japan independent	28 °C	27 %	777

A.3.4.2 Corrected efficiency

For the purpose of comparison with the flat-plate module, efficiency was measured by the spectrum condition closed to AM 1.5 G and corrected to the operation under 25 °C cell temperature

First, the temperature coefficient was frequently evaluated with non-concentration measurements. However, with the increase of the concentration ratio and short-circuit current, the influence of the dark current is supposed to logarithmically decrease, thus leading to a lower temperature coefficient. Temperature correction was achieved by the calculated temperature coefficient from a Syracuse computer model considering influence from short-circuit current density.[20] The absolute value of the coefficient logarithmically drops with the increase of the current density per dark current density.[21] For the purpose of estimating the temperature coefficient, we adopted a simple logarithmic calculated trend.

Next, the temperature of the concentrator cell frequently varies from the module temperature measured by a thermocouple mounted on the module body. In most cases, the concentrator solar cells are electrically isolated from the module body. In contrast to a flat-plate module, the concentrated heat flux flows through the heat-conducting, but electrically insulating, layer to the module body. The cell temperature is raised by the product of heat flux and thermal resistance. The cell temperature with given direct beam irradiance was simulated using forward bias heating. A fine PT100 sensor mounted on an electrically insulated 1 mm diameter stainless steel tube was contacted on a cell surface and a resistance increase was measured with the increase of forward bias current. The measurement was repeated five times at different positions on the cell. The temperature drop between the cell and the enclosure was corrected in proportional to the direct normal irradiance.

Finally, the output power was calculated from I–V curves taken at one-minute intervals (the module temperature), and the direct normal irradiance was smoothed using a moving average method over 20 minutes. This procedure attempted to compensate for the difference between time constants of the solar cells and the pyrheliometers. Without the smoothing effect, the module efficiency will be overestimated by the relatively large time constant of the pyrheliometers, when the direct beam abruptly increases due to the clearing of small clouds. A histogram of the corrected efficiency is shown in Figure A.65. Considering a possible error in measurements, the peak efficiency value in the best sunshine condition (with error) was 31,5 ± 1,7 % and the most frequent efficiency was 28,6 ± 1,7 %.

Again, the aim of this was to compare results with those of the flat-plate module. Considering the fact that it is difficult to control the cell temperature under the illumination of concentrated sunlight, a realistic rating method may be a histogram uncorrected efficiency.[22]

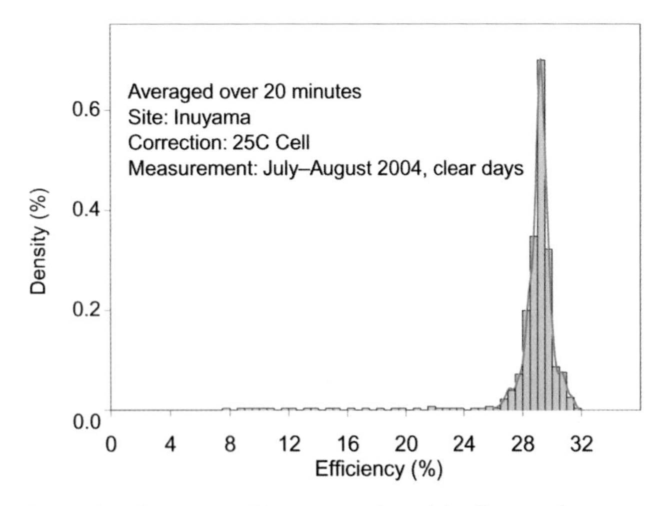

Figure A.65 Histogram of the corrected module efficiency for a clear summer day

A.3.5 Comparison with the crystalline silicon flat-plate module

A concentrator module utilizes only the direct beam irradiance. It has long been said that a concentrator PV was not suitable for the climate in Japan. For this reason, a long-term field test was started to see if the high efficiency of concentrator PV overcomes the reduction in irradiation. A typical flat-plate module was installed at the Inuyama site with 14,06 % rated efficiency multi-crystalline silicon solar cells mounted on a 30° sloped fixture. The rated efficiency is a simple average of the labelled efficiency from eight modules. The flat-plate module was mounted on the ground so that it did not receive reflected sunshine from the

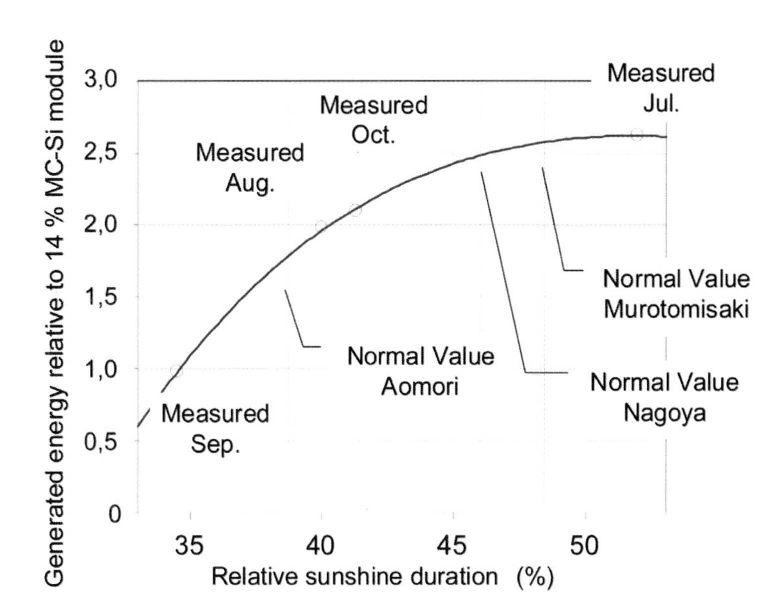

Figure A.66 Comparison to the mc-Si flat-plate module (rated efficiency is 14,06 %)

concrete floor, unlike the PV system on the building or apartment houses. Therefore, the power generation of the flat-plate system in this test is expected to be around 1 000 kWh/kWp. On a clear day, the energy per area from the concentrator module was about three times more than that of the alternative; but as soon as the sun was covered by the clouds, the concentrator module stopped generating power. The amount of the energy relative to the flat-plate module depended upon the relative sunshine duration, defined as the probability that direct normal irradiance (DNI) exceeds more than 120 W/m^2 from dawn to sunset.

Figure A.66 indicates the ratio of energy production from 15 July to 15 October 2004. For the period with a lower percentage of sunshine, such as September, the concentrator module only produced the same amount of energy as the flat-plate module with the same area. However, over a period with many sunny days, the energy per area from the concentrator module was more than 2,5 times that of the flat-plate system. Applying the normal value of the sunshine duration, defined as the averaged sunshine duration from 1960 to 1990, the concentrator module would produce roughly 1,7 to 2,6 times more energy per area per annum than the 14 % multi-crystalline silicon module in most of Japan's cities. The sunshine duration data was collected in many cities. Comparing sunshine duration is convenient when anticipating energy production by the concentrator PV. Nevertheless, the analysis is crude and will lead to some inaccuracy in the result. In the future, a more rigorous estimate will be made and compared against long-term data.

A.3.6 Land utilization for PV generation

Depending upon the land price and land preparation cost, it is important to think about the packing density of the PV modules in the designated area. One of the well-known disadvantages for the concentrator system is shading by adjacent modules. This corresponds to the fact that direct horizontal irradiance is always less than the global irradiance on a horizontal surface. This also applies to the concentrator system using single-junction silicon solar cells. However, for the system using high efficiency multi-junction solar cells, the issue is not as simple as it once was. This section is devoted to discussing the space factor of concentrator systems using spectrum-sensitive multi-junction solar cells.

The concentrator system using multi-junction solar cells boosts power generation and is expected to reduce the PV cost. One of the disadvantages of the multi-junction solar cell is its sensitivity to the spectrum change. Each junction of the multi-junction solar cell is connected in a series, and spectrum change leads to an imbalance of spectrum matching. One of the main reasons for spectrum change is sun height and corresponding air mass.

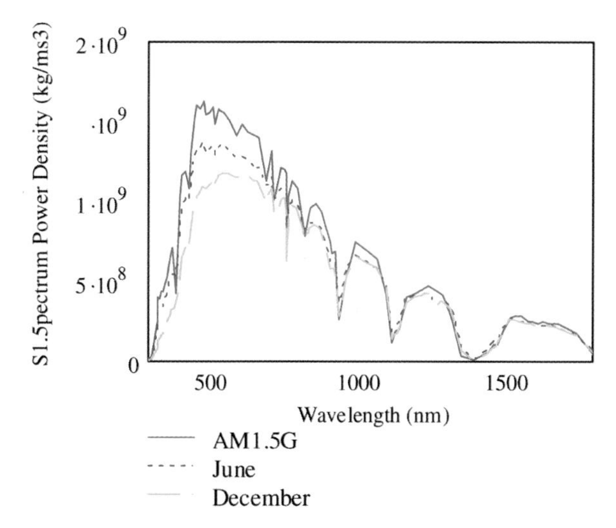

Figure A.67 Spectrum calculated by the sunshine model used at solstice in June and December

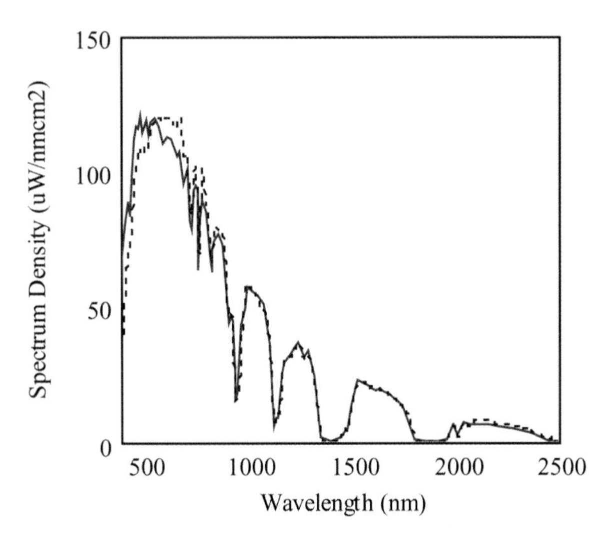

Figure A.68 Direct beam spectrum: Measured versus model

A.3.6.1 Spectrum sensitivity of multi-junction solar cell arrays

The idea of modified and reduced turbidity is generally called low aerosol density (AOD); but the value of aerosol density is not always identical.[23] For the sake of experimentation, the aerosol density value we chose is the fitted value from the data of DNI at 14 sites, measured by the Japanese Meteorological Agency. The calculated spectrum by the sunshine model at solstice in June and December is shown in Figure A.67. The overall spectrum is revealed to be close to the AM 1.5 G spectrum, rather than the AM 1.5 D spectrum. However, when compared to the recently proposed standards in the low AOD spectrum SMART2,[24] it is not as close as the AM 1.5 G spectrum. This is possibly because the Japanese atmosphere contains more moisture and the aerosol density is thicker than in the US. The spectrum obtained by the model was compared to the spectrum measurement and showed a good agreement, as shown in FigureA.68.

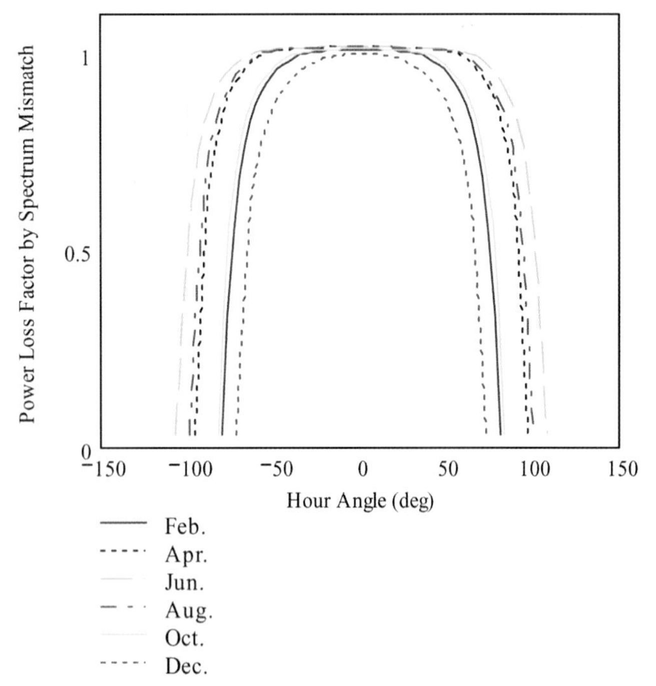

Figure A.69 Relative efficiency by the change of spectrum of direct normal irradiance: Ge-based spectrum-matched three-junction cell

Supposing that a DNI spectrum is calculated by the most recent reduced-aerosol optical density model,[23,26] and a three-junction concentrator solar cell is completely matched to air mass 1,5 calculated from the low AOD model, the power loss factor in arbitrary months and hour angle would be calculated as in Figure A.69. This is derived from the results of the following two steps:

- the numerical model for the spectrum in given time angle and day number;
- calculation of short-circuit current (Isc), fill factor (FF) and efficiency of spectrum-sensitive multi-junction solar cells and concentrator modules under the given spectrum.

A.3.6.2 Experimental proof

The model introduced above was proved by a spectrum-sensitive concentrator module with 32 series-connected InGaP/InGaAs/Ge three-junction solar cell. The module only converts DNI into electricity and is directly influenced by the spectrum change.

Assuming that the exoatmosphere spectrum and energy density is known by day number, and the distance between sun and Earth, the direct normal irradiance will be given by the subtraction of absorbed and diffused components of the atmosphere. The absorption spectrum of the air, including ozone, water and other uniformly mixed gases, is relatively sharp and would not significantly affect the balance of the short-circuit current among the top, middle and bottom cells. Therefore, the value of direct normal irradiance after

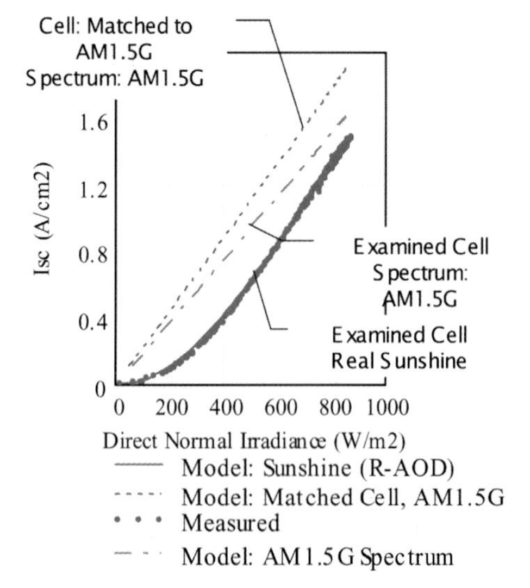

Figure A.70 Short-circuit current of concentrator module versus direct normal irradiance (DNI)

Notes: dot points = measured; solid line = calculated by the model.

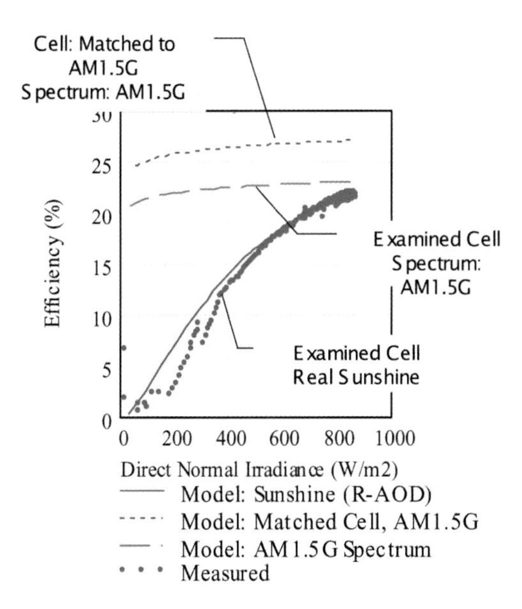

Figure A.71 Efficiency of concentrator module versus DNI

Notes: dot points = measured; solid line = calculated by the model.

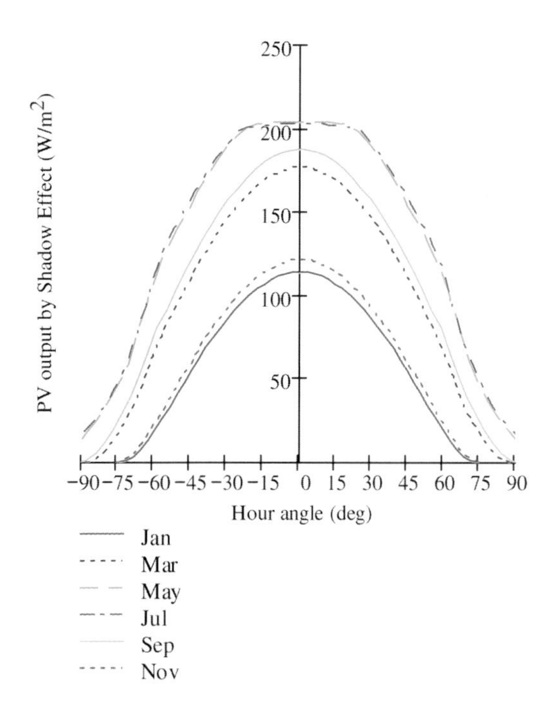

Figure A.72 Power generation of spectrum-sensitive concentrator array with a space factor of 0,9 in an east–west direction and a space factor of 0,5 in a north–south direction

compensation for Sun–Earth distance and function of day number would be a good indication of how much the direct normal spectrum is modified from the original blue-rich standard spectrum – for example, AM 1,5 G. It is also convenient to use direct normal irradiance as an explanatory variable because it is a typical measured variable in a concentrator or multi-junction solar cell evaluation and no additional measurement is necessary.

Figure A.70 compares Isc from the model and measured data. The solid line corresponds to the Isc anticipated by the model. The dot points correspond to measured ones. The measured Isc was distributed along the anticipated Isc curved line. The examined module was a concentrator module; therefore, Isc was supposed to increase linearly with DNI, on the X-axis, unless the spectrum changes. Because of influences of the spectrum change, Isc did not increase linearly with irradiance since the calculation is based on a constant spectrum, represented as a dash line and dash–dot line. The model explained the influence of the spectrum change and the trend of measured Isc.

The same comparison was made for the efficiency of the concentrator module. Figure A.71 compares conversion efficiency from the model and measured data. The solid line corresponds to the conversion efficiency anticipated by the model. The dot points correspond to measured ones. The measured efficiency was also distributed along the anticipated trend. There were some discrepant measured data points in the low DNI region. The causes are not clear; but one possible explanation is the influence of shading from a handrail on the building's roof or some strange tracking error during the low sun height period. This unusual trend in the low DNI region was not seen after the roof handrail was removed.

A.3.6.3 Space factor calculation and optimization

Finally, the influence of shading was analysed considering the above-mentioned characteristics of spectrum-sensitive multi-junction cells. Different from the characteristics of flat-plate systems, the calculation of the space factor for concentrator modules (especially with spectrum-sensitive multi-junction solar cells) must consider the following points:

- Each module moves while tracking.
- Shading influence occurs in an east–west direction, as well as a north–south direction.
- Conversion efficiency changes according to sun height due to spectrum change.
- Partially shaded cells are bypassed by diodes. It is necessary that individual multi-junction cells have bypass diodes for protection of reverse bias voltage.

Some terms needed to be clarified:

- Space factor in an east–west direction:
  - module width/module width plus gap between modules lined in an east–west direction.
- Space factor in a north–south direction:
  - module length/module length plus gap between modules in a north–south direction.
- Area space factor:
  - product of space factor in an east–west direction and space factor in a north–south direction.
- Area efficiency:
  - (annual power generation at a given area space factor × total module area)/(total power generation of isolated module × land area).

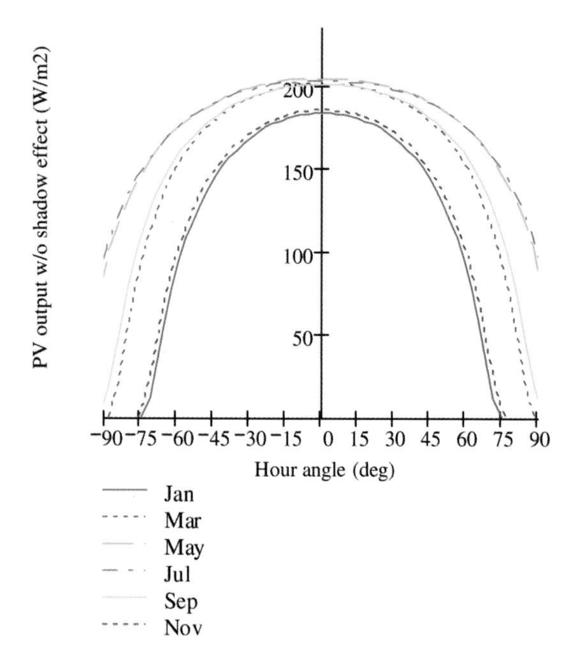

Figure A.73 Power generation of isolated spectrum-sensitive concentrator module without shading effect

The idea of area efficiency came from an optimization problem that maximizes the annual power generation for a given land area. The maximum area efficiency corresponds to the maximum power generation of the given land area.

The array was supposed to be implemented in a rectangular arrangement along exactly an east–west and north–south direction in an infinitely large area.

Figure A.72 shows the calculated power generation of the three-junction solar cells concentrator array. Modules are densely packed in east–west line (space factor of 0,9 along this direction) and modestly packed in a north–south line (space factor of 0,5 along this direction). It was assumed that the power generation of a partially shaded module would generate power proportional to the unshaded area. Compared to the power generation from a stand-alone module described in Figure A.73, power generation for an hour angle of more than 60° (summer) or 30° (winter) was decreased due to shading effects. However, the power generation outside that hour angle region was found to drop rather quickly by spectrum mismatching effect. The overall influence of the shading effect of adjacent modules was not significant, even in a moderately densely packed array, compared to a flat-plate array.

Next, the best space factor was considered either in an east–west direction or a north–south direction in light of land utilization. A contour plot of the area efficiency was made as a function of the space factor in an east–west direction or north–south direction (see Figure A.74). The calculated optimum space factor in an east–west direction and north–south direction was 0,92 and 0,78, respectively. The ratio of the total module and land area was 0,71. It was found that the concentrator arrays using spectrum-sensitive multi-junction solar cells would be able to be packed more densely than a typical flat-plate array without an internal bypass diode. In this situation the total annual power generation in an optimum packed array (area space factor of 0,71) was 53 % of the power of the same number of stand-alone modules placed in an infinitely large area.

Finally, the use of a spectrum model was confirmed by the observed DNI data in Japan (see Figure A.75). It was shown that the overall contour trend matched the calculated trend. It was thought that the optimum space factor in a three-junction concentrator solar cell was packed more densely than single-junction cells, such as crystalline silicon solar cells. It was also thought that the area efficiency in three-junction solar cells was greater than the single-junction cells because the shadow loss in three-junction solar cells that appeared in low sun height had a relatively lower impact on annual power generation. This result also implies that

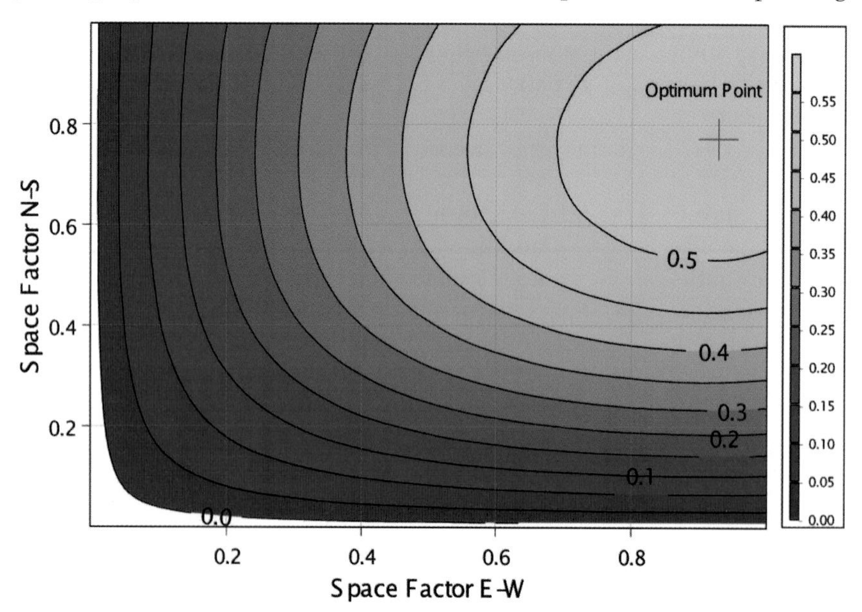

Figure A.74 Contour plot of area efficiency as a function of space factor in east–west and north–south directions

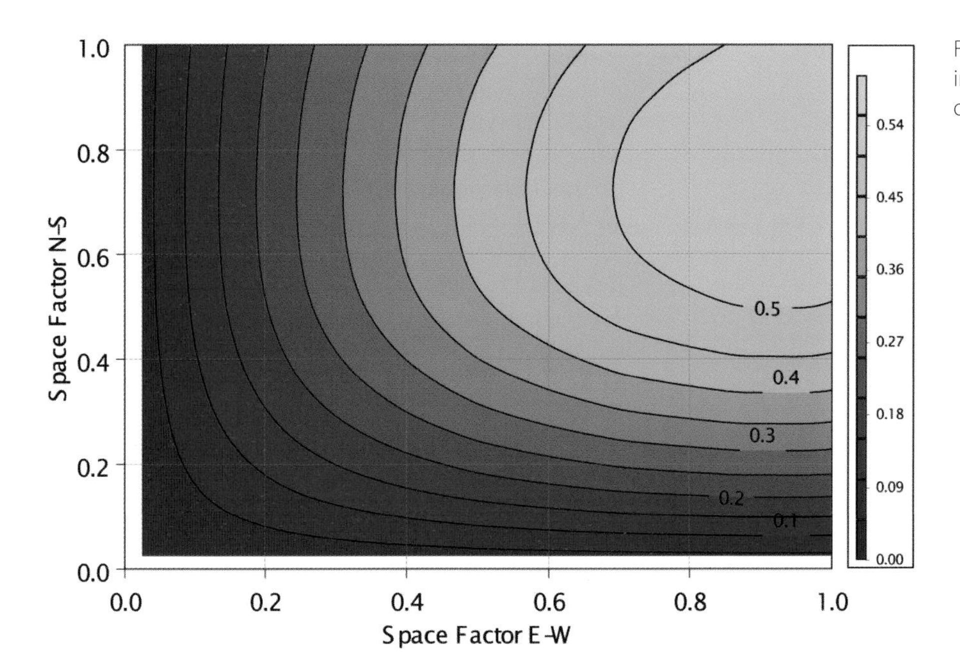

Figure A.75 Contour plot of the ratio of integrated irradiance based on the observed DNI in Matsumoto, Japan

the calculated optimum space factor and the maximum area efficiency in three-junction cell were almost equal to those of single-junction cells.

It was believed that a huge expanse of land would be necessary for the concentrator array in order to avoid shading loss from adjacent modules. This is true for classical single-junction concentrator solar cells; but the situation will be different for the next generation of super high efficiency multi-junction solar cells. The multi-junction solar cells are sensitive to spectrum change. Therefore, the performance in the period of low sun height when shading influence is dominant, cannot be compared to the period of no shading. In other words, it is possible to suppose that shading loss for spectrum-sensitive multi-junction solar cells will have less impact than that of conventional concentrator modules and the new concentrator array with multi-junction solar cells may be packed more densely than the alternatives.

The concentrator array's space factor with spectrum-sensitive multi-junction solar cells was examined. First, the power generation results from daily and yearly spectrum change were modelled and proved through a series of experiments.

Second, the array's annual power generation was calculated considering efficiency drop by spectrum mismatch and recovery of partial shading by internal bypass diodes. The optimum space factor in light of land utilization was as large as 0,71. This high packing density resulted from recovery of partial shading by implemented bypass diodes.

Figure A.76 Concentrator photovoltaics (CPV) for 'breeding plants', which collects a direct beam and provides diffused sunlight to plants

Figure A.77 Soft shadow by rejection of direct beam but transmittance of diffused sunlight

A.3.7 Possibility of 'plant breeding' PV

Another interesting application is what we call the tree planting PV (see Figure A.76). Concentrator photovoltaics (CPV) only utilize the direct beam of the sunlight, which is often harmful for tree planting. The CPV system without the back cover is transparent to diffused sunlight. The CPV module shades the strong direct beam and provides rich diffused sunlight to plants, as shown in Figure A.77. The rejection of the direct beam to the ground is also helpful in controlling irrigation. With this transparent module, the area under the module is no longer a 'dead' area. Different from a 'see-through' flat-plate module, power generation is not compromised at all.

A.3.8 Conclusions

A new 550X (geometrical concentration ratio) module is being developed and shows the highest efficiency of any type of PV module, as well as more than 20 years of accelerated lifetime (receiver section) in desert areas. This achievement is blessed with new innovative concentrator technologies. The new concentrator system is expected to open a door to a new age of high efficiency PV.

In light of applications, there are two original aspects of CPV modules that are advantages compared to the alternative flat-plate modules. The spectrum sensitivity of multi-junction cells is expected to allow the CPV modules to be densely packed in the desert. The see-through structure of CPV modules also shades the strong direct beam and provides rich diffused sunlight to plants without compromising power generation.

REFERENCES

1. Green, M. A., Emery, K., King, D. L., Igari, S. and Warta, W. (2003) 'Solar cell efficiency table version 22', *Progress in Photovoltaics: Research and Applications*, vol 11, pp347–352
2. Zhao, J., Wang, A., Yun, F., Zhang, G., Roche, D. M., Wenham, S. R. and Green, M. A. (1997) '20,000 PERL silicon cells for the 1996 World Solar Challenge solar car race', *Progress in Photovoltaics: Research and Applications* (1997) vol 5, pp269–276
3. Yamaguchi, M. and Luque, A. (1999) 'High efficiency and high concentration in photovoltaics', *IEEE Transactions on Electron Devices*, vol 46, no 10, p2199
4. See www.syracuse-pv.webhop.org
5. Nishioka, K., Takamoto, T., Nakajima, W., Agui, T., Kaneiwa, M., Uraoka, Y., and Fuyuki T. (2003) 'Analysis of triple-junction solar cell under concentration by SPICE', *Proceedings of the Third World Conference on Photovoltaic Energy Conversion*, Osaka, Japan, p869
6. Nishioka, K., Takamoto, T., Agui, T., Kaneiwa, M., Uraoka, Y. and Fuyuki, T. (2004) 'Evaluation of InGaP/InGaAs/Ge triple-junction solar cell under concentrated light by simulation program with integrated circuit emphasis', *Japanese Journal of Applied Physics*, vol 43, no 3, p882
7. Takamoto, T., Agui, T., Kamimura, K., Kaneiwa, M., Imaizumi M., Matsuda, S. and Yamaguchi, M. (2003) ' Multijunction solar cell technologies, high efficiency radiation resistance and concentration applications', *Proceedings of the Third World Conference on Photovoltaic Energy Conversion*, Osaka, Japan, p581
8. Araki, K., Kondo, M., Uozumi, H., Kemmoku, Y., Ekins-Daukes, N. J. and Yamaguchi, M. (2004) 'Packaging III-V tandem solar cells for practical terrestrial application achievable to 28 % of module efficiency by conventional machine assemble technology and possibility of 500X and low weight HCPV for space', *Proceedings of the 19th EU-PVSEC*, Paris, France, p2451
9. Araki, K., Uozumi, H. and Yamaguchi, M. (2002) 'A simple passive cooling structure and its heat analysis for 500X concentrator PV module', *Proceedings of the 29th IEEE PVSC*, New Orleans, LA, p1568
10. Araki, K., Uozumi, H., Egami, T., Hiramatsu, M., Miyazaki, Y., Kemmoku, Y., Akisawa, A., Ekins-Daukes, N. J., Lee, H. S. and Yamaguchi, M. (forthcoming) 'Development of concentrator modules with dome-shaped Fresnel lenses and 3J concentrator cells', *Progress In Photovoltaics: Research and Applications*, vol 13, no 6, September
11. Araki, K., Kondo, M., Uozumi, H., Kemmoku, Y., Egami, T., Hiramatsu, M., Miyazaki, Y., Ekins-Daukes, N. J., Yamaguchi, M., Siefer, G. and Bett, A. W. (2004) 'A 28 % efficient, 400X and 200 Wp concentrator module', *Proceedings of the 29th IEEE PVSC*, New Orleans, LA, p2495
12. Araki, K., Leutz, R., Kondo, M., Akisawa, A., Kashiwagi, T. and Yamaguchi, M. (2002) 'Development of a metal homogenizer for concentrator monolithic multi-junction cells', *Proceedings of the 29th IEEE PVSC*, New Orleans, LA, p1572
13. Hiramatsu, M., Miyazaki, Y., Egami, T., Akisawa, A. and Mizuta, Y. (2003) 'Development of non-imaging Fresnel lens and sun-tracking device', *Proceedings of the Third World Conference on Photovoltaic Energy Conversion*, Osaka, Japan, p2379
14. Yamaguchi, M. and Araki, K. (2002) 'Japanese R&D activities of multi-junction and concentrator solar cells', *Proceedings of the 29th IEEE PVSC*, New Orleans, LA, p1568
15. Araki, K., Kondo, M., Uozumi, H. and Yamaguchi, M. (2003) 'Development of a robust and high efficiency concentrator receiver', *Proceedings of the 29th IEEE PVSC*, New Orleans, LA, p630
16. Araki, K., Kondo, M., Uozumi, H. and Yamaguchi, M. (2003) 'Material study for the solar module under high concentrator UV exposure', *Proceedings of the Third

World Conference on Photovoltaic Energy Conversion, Osaka, Japan, p80
17. Leutz, R., Suzuki, A., Akisawa, A. and Kashiwagi, T. (2001) 'Flux uniformity and spectral reproduction in solar concentrators using secondary optics', *Proceedings of the ISES*, Adelaide, South Australia
18. IEEE (2001) *IEEE Standards*, 1513-2001, Institute of Electrical and Electronics Engineers
19. van Riesen, S., Bett, A. W. and Willeke, G. P. (2003) 'Accelerated aging tests on III-V solar cells', *Proceedings of the Third World Conference on Photovoltaic Energy Conversion*, Osaka, Japan, p837
20. Ekins-Daukes, N. J., Kemmoku, Y., Araki, K., Betts, T. R., Gottschalg, R., Infield, D. G. and Yamaguchi, M. (2004) 'The design specification of Syracuse: A multi-junction concentrator system computer model', *Proceedings of the 19th EU-PVSEC*, Paris, France, p2466
21. Anton, I., Sala, G. and Pachon, D. (2001) 'Correction of the voc v. temperature dependence under non-uniform concentrated illumination', *Proceedings of the 17th EU-PVSEC*, Munich, Germany, p156
22. Siefer, G., Bett, A. W. and Emery, K. (2004) 'One year outdoor evaluation of FLAT-CON concentrator module', *Proceedings of the 19th EU-PVSEC*, Paris, France, p2078
23. Myers, D. R., Kurtz, S. R., Emery, K., Whitaker, C. and Townsend, T. (2001) 'Outdoor meteorological broad-band and spectral conditions for evaluating photovoltaic modules', *Proceedings of the 28th IEEE PVSC*, Anchorage, AK
24. Myers, D. R., Emery, K. and Gueymard, C. (2002) 'Terrestrial solar modeling tools and applications for photovoltaic devices', *Proceedings of the 29th IEEE PVSC*, New Orleans, LA
25. Myers, D. R., Emery, K. and Gueymard, C. (2002) 'Proposed reference spectral irradiance standards to improve concentrating photovoltaic system design and performance evaluations', *Proceedings of the 29th IEEE PVSC*, New Orleans, LA
26. Emery, K., DelCueto, J. and Zaaiman, W. (2002) 'Spectral corrections based on optical air mass', *Proceedings of the 29th IEEE PVSC*, New Orleans, LA

A.4 RECENT DEVELOPMENTS IN LOW-CONCENTRATION PHOTOVOLTAICS (LCPV)

A.4.1 Introduction

The challenge to developing very large scale photovoltaic systems as a mainstream source of electrical generation is economics. Photovoltaic technology, while technically capable, is now too expensive to serve as a large scale source of generation without significant subsidies. The dominant cost in current photovoltaic systems is the cost of the semiconductor material. In a typical solar panel, the cost of the silicon cells exceeds 50 % of the total cost and is therefore the principal economic barrier to large scale photovoltaic generation using conventional flat-panel technology. An obvious solution to this problem is to produce more power per unit area of silicon by concentrating the sun's light onto the silicon through an inexpensive optical system. Most of the recent developments in concentrator-based systems have utilized high concentration photovoltaic (HCPV) system approaches using advanced and more efficient cell technology and high concentration ratios in the range of 200 to 1 000. Thus far, this approach, while successful technically, has not been successful from a financial standpoint. It has not yet led to lower cost PV systems so far; but with suitable capital investments it is possible that the HCPV approach can yield a satisfactory solution as discussed in the previous section. In this section we examine the potential of recent developments in LCPV systems that offer the potential to break through the cost barrier and lead to practical VLS-PV systems based on currently available commercial PV cell technology.

The concept of concentrator-based PV systems has been an active field of investigation since the beginning of terrestrial photovoltaics. A significant stumbling block has been that conventional low-cost mass-produced PV cells have never been able to function efficiently under concentrated illumination. The principal reason is that the series resistance of the screen-printed metallization used on conventional cells is too high, resulting in a decrease in the output voltage of the cell as the current produced by the cell rises under increased illumination. The result is a marked decrease in cell efficiency as illumination is increased, which effectively neutralizes the economic gains sought through optical concentration. The recent development of a special electrode for PV cells has changed this picture. This electrode, the Day4™ electrode, when applied to the front surface of a conventional PV cell, decreases the series resistance of the electron harvesting component of the cell by a factor greater than ten. This order-of-magnitude decrease in resistance enables a conventional commercially available mass-produced cell to retain its efficiency under concentrated illumination,[1] which enables that cell to function satisfactorily

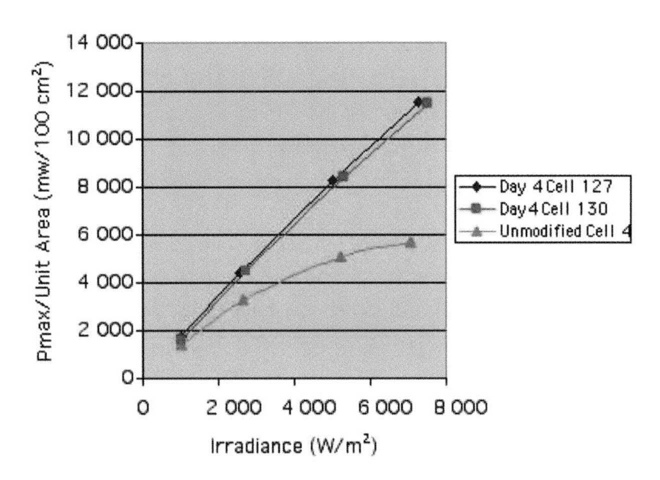

Figure A.78 Measurement results, Day4™ Cell and unmodified cell

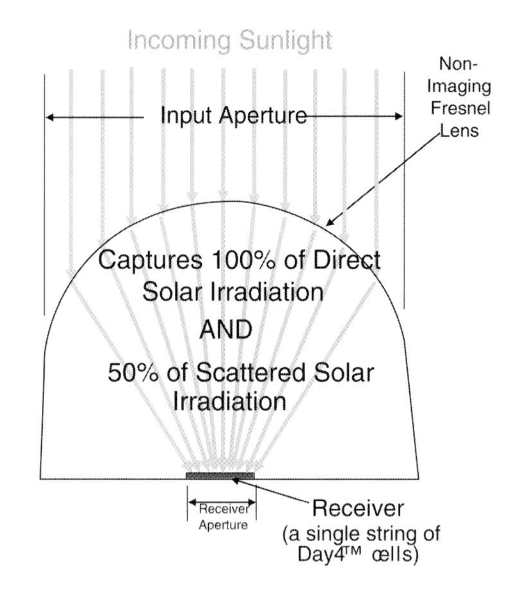

Figure A.79 Non-imaging Fresnel lens with receiver plane

in a solar concentrator. This can potentially lead to a solution for large-scale systems based on low-cost cells and existing production capacity.

A.4.2 The Day4™ Electrode

The Day4™ Electrode is a patented electrode structure consisting of specially coated wires embedded in a polymer film that create a modified cell with an electrode resistance less than one tenth that of the original cell when laminated onto the front face of a conventional mass-produced mono- or poly-crystalline PV cell. The incremental cost of the electrode and its application to the original cell are a very small fraction of the cell's original cost. The resulting Day4™ Cell is capable of performing in a solar concentrator with a concentration factor between 6 and 10 without significantly decreasing the overall efficiency of the cell. This type of cell, when employed in an LCPV concentrator receiver, is therefore capable of delivering several times more power per unit area of silicon material than the original conventional cell upon which it is based. Figure A.78 shows comparative plots of the performance of Day4™ Cells versus an unmodified conventional cell under increasing illumination of up to seven times concentration. In this case, the Day4™ Cells and the unmodified cell are from the same batch of original cells.

A.4.3 Concentrator optics

In order to achieve the objective of cost-competitive VLS-PV systems using the LCPV approach, it is necessary to create a low-cost optical system capable of capturing as much of the available sunlight as possible and concentrating this light onto the PV receiver. A non-imaging Fresnel lens system with relatively wide acceptance angles in both the north–south and east–west directions is capable of capturing a significant fraction of the diffuse illumination, as well as 100 % of the direct solar illumination.[1] Such a system performs well under single-axis tracking of the sun's motion. In this case, tracking accuracy requirements are not stringent and it is possible to realize a low-cost tracking concentrator system employing conventional PV cells with Day4™ Electrodes applied and an acrylic non-imaging Fresnel lens in a single-axis tracking mode.

A non-imaging Fresnel lens is illustrated in Figure A.79. It is made of prisms of minimum dispersion and deviation.[2] Both the outer and the inner surfaces of the lens are optically active in refracting the incoming light onto a receiver of defined width. The lens is designed to accept radiation incident with angles $\theta = \pm 6°$ in the cross-sectional plane (the plane of the paper in Figure A.79) and $\varphi = \pm 18°$ in the plane perpendicular. The seasonal variation in the sun's declination is $\delta = \pm 23{,}45°$. The focal foreshortening induced by the secondary acceptance angle ϕ leads to the acceptance of most radiation without having to seasonally adjust the tilt of the concentrator. The concentrator is installed at latitude tilt, in the north–south direction, tracking the sun's azimuth. A simulation of the acceptance angles for incident radiation from the angles θ and φ is shown in Figure A.80.

The entire concentrator concept has undergone preliminary tests by Day4Energy Inc using the small (25 W) pilot concentrator shown in Figure A.81. A conceptual view of a PV power generation system based on this concentrator concept is illustrated in Figure A.82.

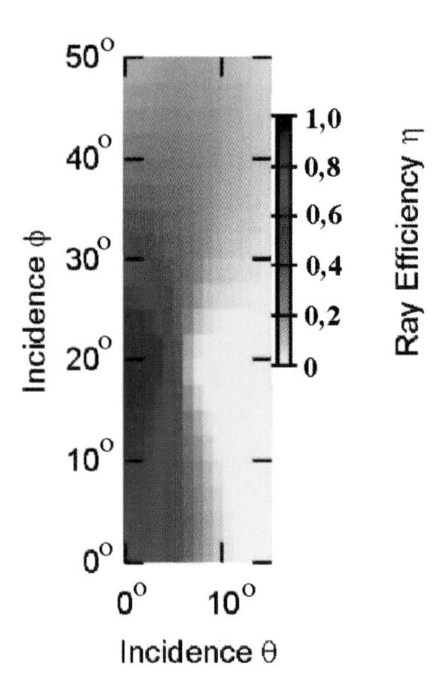

Figure A.80 Simulation of Fresnel lens acceptance angles for incident radiation from the angles θ and φ

Figure A.81 Pilot concentrator

Figure A.82 Concentrator-based power system: Conceptual view

A.4.4 Cost estimates

Table A.39 shows rough estimates of the capital costs of one-axis tracking LCPV systems based on the technology described above under the following assumptions:

- PV receiver efficiency = 18 %;
- optical system efficiency = 80 %;
- overall AC efficiency = 13,68 %;
- scattered light sensitivity = 50 %;
- system expected lifetime = 30 years;
- financing cost = 3 %.

Installed cost is expressed in USD/W for an optical concentration factor of 6. Large system costs are based on economies of scale derived from bulk purchase of components. Small system costs assume modest volume purchases.

An example of a suitable climate for large scale solar power generation is the south-west desert of the US. Table A.40 shows the range of average annual irradiation for the state of Arizona.

Based on the cost data in Table A.39 and the irradiation data in Table A.40, assuming the system captures 100 % of the direct irradiation and 50 % of the scattered irradiation, power costs for the climate found in the state of Arizona in the south-west US desert are estimated to range from 0,053 USD to 0,061 USD/kWh for relatively small systems with capacities of a few megawatts. These prices are in the range of current electricity prices in the south-west US. Assuming the economies of scale that can be realized for larger systems, as shown in Table A.39, power costs are estimated to be reduced to a range of 0,036 to 0,041 USD/kWh.

A.4.5 Leveraging current production

Since the LCPV technology described in this section is based on the currently mass-produced conventional PV cell technology, it has the potential for realizing VLS-PV systems today by utilizing the output of the current

Table A.39 Low-concentrator photovoltaics (LCPV) sun concentrator system cost estimates

	Small system costs	Large system costs
System component		
Concentrator receiver (USD/m^2)	145,00	115,00
Non-imaging Fresnel lens (USD/m^2)	30,00	16,00
Tracker (USD/m^2)	100,00	75,00
Balance of system (BOS) (USD/m^2)	95,00	75,00
DC/AC inverter (USD/kW)	400,00	40,00
System efficiency (at the AC level)		
Direct illumination	13,68 %	13,68 %
Scattered illumination (average)	6,84 %	6,84 %
System-installed cost per watt		
Capital cost (USD/m^2 of aperture)	424,72	286,47
Total power produced (WpAC/m^2)	136,8	136,8
Total installed cost (USD/WpAC)	3,10	2,09

Table A.40 Range of solar irradiation, state of Arizona

Total solar irradiance	9,1–8,1 kWh/m^2/day
Direct solar irradiance	7,2–6,2 kWh/m^2/day
Scattered solar irradiance	1,9 kWh/m^2/day

Source: US Energy Information Agency

industry. In 2004, the total world production of PV cells was approximately 1,2 GW. To produce a 1 GW VLS-PV system utilizing LCPV technology with a concentration factor of 6 would require approximately 160 MW of conventional cell production, or less than 15 % of current production levels. The LCPV technology described here is currently entering its pilot demonstration phase. If it proves to meet its cost expectations, realization of VLS-PV systems will become a practical reality in the near future through leveraging the existing industrial production capability without needing large capital investments in new production technology.

REFERENCES

1. Rubin, L. B, Rubin, G. L. and Lutz, F. (2005) *One-axis PV Concentrator Based on Linear Non-imaging Fresnel Lens*, International Conference on Solar Concentrators for the Generation of Electricity or Hydrogen, Scottsdale, Arizona, 1–5 May 2005
2. Leutz, R. and Suzuki, A. (2001) *Non-imaging Fresnel Lenses: Design and Performance of Solar Concentrators*, Springer, Germany

A.5 APPLICATION OF BIFACIAL PV SYSTEMS TO VLS-PV

A.5.1 Introduction

In October 2003, Hitachi Ltd began to manufacture and market production-model bifacial PV cells. When the bifacial PV arrays are installed vertically, they can generate almost the same annual power as mono-facial PV arrays installed in the optimal azimuth at the optimal inclination. The power generated by Hitachi's bifacial PV arrays stays almost constant, independent of the azimuth of their installation. Thus, when the vertical installation of bifacial PV arrays is applied to the construction of the VLS-PV, the following features can be expected:

- Reflected irradiation from desert surfaces can be utilized.
- A particular combination of arrays can level the distribution of daily power generation output and optimize (level) the seasonal fluctuation of generated power.
- Vertically installed arrays can be used for windbreak walls, sand-break walls and other fences.

This section introduces the basic characteristics of Hitachi's bifacial PV cells and modules, and refers to the results of a preliminary examination of bifacial PV cells that were used to build a VLS-PV system – in particular, the simulation results of power generated.

A.5.2 Characteristics of bifacial PV cells and modules

A.5.2.1 Structure of bifacial PV cells

The structure of the Hitachi PV cell is shown in Figure A.83. This cell has a pyramid-like texture on both sides of the thin mono-crystal silicon substrate (p-type, 125 mm chip, 220 μm of thickness) whose front surface

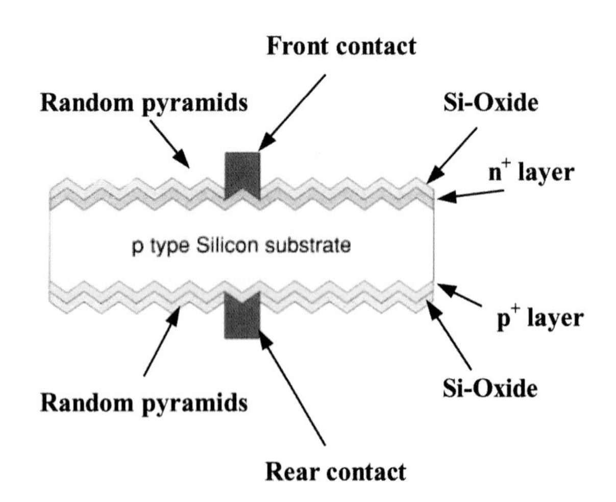

Figure A.83 Cross-section of a bifacial cell

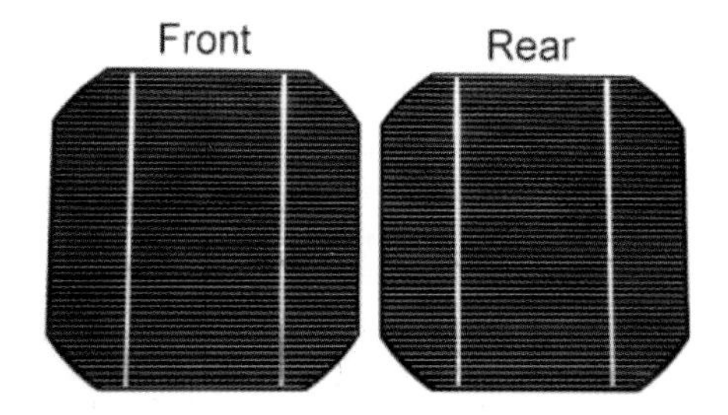

Figure A.84 Photograph of a bifacial cell

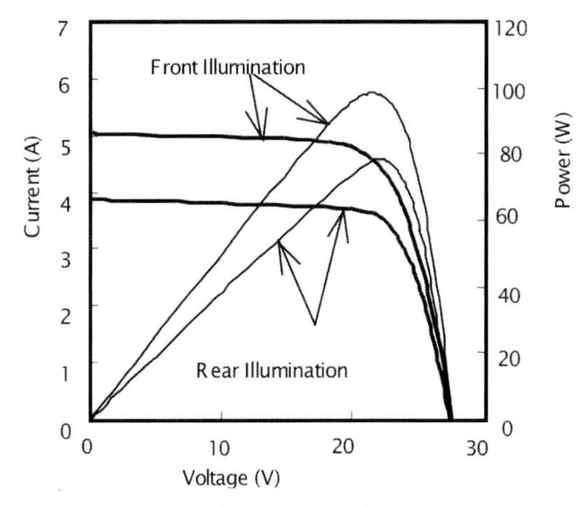

Figure A.85 Typical I–V and P–V curves of a bifacial PV module HB3M-48

proximity forms n+p junctions. The entire rear surface has a back surface field (BSF) as the result of boron diffusion. As shown in Figure A.84, the appearances of both faces are the same because their electrodes are constructed as grids.

A.5.2.2 Specifications of a bifacial PV module

Table A.41 lists the specifications of a typical bifacial module (type HB3M-48) and Figure A.85 shows an example of its electrical characteristics. Electricity outputs from the front and rear faces are separately measured. The bifaciality (ratio of the rear face output to the front face output) of this module is about 0,8 (74 W/96 W).

Table A.41 Specifications of a bifacial PV module HB3M-48

Module type		HB3M-48
Cells	Type Size	Single-crystalline (SOG) 125 x 125 x 0,22 mm
Surface materials		Double glass
Series number of cells		48 (= 8 x 6)
Size		Width: 1 120 mm Length: 868 mm Thickness: 8 mm
Electrical performance	Voc (V)	Front: 27,5 Rear: 27,2
	Isc (A)	5,0 f/3.7 r
	Pmax (W)	96 f/74 r
	Vpm (V)	21,3 f/21.8 r
	Ipm (A)	4.5 f/3.4 r
Temperature coefficient of efficiency (%/°C)		–6.0 f/–2.7 r

Figure A.86 Photograph of PV module HB3M-48

Figure A.86 shows Hitachi's bifacial PV module. The module is called a lighting double-glass module and therefore has the same appearance on the front and rear faces. Compared with the single-glass module, Hitachi's PV module could offer a substantial improvement in the length of the useful life. In addition, as shown in Figure A.87, the module could be applied to fence-integrated vertical installation.

A.5.2.3 Characteristics of generated power of bifacial PV system with vertical installation

Distribution of daily power output

Figure A.88 shows the distribution of daily power output (annual average) of four types of installation systems: bifacial PV south–north vertical; bifacial PV east–west vertical; mono-facial PV south optimal inclination; mono-facial PV south vertical. The solar irradiance and temperature information of the Meteorological Test Data for Photovoltaic System (METPV) in Mito city (Ibaraki prefecture) is used to calculate the distribution:

Figure A.87 Photograph of a bifacial vertical fence-integrated PV array

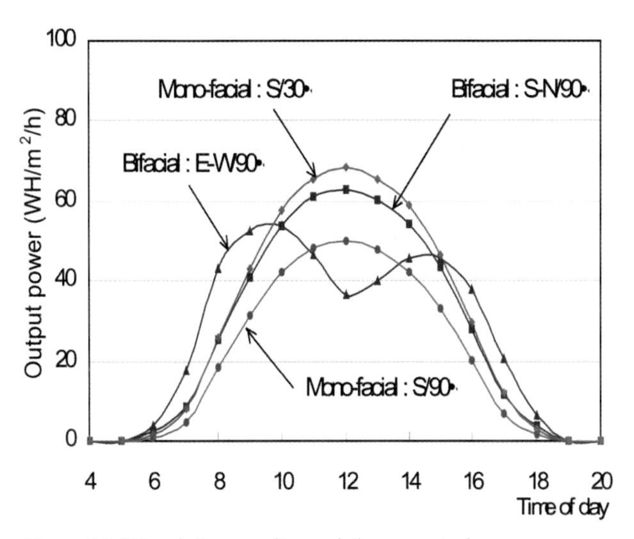

Figure A.88 Simulation results on daily generated power distribution

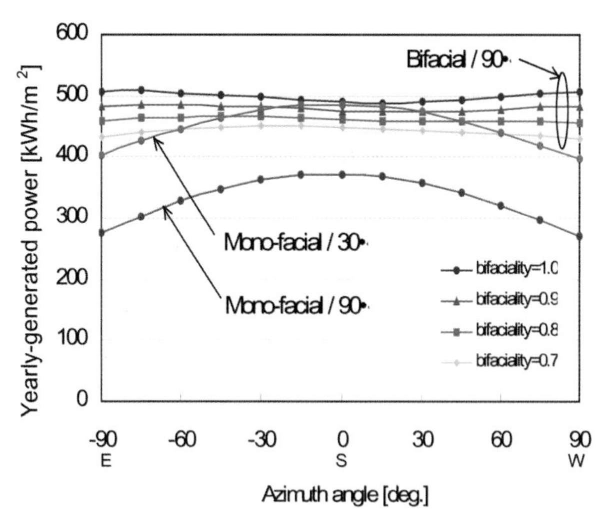

Figure A.89 Simulation results on annually generated power with various azimuth angles and bifacialities

- Bifacial PV installed vertically facing east–west generates electricity from early in the morning until late in the evening. This generates power peaks in the morning and the evening. The generated power of bifacial PV installed vertically facing south–north peaks around noon.
- A combination of east–west and north–south facing installation could generate a flat distribution of generated power throughout the day.

Azimuth angle dependency

Figure A.89 shows the result of calculations of annually generated power when the azimuth of bifacial PV installed vertically (90°) is changed by using the METPV of Mito city, compared with cases of mono-facial PV installed at an optimal inclination (30°) and installed vertically (90°). Figure A.89 also evaluates bifaciality dependency:

- Since mono-facial PV is dislocated from azimuthal angle 0° (facing south), yearly generated power falls greatly. However, the annual generated power of bifacial PV installed vertically remains constant regardless of the azimuthal angles.
- Since the bifaciality of bifacial PV installed vertically is greater, the ratio of rear-face output power increases, leading to increasing annually generated power.

Application of a bifacial PV system with a vertical installation to snow areas

The vertical installation system of bifacial PV has no snow accumulation on the solar cell module and can utilize reflected light from snow on the ground. For a demonstrative experiment in a cold region with snow, we installed bifacial fence-type PV arrays that combine sub-arrays facing south–north and sub-arrays facing east–west on the campus of Chitose Institute of Science and Technology (Chitose city) in January 2003 and started measuring. Figures A.90 to A.92 show the installation, layout and single-line diagram. Table A.42 indicates electrical characteristics by the sub-arrays.

Figure A.90 Overview of the PV array on the campus of the Chitose Institute of Science and Technology

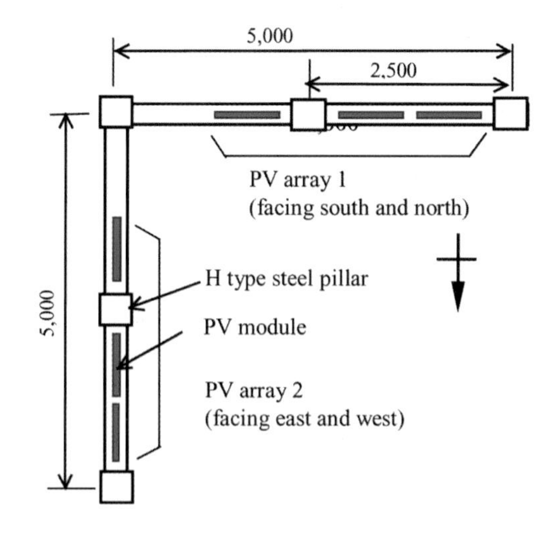

Figure A.91 Layout of bifacial fence-type PV array

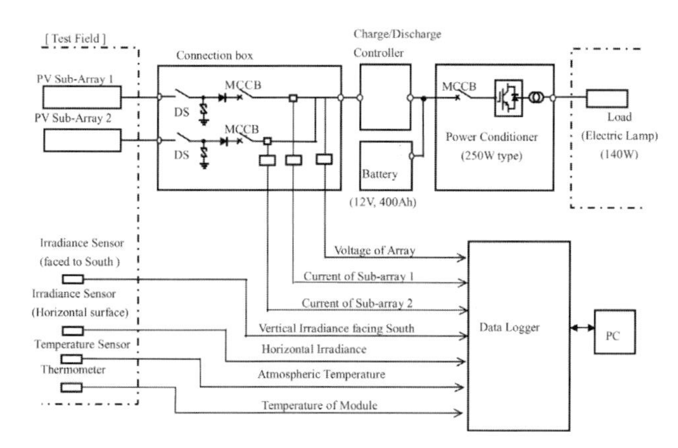

Figure A.92 Electrical diagram of the PV system

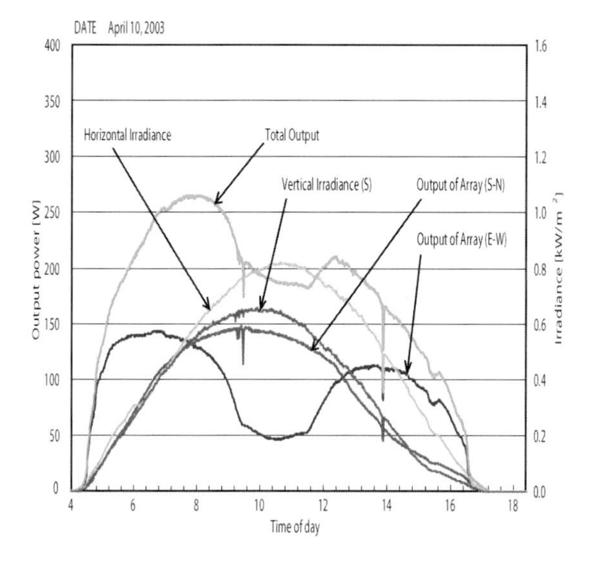

Figure A.93 Measured generated power on a fine day without snow on the ground

Figure A.93 shows the daily distribution of generated power output during a fine day without snow on the ground (10 April 2003). As shown in Figure A.88, the total power output, which combines the power generation pattern of sub-array 1 that peaks at noon (facing north–south) and the power generation patterns of sub-array 2 that peaks in the morning and afternoon (facing east–west), shows a relatively flat output from dawn to sunset.

Figure A.94 shows the distribution of generated power output on a fine day with snow on the ground (9 March 2003). Despite the fine days being in almost the same season, power generation on 9 March 2003, which has snow on the ground, is far larger than on 10 April 2003 when there is no snow on the ground.

The large power generation seems attributable to the reflected light from snow on the ground. As indicated in Table A.43, we compared the daily total generated power and the total irradiation of these two days. Table A.43 reveals that the daily generated power on the day with snow on the ground is 1,4 times larger than the daily generated power on the day without snow on the

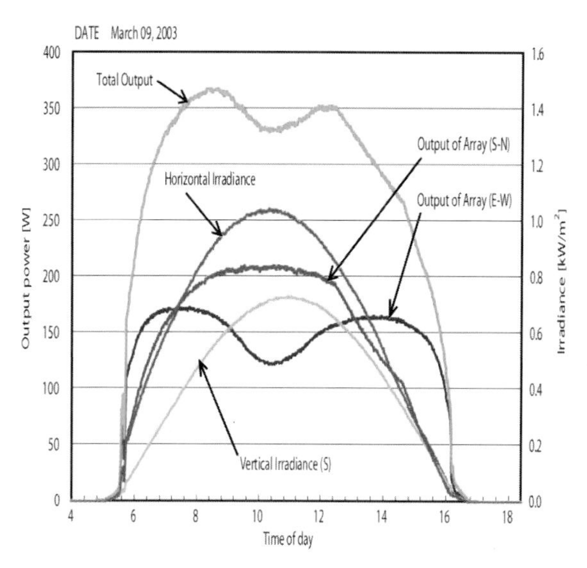

Figure A.94 Measured generated power on a fine day with snow on the ground

ground. The irradiation measured at vertical south face is 1,6 times greater, so this seems attributable to reflected light from snow on the ground. Note that no snow was observed on the front surface of the PV cell module.

A.5.3 Preliminary examination of application of bifacial PV cells with vertical installation to VLS-PV

We conducted a preliminary experiment with a bifacial PV system installed vertically and applied to VLS-PV to learn the characteristics of generated power as the arrangement of arrays is changed.

A.5.3.1 VLS-PV system specifications

The VLS-PV system specifications that we examined are as follows:

Table A.42 Electrical performance of the sub-arrays

	Front illumination			Rear illumination		
	P_{max} (W)	V_{pm} (V)	I_{pm} (A)	P_{max} (W)	V_{pm} (V)	I_{pm} (A)
PV array 1	231,9	17,57	13,20	179,7	18,24	9,84
PV array 2	228,3	17,45	13,08	170,1	18,39	9,24
System	460,2	17,51	26,28	349,8	18,32	19,08

Table A.43 Comparison of irradiances and generated energies

		Figure A.93	Figure A.94	Figure A.94/ Figure A.93
Horizontal irradiance (kWh/m²/day)		6,10	4,96	0,81
Vertical irradiance south (kWh/m²/day)		4,60	7,27	1,58
Generated energy (kWh/day)	Sub-array 1 (south–north)	1,082	1,636	1,51
	Sub-array 2 (east–west)	1,164	1,577	1,35
	Total	2,246	3,213	1,43

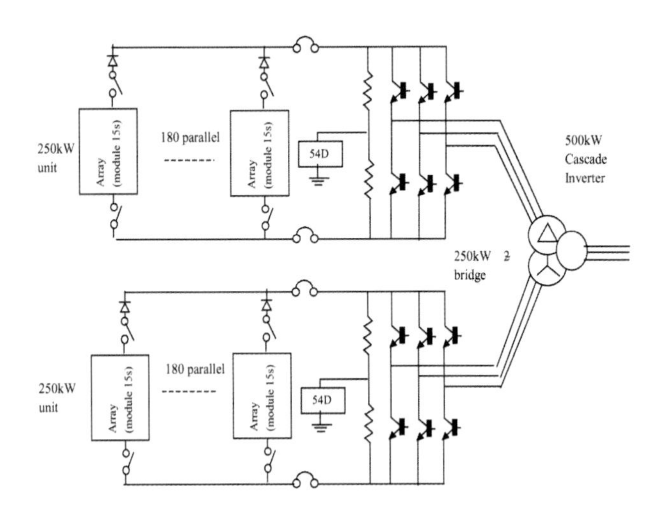

Figure A.95 Electrical diagram of a VLS-PV system with bifacial PV modules

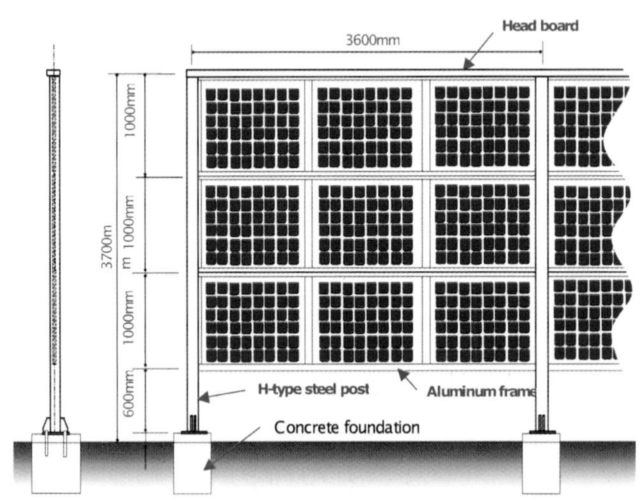

Figure A.96 A schematic image of a bifacial PV fence-type installation

- bifacial PV module type HB3M-48 (the specifications are provided in Section A.5.2.2);
- system configuration of 500 kW (nominal front surface output of bifacial PV);
- planned place to install: Gobi Desert.

Figure A.95 shows the system configuration. A 500 kW system has two 250 kW units in parallel structure. One 250 kW unit (250 kW) has 180 arrays (1,44 kW/array) in parallel structure. One array (1,44 kW) has 15 modules (HB3M-48) in a serial structure. The total number of PV modules is 5 400.

The metrological data of Huhhot (Inner Mongolia) was adopted for power generation simulation.

A.5.3.2 Structure of bifacial PV arrays with vertical installation

Rough structure of fence-type bifacial PV arrays

Figure A.96 shows a sketch of bifacial PV arrays installed vertically. We fixed H-type steel posts at intervals of 3,6 m, where triple-layer PV modules were placed after three PV modules were integrated into one unit with an aluminium frame. The top of the fence is a head board in which a direct current cable was fixed. The fence has a height of 3,7 m, and the fence bottom has a 600 mm high air space to prevent the effects of shadows from weeds and other plants.

How to assemble fence-type bifacial PV arrays on site

Figure A.97 shows how to assemble fence-type bifacial PV arrays on site. The specific steps to assemble the arrays are as follows:

1. Fix H-type steel posts on the foundation at planned intervals.
2. Insert PV modules between the steel posts from above. Similarly, insert modules on the second and third stages from above. With regard to the PV modules, three sheets of modules are integrated within one unit with an aluminium frame in advance.
3. Fix the head board on the top.

As mentioned earlier, since the PV modules do not need to be screwed on for support, the number of assembly tasks on site can be cut substantially.

A.5.3.3 Evaluating the strength of bifacial PV arrays with vertical installation

According to JIS TR C 0006 (design standard for solar

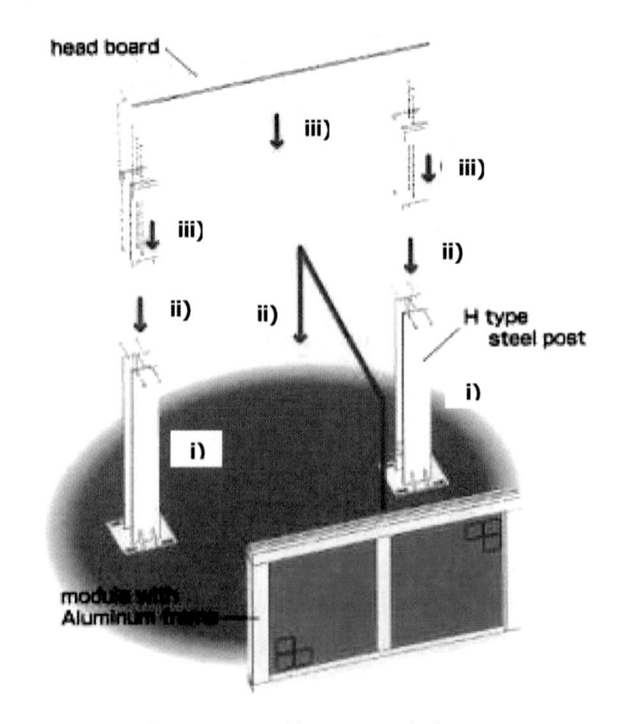

Figure A.97 A schematic assembly process of a fence-type bifacial PV

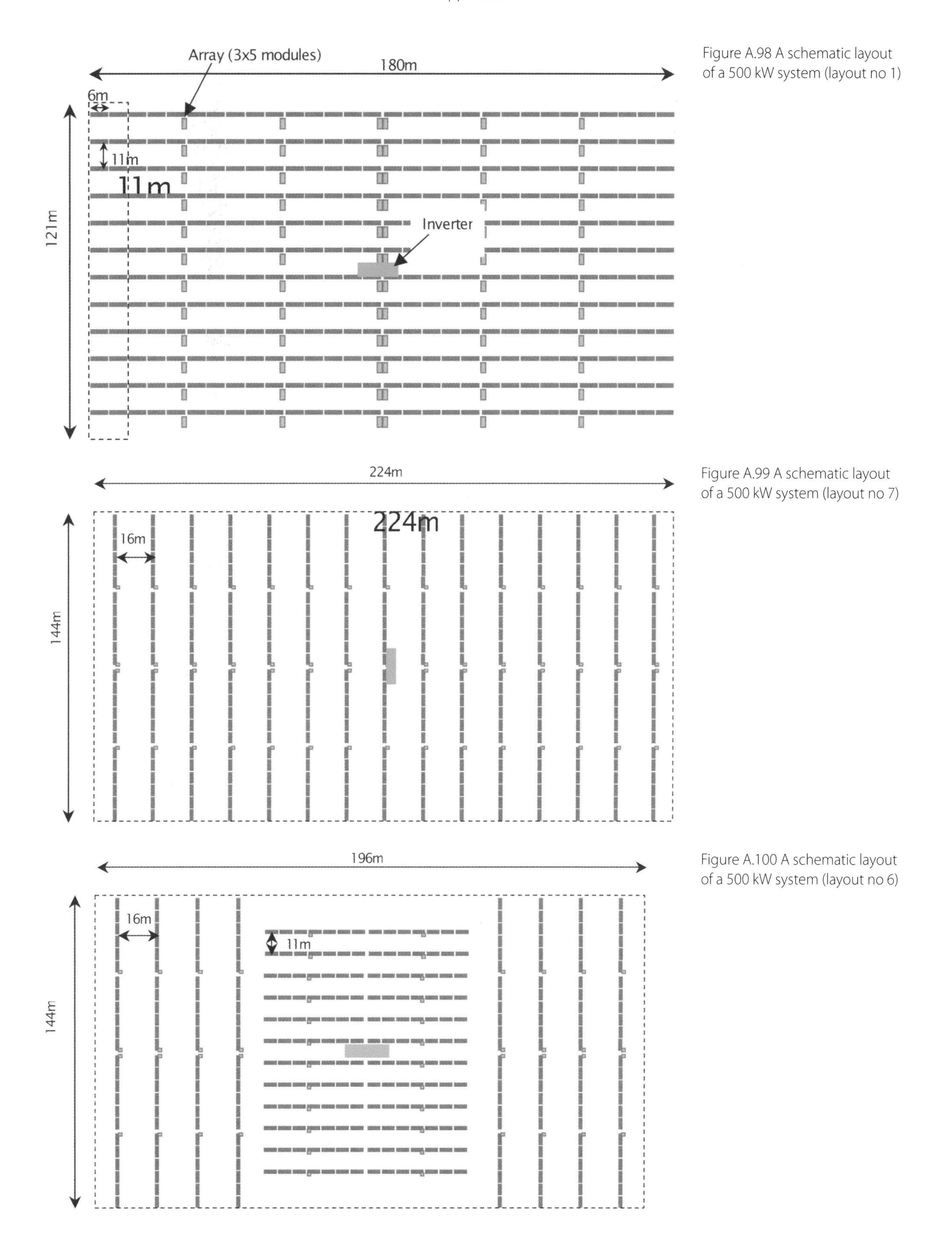

Figure A.98 A schematic layout of a 500 kW system (layout no 1)

Figure A.99 A schematic layout of a 500 kW system (layout no 7)

Figure A.100 A schematic layout of a 500 kW system (layout no 6)

Table A.44 Evaluation results of wind pressure resistance of fence-type PV arrays

Calculation of wind pressure	Height of array	3,611 m
	Height of panel	2,911 m
	Air density	1,274 Ns^2/m^4
	Reference wind velocity for design	42 m/s
	Reference velocity pressure	1 123,7 N/m^2
	Height correction coefficient	0,82 –
	Application coefficient (return period: 50 years)	1,00 –
	Environmental coefficient (flat land)	1,15 –
	Velocity pressure for design	1 054,1 N/m^2
	Wind factor	1,2 –
	Wind pressure	1 260 N/m^2
Evaluation of glass surface stress	Thickness of glass used	3,2 mm
	Thickness of equivalence glass	5,27 mm
	Length of short side of glass	867 mm
	Length of long side of glass	1 118 mm
	Coefficient,	0,4
	Glass surface stress (allowance: 88 N/mm^2)	13,6 N/mm^2
Evaluation of frame stress	Height of frame	950 mm
	Width of frame	3576 mm
	Part A load W_A	284,3 N
	Part B load W_B	1 855,9 N
	λ	0,13
	Frame bend moment	9,32E+05 N.mm
	Section modulus of frame (upper side)	1,20E+04 mm^3
	Section modulus of frame (lower side)	9,00E+03 mm^3
	Frame (upper side) stress (allowance : 107 N/mm^2)	77,7 N/mm^2
	Frame (lower side) stress (allowance : 107 N/mm^2)	103,6 N/mm^2
Evaluation of pillar base stress	Pillar pitch	3 600 mm
	Load on the pillars	13 204 N
	Pillar base to the height of panel centre	1060 mm
	Pillar base bending moment	1,40E+07 N.mm
	Section modulus of pillar (H type steel 100 × 100)	7,56E+04 mm^3
	Pillar base stress (allowance: 240 N/mm^2)	185,2 N/mm^2

array supports), we evaluated the wind pressure resistance of the fence-type PV arrays shown in Figure A.96. Table A.44 shows the evaluation conditions and results, indicating that the glass surface stress, frame stress and pillar base stress seem to have no problems.

A.5.3.4 Azimuth angles to install arrays

We analysed seven installation layouts, which are in the directions of south–north and east–west, as well as their combined directions, and we conducted simulations of these seven types of power generation.

Table A.45 lists these installation layouts. Among these cases, Figure A.98 shows an example with all arrays facing north–south, while Figure A.99 shows an example with all arrays facing east–west. Figure A.100 shows an example with total arrays facing both directions.

Table A.45 Layout summary of fence-type bifacial PV

Layout number	South–North Arrays	South–North Output (kW) (front/rear)	East–West Arrays	East–West Output (kW) (front/rear)	Space requirement (m x m)	Space requirement (m^2)
1	360	518/400	0	–/–	132 x 180	23 760
2	312	449/346	48	69/53	144 x 172	24 768
3	264	380/293	96	138/107	144 x 180	25 920
4	216	311/240	144	207/160	144 x 188	27 072
5	180	259/200	188	259/200	108 x 252	27 216
6	168	242/187	192	277/213	144 x 196	28 224
7	0	–/–	360	518/400	144 x 224	32 226

A.5.3.5 Generated power simulation

Estimating irradiation on the inclined plane

We adopted a formula in estimating the global irradiation (inclination angle of 90°) on an inclined plane.

The global irradiation on the inclined plane =

direct irradiation constituent on inclined plane;

+

diffuse sky irradiation constituent on inclined plane;

+

constituent of reflected irradiation from the ground on the inclined plane.

[A.15]

We used the RB model to estimate the direct constituent and the isotropic reflection model to estimate the constituent of reflected irradiation from the ground. In estimating the diffuse sky irradiation constituent, we used two models, the Hay model (anisotropic diffused model) and the global isotropic diffused model. We refer to the anisotropic diffuse model as case A and the isotropic diffuse model as case B.

We used the above calculation models to estimate monthly global irradiation on the inclined plane (inclination angle of 90°). The result of case A is shown in Figure A.101 and that of case B is shown in Figure A.102. There are large differences of irradiation evaluation values between cases A and B, which will be discussed in future. A combination of arrays facing south–north and east–west could level seasonal fluctuations of irradiation, leading to a levelling of generated power.

Table A.46 indicates the annual global irradiation on the daily average that is converted from monthly irradiation data. Naturally, the irradiation in case B is evaluated as greater than in case A. For vertical installation, arrays facing east–west can obtain more irradiation than those facing north–south.

Table A.46 Global irradiation on the inclined plane for vertical installation with an inclination angle of 90° ($kW/m^2/day$)

Face direction	Case A	Case B
South	3,89	4,23
North	0,95	2,11
South + north	4,84	6,34
East	2,70	3,43
West	2,71	3,44
East + west	5,41	6,87

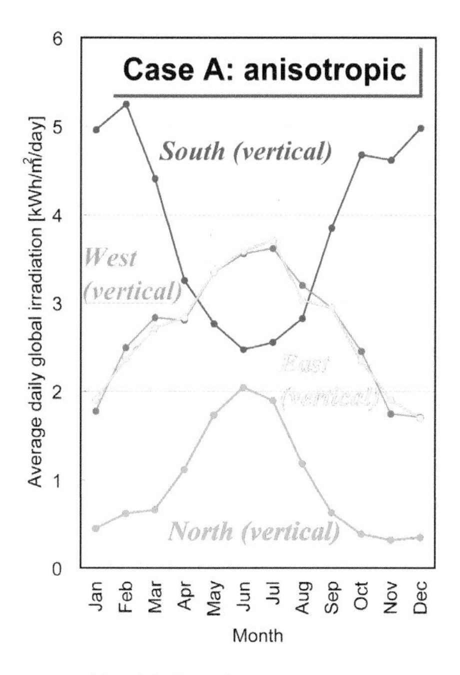

Figure A.101 Monthly global irradiation in case A

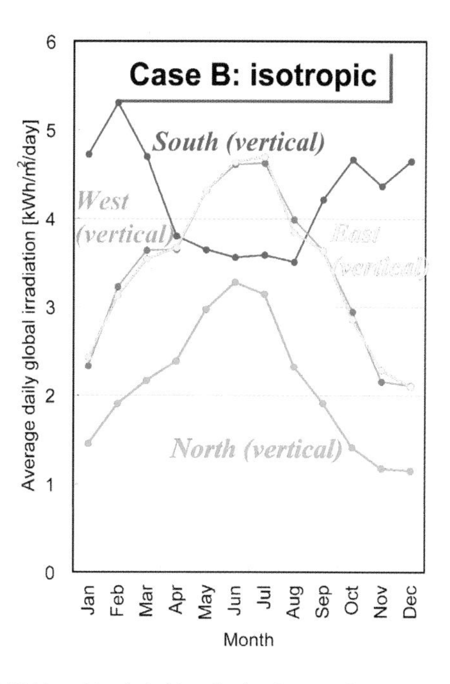

Figure A.102 Monthly global irradiation in case B

Simulation results of power generated

Table A.47 shows the calculation conditions used for simulations of power generated. We adopted the database for the Huhhot point for the monthly average temperature.

With these calculation conditions, we calculated the annual total generated power for the seven types of layouts, numbering 1 to 7, for cases A and B based on the irradiation evaluation described in Section A.5.3.3. Figure A.103 shows the results. For information, Figure A.103 also refers to generated power from a 504 kW system that has mono-facial PV arrays, installed at an inclination angle of 30° facing south.

According to Figure A.103, the following observations were made:

- Any combination of arrays facing south–north and east–west produces the same annual total of power output. Our evaluation of irradiation on the inclined plane alone predicted that the arrays facing

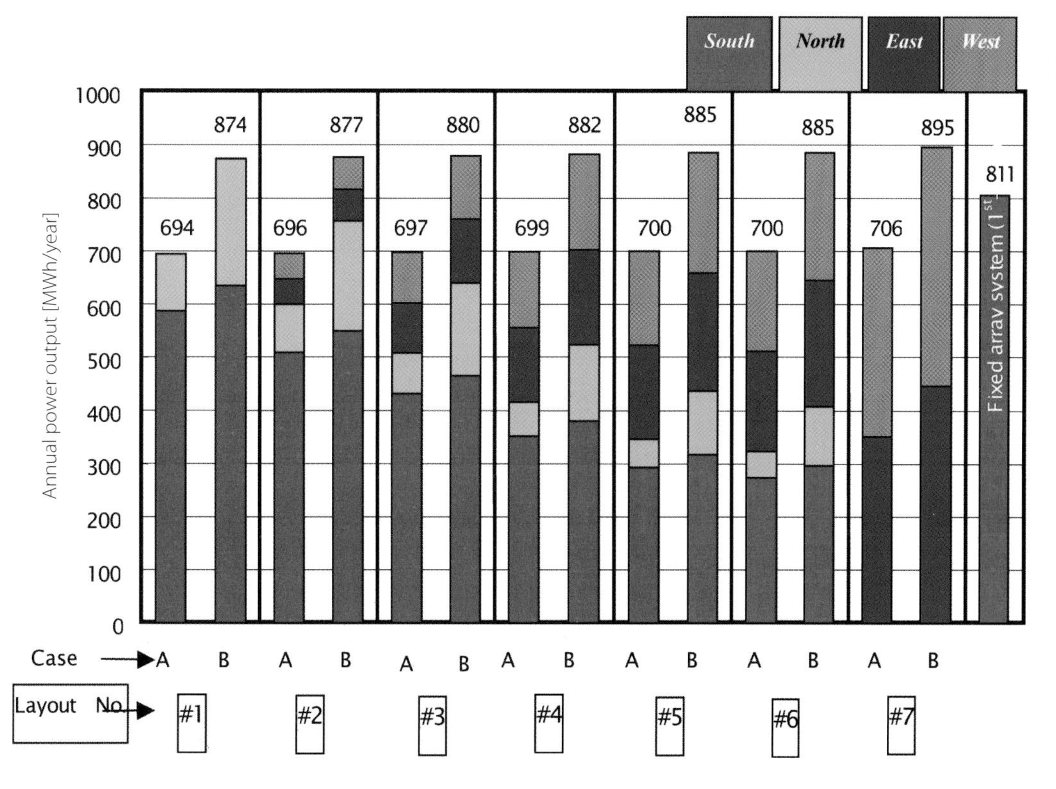

Figure A.103 Annual power output of bifacial PV system (simulation)

Table A.47 Calculation conditions used for simulations of power generated

Month	January	February	March	April	May	June	July	August	September	October	November	December	Average
Monthly average (°C)	−13,1	−9,0	−0,3	7,9	15,3	20,1	21,9	20,1	13,8	6,5	−2,7	−11,0	5,8
Rise in module temperature (°C)	13	13	13	13	13	13	13	13	13	13	13	13	13
Temperature correction coefficient	1,126	1,105	1,062	1,021	0,984	0,960	0,951	0,960	0,991	1,028	1,074	1,115	1,031
Stain correction coefficient	0,950	0,950	0,950	0,950	0,950	0,950	0,950	0,950	0,950	0,950	0,950	0,950	0,950
Annual average degradation rate	0,931	0,931	0,931	0,931	0,931	0,931	0,931	0,931	0,931	0,931	0,931	0,931	0,931
Array circuit correction coefficient	0,980	0,980	0,980	0,980	0,980	0,980	0,980	0,980	0,980	0,980	0,980	0,980	0,980
Coefficient of array load matched correction	0,980	0,980	0,980	0,980	0,980	0,980	0,980	0,980	0,980	0,980	0,980	0,980	0,980
Inverter correction coefficient	0,900	0,900	0,900	0,900	0,900	0,900	0,900	0,900	0,900	0,900	0,900	0,900	0,900
System output coefficient	0,860	0,845	0,811	0,780	0,752	0,734	0,727	0,734	0,758	0,786	0,821	0,852	0,788

east–west would be more advantageous. However, in reality, the difference of generation efficiencies between the front and rear sides of the PV modules seems to have brought about the result.

- There is a clear difference of annual generated power between case A (assumed as the anisotropic diffuse model) and case B (assumed as the isotropic diffuse model). The generated power needs to be corrected according to the measurements of irradiation on sites in the future.

A.5.4 Conclusions

Hitachi's bifacial PV with vertical installation can level the distribution of daily output of power generation. Its annual generated power will remain unchanged despite the direction used for the installation. Moreover, this PV system can effectively utilize reflected light from snow on the ground. Furthermore, a fence-integrated PV structure can cut the number of assembly tasks at local sites.

We conducted a preliminary examination of a system design of this bifacial PV system with vertical installation for possible application to a 500 kW VLS-PV. Our simulation of power generated confirmed that any combination of layouts of arrays facing south–north and east–west resulted in the almost identical annual total power generation. There was a large difference between the anisotropic diffuse model (Hey model, case A) and the isotropic diffuse model (case B) concerning the evaluation of the constituent of diffuse sky irradiation on the inclined plane. The irradiation needs to be measured at sites for confirmation in the future.

A.6 LARGE SCALE PV PLANT EXPERIENCE IN THE ARIZONA DESERT AND EVALUATION OF THE ENERGY PAYBACK TIME

A.6.1 Introduction

The Arizona Corporation Commission Environmental Portfolio Standard (EPS) programme has provided a significant stimulus for the construction and operation of renewable resource energy-generating capacity, particularly photovoltaic systems, in the state of Arizona.[1] Through the EPS, over 10 MW DC of grid-connected PV are currently installed in Arizona, primarily by the state's two largest investor-owned utilities, Arizona Public Service (APS) and Tucson Electric Power (TEP). The EPS programme provides for multi-year, pay-as-you-go development of renewable energy, with kWh AC energy production as a key programme measurement. The programme has established a goal for Arizona's utilities that 1,1 % of the energy generation in the state must be derived from renewable energy sources, with 60 % of that from solar by 2007. To achieve that goal, upwards of 50 MW AC of solar, power systems will need to be installed in the state in addition to large quantities of non-solar renewable plants.

The vast majority of the state's installed generating capacity of utility-scale PV systems (100 kW and larger) utilizes flat-plate, crystalline-silicon collector technology. The APS experience has focused on one-axis, north–south oriented, horizontal-tracking arrays.[2] The TEP systems incorporate standardized fixed arrays. The TEP experience with these systems, including performance, cost, maintenance, installation and design, is introduced in this section.

A life-cycle analysis of the energy requirements and greenhouse gas emissions of the balance of system (BOS) components at TEP's Springerville PV plant is also discussed. The Springerville PV plant, located in eastern Arizona, US, currently has 4,6 MWp of installed PV modules, of which 3,5 MW are mc-Si PV modules. Electricity from the PV modules is used to power the 10 MW AC-el water pump load at the Springerville coal-fired electricity-generating plant. When the water pumps are not operating, the PV electricity is distributed over the transmission grid for general consumption. While PV plants produce fossil fuel-free and non-polluting electricity, the life cycle of PV plant components from production through to disposal does consume fossil fuels, which causes the release of greenhouse gases (GHGs). With the growing need to conserve fossil fuel and mitigate GHG emissions, this evaluation of a large field PV plant in terms of its potential to achieve such savings and lower emissions is timely. Previous life-cycle assessments of field and rooftop PV systems indicated that the energy embodied in the BOS components and their installation in ground-mounted utility plants is much greater than the energy requirements in rooftop and façade installations.[3–6] These assessments were based on a single plant, the Serre plant in Italy, whose BOS life-cycle energy was estimated to be 1850 MJ of primary energy per m^2 of installed PV modules, and the energy payback time (EPT) was around 2 to 2,2 years. By comparison, several studies of the BOS in rooftop and façade PV installations showed their energy requirements to be only about 600 MJ/m^2. The much higher energy requirement of field PV plants compared to residential PV systems was attributed to the need for concrete foundations and metal support structures. Furthermore, projections of the energy requirements for BOS of central plants predicted only small reductions with time (for example, 1700 MJ/m^2 by 2010, and 1500 MJ/m^2 by 2020).[6] This study discusses an actual installation where the life-cycle energy requirements are drastically lower than those predicted numbers.

A.6.2 Photovoltaic power plant experience at Tucson Electric Power

A.6.2.1 Tucson Electric Power (TEP)

TEP is the second-largest investor-owned utility in Arizona, providing electricity to nearly 370 000 residential, commercial and industrial customers in Tucson and surrounding areas in south-eastern Arizona.[7] With about 2 000 MW of net generating capacity (primarily coal fired), TEP supplies most of the power that it distributes. The company operates nearly 15 000 miles of transmission and distribution lines throughout its service territory of 1 155 square miles. The utility is involved in a very active renewable energy programme that directly supports the EPS initiative. Primarily focused on landfill gas and PV, the programme also includes solar thermal electric, wind, biomass and geothermal.[8]

The utility-scale PV generation effort is centred at the Springerville Generating Station Solar System in eastern Arizona. As shown in Figure A.104, this facility, one of the largest PV generating plants in the world, includes 4,6 MW DC of installed PV systems. Covering 44 acres, this PV generating plant is grid inter-tied with a 34,5 kV TEP transmission line. Although the Springerville plant includes other collector technologies, including amorphous silicon and cadmium telluride, crystalline silicon accounts for nearly 80 % of the plant's capacity and is the focus of this section. The field experiences with these systems provide a treasure of information that not only establishes a baseline for today's state-of-the-art system capabilities, but also can help guide the development of PV system technology for the future. These are the reasons that TEP and Sandia National Laboratories entered into a collaborative effort to track, analyse and document the cost and field performance, as well as operations and maintenance (O&M) experience associated with these systems.

Figure A.104 Springerville PV generating plant

A.6.2.2 Design and installation experience

The 26 crystalline silicon systems at the Springerville plant are listed in Table A.48. Each of these systems is an identical copy of a standardized array field configuration that utilizes the same hardware components, wiring topology and structural mounting plan. The standard system configuration includes ASE Americas (now RWE Schott Solar) ASE-300-DG/50 modules and a Xantrex PV-150 inverter. The arrays are mounted at a fixed tilt of 34° facing due south with 450 modules per array. Based on each system's areal footprint of 300 feet north–south and 140 feet east–west, the system power density is 110,6 kW AC per acre of ground. Each array string includes nine modules with two strings per row. The power per string is 2,7 kW and the maximum string design voltage is 595 V at –22 °F. The operating voltage of each string is 380 to 430 V. The Xantrex PV-150 inverter converts the variable voltage DC power to 208 V three-phase AC power. The inverters have a maximum rating of 157 kVA, at which point they will limit output or come off line, followed by an automatic restart. Each unit has a DC disconnect, 208 to 480 V step-up/isolation high efficiency transformer, revenue meter and AC disconnect. Groups of four units are connected in parallel to each of 11 500 kVA 480 V to 34 500 V high efficiency step-up transformers. Each transformer has a continuous rating of 500 kVA and can accommodate up to 650 kVA for brief intervals. The high voltage sides of the transformers are connected in parallel in a daisy chain configuration to a 34,5 kV underground distribution line that connects to the overhead 34,5 kV distribution line that feeds the well field pumps of the nearby 760 MW coal-fired Springerville Generating Station. The pumps operate continuously with an average load of about 6 000 kW.

The array configuration is designed to minimize the array field balance of system (BOS) cost. The dual-stanchion array structural supports are fabricated steel with a powder coating to minimize corrosion. The steel supports are staked to the ground to prevent wind-induced uplift and sliding. The site preparation includes minimal surface disturbance/levelling of the natural terrain while retaining the native vegetation as much as possible to reduce surface erosion and to minimize dirt splash on the modules. Mounting of the arrays to the terrain may result in a slightly jagged array appearance along the row due to surface variations, but the PV

Table A.48 List of Springerville crystalline silicon systems

System	Array size (kW DC)	Installation date
SGS-135C-1	135	13 July 2001
SGS-135C-2	135	13 July 2001
SGS-135C-3	135	17 August 2001
SGS-135C-4	135	2 October 2001
SGS-135C-5	135	23 October 2001
SGS-135C-6	135	14 December 2001
SGS-135C-12	135	30 May 2002
SGS-135C-7	135	1 August 2002
SGS-135C-8	135	1 August 2002
SGS-135C-9	135	1 August 2002
SGS-135C-10	135	17 September 2002
SGS-135C-11	135	24 June 2002
SGS-135C-13	135	15 June 2003
SGS-135C-14	135	15 July 2003
SGS-135C-15	135	15 July 2003
SGS-135C-16	135	30 July 2003
SGS-135C-29	135	15 October 2003
SGS-135C-30	135	30 October 2003
SGS-135C-31	135	15 August 2003
SGS-135C-32	135	30 August 2003
SGS-135C-26	135	22 June 2004
SGS-135C-27	135	22 June 2004
SGS-135C-28	135	24 June 2004
SGS-135C-23	135	20 July 2004
SGS-135C-25	135	21 July 2004
SGS-135C-24	135	23 July 2004

Figure A.105 Typical 135 kW DC system

Figure A.106 Xantrex PV-150 system inverter

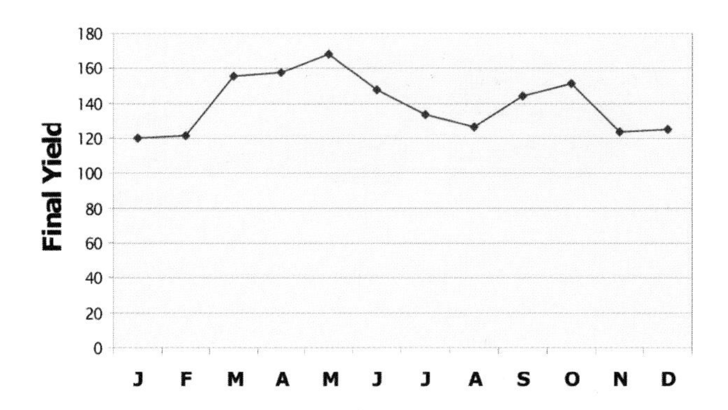

Figure A.107 Average monthly final yield for all systems over operating history

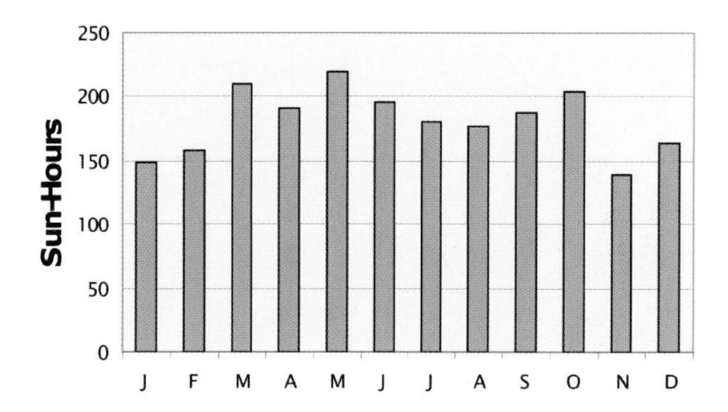

Figure A.108 Monthly reference yield (sun hours) for 2004

output effects are near zero. The array electrical interconnection uses 600 V-rated DC equipment and underground power distribution to minimize cost. Each system is installed in exactly the same way using a trained local labour pool. This standardized approach has resulted in a total system BOS cost of less than 1,00 USD/W DC.[9]

A Springerville system is shown in Figure A.105. Note the white inverter enclosure at the back of the arrays near the centre of the picture. A close-up photo of the Xantrex PV-150 inverter and enclosure is shown in Figure A.106.

A.6.2.3 System performance

To describe the system performance of the Springerville systems, the authors have chosen to utilize PV energy parameters that have been established by the International Energy Agency (IEA) Photovoltaic Power Systems Programme as described in the IEC standard 61724.[10] Three of the IEC standard 61724 system performance parameters – final yield, reference yield and performance ratio – define the system field performance in terms of energy production, solar resource and system losses. These provide an easily understood method to not only compare system performance with other system options, but also to permit system owners/customers to determine if system performance is meeting expectations. This process has been proposed for widespread adoption in the US, and the authors certainly support this effort.[11]

The final yield is the net AC energy output of the system divided by the aggregate nameplate power of the installed PV array at standard test conditions (STC) of 1 000 W/m² solar irradiance and 25 °C cell temperature:

$$Final\ yield = \frac{kWh\ AC}{kW\ DC} \quad [A.16]$$

It represents the number of hours that the PV array would need to operate at its rated power to provide the same energy. All UL-listed modules require a nameplate on the back of the module that identifies the STC rated DC power. The aggregate array power can easily be determined by summing the nameplate power ratings for the array. The average monthly final yield for all 26 Springerville systems over the 3,5-year operating history is shown in Figure A.107. The average annual final yield is 1 673 kWh AC/kW DC. The average final yield for 2004 is 1 720 kWh AC/kW DC.

The reference yield is the total in-plane solar insolation (kWh/m²) divided by the array reference irradiance. It represents an equivalent number of hours at the reference irradiance. The reference irradiance is typically equal to 1 kW/m²; therefore, the reference yield is the number of peak sun hours:

$$Reference\ yield = \frac{total\ plane\ of\ array\ insolation}{1\ kW\ /\ m^2} \quad [A.17]$$

The monthly reference yield for the Springerville arrays in 2004 is shown in Figure A.108. The annual reference yield for 2004 is 2 175 sun hours.

The performance ratio is the final yield divided by the reference yield and is dimensionless. It represents the total losses in the system when converting from nameplate DC rating to AC output. Typical system losses include DC wiring, module mismatch, bypass diodes, module temperature effects, inverter conversion efficiency, as well as others:[12]

$$Performance\ ratio = \frac{final\ yield}{reference\ yield} \quad [A.18]$$

The average monthly performance ratio for all the systems in 2004 is shown in Figure A.109. The average annual performance ratio for all systems in 2004 is 0,79.

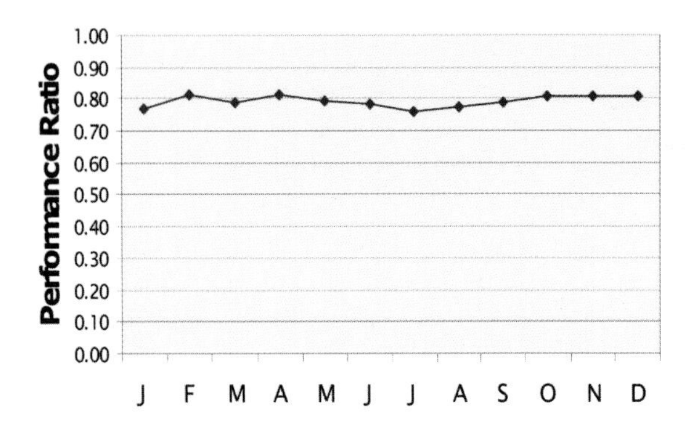

Figure A.109 Average monthly performance ratio for all systems in 2004

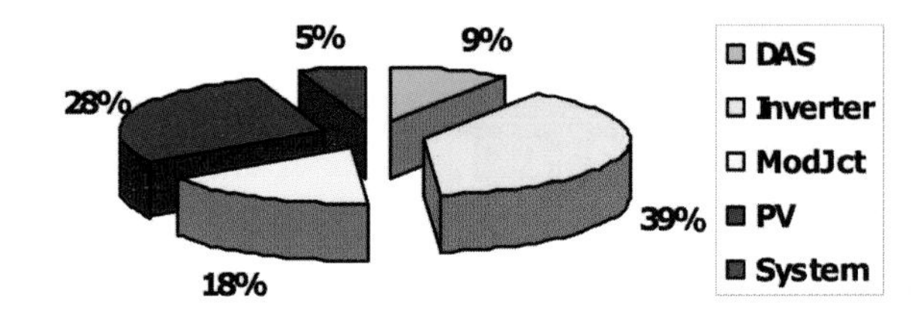

Figure A.110 Unscheduled maintenance events by component

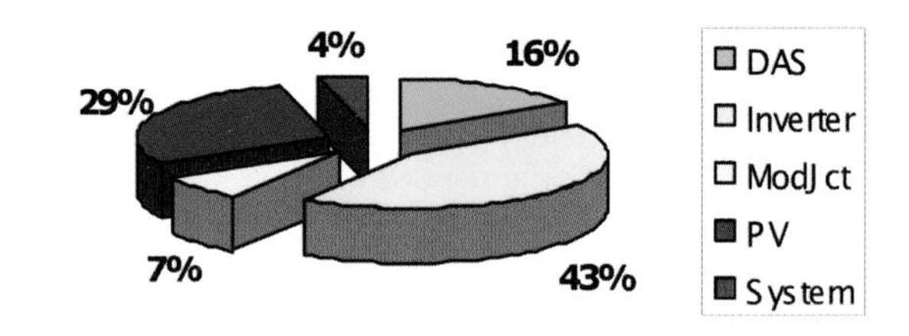

Figure A.111 Unscheduled maintenance costs by component

A.6.2.4 System maintenance experience

For the past five years, Sandia has been working to develop a comprehensive database model to track the life-cycle costs of PV systems.[13] This database, which continues to undergo improvements, was utilized to capture, document and track scheduled and unscheduled maintenance service, repairs, replacements, and labour and travel costs associated with maintenance activities for these systems. Based on Microsoft Access, the database architecture is modular to support future additions, allows associations at the component level, allows multiple components to be tracked with a system, and provides for multiple failures to be documented as a result of a maintenance visit. Failure modes (what and why), activity dates (failure and repair) and costs (labour, parts and travel) were captured and analysed from system maintenance activity logs covering the period of mid 2001 through to 2004. From this data, analyses of failure modes and O&M costs were made.

The Springerville systems provide a significant database for assessing the reliability and maintenance needs for a utility-scale generating plant operating in a utility environment. Altogether, a total of 11 700 identical PV modules and 26 identical inverters have been installed since mid 2001.

Over the operating history from mid 2001 through to 2004, a total of 85 unscheduled maintenance events were recorded for Springerville systems. The events are grouped by categories, including data acquisition systems (DAS), inverters, junction boxes, modules (PV) and systems. Figure A.110 presents the breakdown of these events by component as a percentage of the total number of events.

The unscheduled events resulted in a loss of generating capacity that affected one or more systems and required human intervention to restore the system(s) to full operational capacity. These events could be as simple as a manual restart of a tripped inverter or considerably more complex, such as the repair of damage resulting from a lighting strike (the plant experienced strikes in 2003 and 2004). Figure A.111 presents a breakdown of unscheduled costs by component as a percentage of the total unscheduled repair costs. Up to 1 January 2005, the 26 crystalline silicon Springerville systems had provided 582 system months of continuous operation since installation. Over that same period, a total of 85 unscheduled maintenance events were recorded, which provides a mean time between unscheduled service per system of 6,9 months of operation.

Scheduled maintenance was conducted on the plant each year. This included mowing the native vegetation, as well as visual inspections of the arrays and power handling equipment. In 2004 the bolt torque settings on all current carrying connections were also checked. Table A.49 lists the annual maintenance costs, both scheduled and unscheduled, as a percentage of the cumulative capital investment by year. The average annual maintenance costs since the initial Springerville installations are 0,16 %.

While the above maintenance costs include unscheduled repair/service on the inverters, the costs of inverter rebuild (anticipated every ten years) are not included. Including this expense on an amortized basis is estimated to increase the annual maintenance cost by an additional 0,1 %. Prior to 2004, software revisions were made allowing for the remote resolution of most fault conditions, resulting in fewer unscheduled site visits in 2004. Daily performance analysis tools pinpoint underperforming units, allowing for timely resolution of problems with minimal lost energy production. Consequently, overall system effective availability was 99,78 % in 2003 and 99,72 % in 2004.

Table A.49 Maintenance costs as a percentage of capital investment

Year	Scheduled (%)	Unscheduled (%)	Total (%)
2002	0,08	0,01	0,09
2003	0,07	0,22	0,29
2004	0,06	0,05	0,11

Table A.50 Cost breakdown for Springerville systems

System component	USD/W DC	USD/W AC
Modules	3,33	4,22
Array field BOS	0,56	0,71
Site preparation (0.10 USD/W DC)		
Structure (0.15 USD/W DC)		
Electrical (0.30 USD/W DC)		
AC Inter-tie (0.01 USD/W DC)		
Inverter/transformers	0,40	0,51
Indirect/overhead/profit	1,11	1,40
Total	5,40	6,84

A.6.2.5 System cost experience

Tucson Electric Power is realizing significant cost benefits by incorporating standardized products, volume purchasing and efficient array field design and installation. The Springerville experience has documented some of the lowest installed system costs ever reported, thereby establishing a benchmark for state-of-the-art utility-scale systems. A cost breakdown for systems installed in 2004 is presented in Table A.50.

Modules

The module price reflects a bulk purchase from the module manufacturer.

Array field balance of system

The site preparation cost includes ground levelling, fencing and underground wiring. Structure cost includes mechanical mounting of the modules, support structure hardware and foundation staking. The electrical work includes module interconnect wiring, conduit, junction boxes for both the string and row buses, disconnect switches, system protection and wiring on the AC side of the inverter to the 480 V transformer and the DAS. The AC inter-tie cost includes the wiring and installation labour from the 480 V transformer to the 34,5 kV transformer.

Inverter/transformers

This cost includes the purchase price of the Xantrex PV-150 inverter, the 208/480 V transformer for each system, and one quarter of the 480/34,5 kV transformer cost (each 34,5 transformer gathers four of the systems). Installation labour for these components is included.

Indirect/overhead/profit

Indirect costs include system design, procurement, construction management and project engineering. The overall project management for the Springerville installations is provided via contract by Tucson-based Global Solar Energy.

Energy cost figure-of-merit

The true measure for comparing different PV system options is the cost of delivered kWh AC energy. To put the Springerville cost experience in perspective, the authors have utilized an energy cost figure-of-merit defined as the average installed system cost (USD/kW DC) divided by the energy output (kWh AC/kW DC) expected over a 30-year period. Although the resulting cost figure represents USD/kWh AC, this figure does not include financing costs, the cost of capital, O&M costs or any tax considerations and, thus, is not a levelled energy cost and is not portrayed as such (note that levelled energy cost for TEP is addressed in the following section). For 2004, this energy cost figure-of-merit is 0.10 USD/kWh AC for the Springerville systems. Interestingly, the Springerville energy cost figure-of-merit for fixed flat-plate systems is nearly identical to the energy cost figure-of-merit reported for one-axis, tracking horizontal flat-plate systems installed at Prescott, Arizona.[2]

A.6.2.6 Economic perspective

The experience at Springerville provides a valuable utility perspective on the future use and needs of PV technology. These include actual utility-based energy-generating costs, capacity factors and emission benefits associated with solar electric generation.

Energy cost

The Solar Energy Industries Association (SEIA) recently announced a roadmap with established goals for expanding the use of solar power-generating capacity in the US.[14] The roadmap goal over the next decade for PV systems is a selling price of 3,68 USD/W AC in 2015 and a cumulative installed US capacity of 9,6 GW. Coupling the TEP cost experience at Springerville with this SEIA cost goal provides an interesting perspective for the future of PV. Table A.51 presents a comparison of today's benchmark system costs for Springerville and a proposed breakdown of a 2015 utility-scale PV system, meeting the roadmap goal in today's dollars.

Using the Springerville performance ratio of 0,79, the 3,68 USD/W AC future system cost corresponds to an equivalent cost of 2,91 USD/W DC. The 2015 system cost components follow a proposed breakdown developed elsewhere for a crystalline silicon system.[12] The 2015 module cost is based on a manufacturing cost analysis for a crystalline silicon production plant of 25

Table A.51 System costs for the future (USD/W DC)

System component	Springerville system	2015 system
Modules	3,33	1,78
Array field	0,56	0,58
Inverter	0,40	0,25
Fixed	1,11	0,30
Total	5,40	2,91

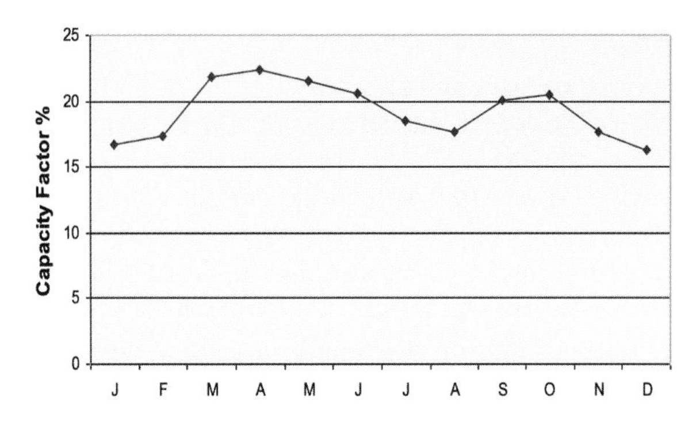

Figure A.112 Average monthly capacity factor for Springerville systems

Source: www.greenwatts.co

MW/year developed by Spire Corporation.[15] The proposed module cost is also consistent with crystalline silicon manufacturing cost projections developed through the US Department of Energy (DOE) Photovoltaic Manufacturing Technology (PVMaT) programme.[16] While module costs and fixed costs require substantial cost reductions to achieve the 2015 goal, this comparison validates the creative system BOS approach developed by TEP at Springerville by already achieving the array field BOS target projected for the next decade. As annual PV installation quantities increase in future years it is expected that the fixed costs will be diluted over larger amounts of installed capacity and will be reduced on a USD per W DC basis.

The industry roadmap goal for 2015 is a levelled energy cost (LEC) of 0,057 USD/kWh AC of PV generation. This compares to the TEP-calculated LEC in 2004 (pay-as-you-go, no-financing costs) of 0,096 USD/kWh AC for the Springerville PV generation.[8] The TEP calculation, which includes both federal income tax credits and state property tax reductions for solar, already meets the roadmap baseline 2015 LEC of 0,115 USD/kWh AC.

Capacity factor

The average monthly capacity factor for the Springerville systems over their operating history is presented in Figure A.112. As depicted here, the capacity factor is defined as the ratio of net electrical generation for the time considered, to the energy that could have been generated if the system were generating at continuous full power during the same period:

$$\text{Annual capacity factor} = \frac{\text{annual final yield}}{8760} \qquad [A.19]$$

The average annual capacity factor for all systems in 2004 was 19,2 %.

A.6.3 Energy payback and life-cycle CO_2 emissions of the balance of system (BOS) in an optimized 3,5 MW PV installation

A.6.3.1 Description and cost of the Springerville PV plant

TEP has designed the Springerville PV plant for 8 MW of field-mounted PV. To date, 4,6 MW of PV modules have been installed, of which 3,5 MW are framed mc-Si PV and 1,1 MW are framed and frameless thin-film PV modules. Figure A.113 shows a close-up view of the PV plant, while an aerial view is as shown in Figure A.104. TEP's philosophy guiding all phases of the PV plant installations is to optimize its design in order to minimize labour, materials and costs. In the design phase of the PV plant, small prototype PV installations were evaluated to assess the appropriate electrical configurations and optimize the performance of standard utility-scale power-conditioning equipment. Having determined the size and design of the PV plant, TEP staged construction in phases to realize benefits from synergies in utilizing labour and equipment. Time and motion studies were conducted to make the best use of construction personnel and equipment. In the site preparation construction stage, TEP installed infrastructure for the electrical connections that is sized to accommodate the addition of planned PV installations until it attains its intended size. Site preparation included levelling the ground, applying soil stabilizers, installing underground conduits, and providing concrete foundations for inverter and transformer pads, high-voltage wiring, high-voltage disconnects, transformers and a grounding system, all to power plant specifications.

The sizing and wiring configuration of the PV arrays is fashioned to produce electricity flows that match the performance characteristics of standard utility electrical equipment. The objective is to maximize the amount of PV capacity per connection point, which minimizes electrical costs per PV W. In addition, the configuration of the PV arrays is designed to maintain operation in defined voltage ranges, enabling the inverters and transformers to operate in their performance 'sweet spot', which also maximizes the amount of AC electricity flowing into the transmission grid. The optimized

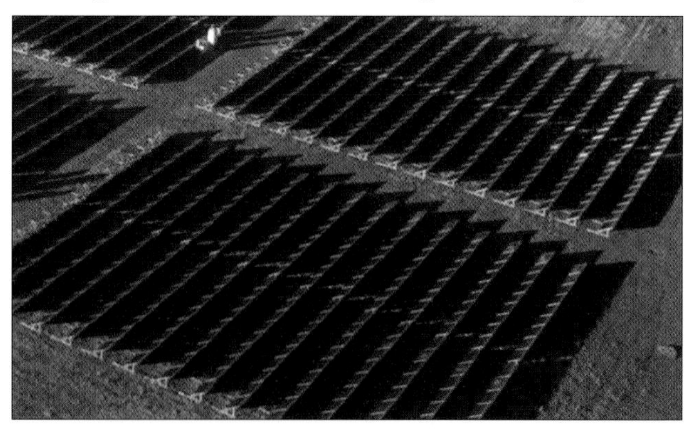

Figure A.113 Photograph of TEP's Springerville PV plant

Source: www.greenwatts.com

layout of the PV installations is 135 kW arrays. Four construction workers installed each of the arrays in 40 hours, from ground level to a standing array. The PV support structures rely on the mass of the PV modules that are anchored into the ground with 1 foot long nails, thereby eliminating the need for concrete foundations. The PV support structures can withstand 193 km/hour (120 mph) winds; to date, they successfully withstood actual sustained winds of 160 km/hour (100 mph).

The performance of the system is monitored with computer software. Detailed records show that the annual average (in 2004) of AC electricity output from the mc-Si section of the plant was 1 730 kWh AC-el/kW (214 kWh AC-el /m^2) of installed PV modules, with an effective system availability greater than 99 %. This electricity output is measured at the grid connection on the 480 V side of the isolation transformer and accounts for all losses. The current real-time day-to-day performance of the plant can be seen at www.GreenWatts.com. The 3,5 MW mc-Si portion of the system was installed in stages between 2001 and 2004 and uses ASE 300DG/50 modules, the specifications of which are the following: 46,6 kg weight (including 5,4 kg of aluminium frame), 2,456 m^2 area and 12,2 % rated efficiency. The thin film modules are being optimized and their performance evaluated.

TEP's optimization of design and forward construction staging enabled them to reduce costs by developing 'cookie-cutter' installation procedures. To minimize the system costs, the timing and sizing of the installations are scheduled to take advantage of volume purchasing and partnership contracts with the manufacturers of the major components. TEP reported that the total installed cost of BOS components was 940 USD/kW,[17] a great improvement from a previous estimate of 1 700 USD/kW[18] (see Figure A.114). These costs do not include financing or end-of-life dismantling and disposal expenses. The end-of-life salvage value of the BOS components is assumed to equal the costs of dismantling and disposal. Inverters and their support software are the most expensive BOS component, with an installed cost of 400 USD/kW. The next largest is the electrical wiring system with an installed cost of 300 USD/kW. TEP states that its detailed attention to designing the DC trunk electrical connections and to the pre-planned construction staging significantly lowered the costs for installing the electrical system. The installed cost of the PV support structures was 150 USD/kW; this low value resulted from simplifying the design, which minimized labour, equipment and materials, and using relatively inexpensive powder-coated angle iron. The site preparation costs were 80 USD/kW.

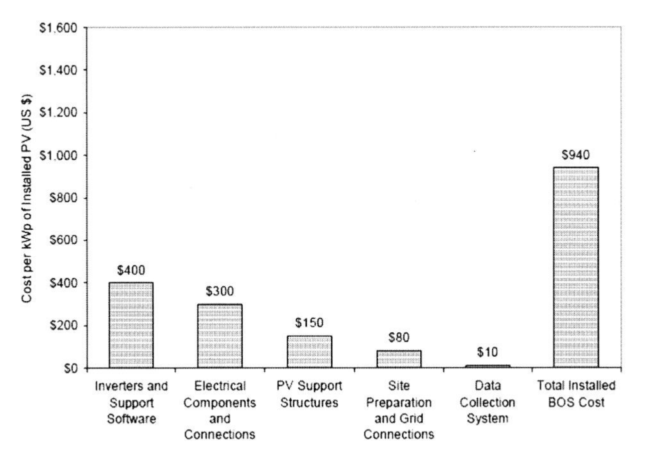

Figure A.114 Installed balance-of-system (BOS) costs for the mc-Si PV installations

Note: Monetary units are in 2003 US dollars.

A.6.3.2 Methodology

BOS life-cycle inventory and boundary conditions

TEP provided an itemized BOS bill of materials for its mc-Si PV installations. Table A.52 is a categorized list of the BOS components. Aluminium frames are shown separately since they are part of the module, not of the BOS inventory, and there are both framed and frameless modules on the market.

The BOS component inventory is scaled to 1 MW of installed PV and a 30-year operating life. The material composition of the BOS components is estimated from information provided by the manufacturers, TEP and published studies. The life expectancy of the PV metal support structures is assumed to be 60 years. Inverters and transformers are considered to have a life of 30 years, but parts must be replaced every 10 years, amounting to 10 % of their total mass according to well-established data from the power industry on transformers and electronic components. The inverters are the utility-scale Xantrec PV-150 models, which have a wide open frame so that a failed part can be easily replaced. We also explored the assumption of total replacement of the inverters every 10 and 15 years as part of the sensitivity analysis.

Although the PV system is currently unmanned, for consistency the material inventory includes an allocation of office facility materials for administrative, maintenance and security staff, as well as staff vehicles for PV plant maintenance.

The boundaries of the life-cycle energy and GHG emissions analysis extend from materials production to product end-of-life disposal. Five life-cycle stages are evaluated: stage 1, BOS part production, including extraction of raw materials, materials processing, manufacturing and assembly; stage 2, BOS part transportation; stage 3, PV plant construction, including BOS installation and office building construction; stage 4, PV plant administrative; and stage 5, product end-of-life management. The geographic system boundary of this study is North America and the time period of the technology and data covers the late 1990s and later.

Table A.52 Material inventory, in kilograms, of the BOS components for a 1 MW field PV plant

BOS components	Steel	Aluminium	Copper	Plastics	Other	Total weight
Frames for PV modules	0	19 861	0	0	0	19 861
Support structure frames	30 906	0	0	0		0 30 906
Support structure hardware	1 333	0	0	0	0	1 333
Bare copper wire	0	0	445	0	0	445
Insulated copper wire	0	0	2 071	323	0	2 394
PVC conduit	0	0	0	3 425	0	3 425
IMC conduit	4 799	0	0	0	0	4 799
Concrete	0	0	0	0	47 405	47 405
Connections	1 296	126	3	36	0	1 462
Inverters	3 036	894	625	485	0	5 040
Transformers	6 756	0	1 652	300	300	9 008
Transformer oil (vegetable)	0	0	0	0	6 001	6 001
Concrete pad foundations	562	0	0	0	18 350	18 912
Grounding and disconnects	178	214	2 721	190	67	3 370
Miscellaneous components	437	0	1	516	0	955
Fence (perimeter)	4 291	0	0	0	10 502	14 793
Water for soil stabilizer	0	0	0	0	60 564	60 564
Vehicles and construction*	330	57	5	23	89	505
Office facilities PV plant staff**	1 968	2	4	532	18 190	20 697
Totals	55 893	21 154	7 527	5 832	161 469	251 876
Energy consumption					Total	Energy (GJ)
Gasoline (litre) staff vehicles					970	312
Diesel (litre) construction					1 472	53
Natural gas (m^3) office					1 480	54
Electricity (kWh) office					5 168	19

Notes: * Automobile material composition data is from Weiss et al (2000)

** The composition of the office building material is estimated from data provided by the LCA software LISA from the case study of a multi-storey office building with LISA default values.

Estimation of life-cycle energy and GHG emissions

The life-cycle energy uses and GHG emissions over the complete life cycle of PV BOS were determined from the commercial Life Cycle Inventory (LCI) databases of Franklin[19] and Ecoinvent,[20] and public domain sources from the National Renewable Energy Laboratory (NREL)[21] and the Aluminium Association.[22] Supplementary data sources include those from the US Energy Information Administration,[23] the US Department of Energy[24] and a European study.[25] The LCA software tool Simapro 6 was used for detailed energy payback and greenhouse emissions analyses. The assumptions for the reference case scenario of the LCI were as follows:

- 33 % of secondary material content in aluminium parts;
- 30 years of inverter lifetime with 10 % of materials' replaced every ten years;
- transport range of BOS components of 1 600 km, with 50 % of the transport made by rail and 50 % by trucks;
- a transport range for disposal of 160 km by trucks.

The following well-established data were used. The US electricity production mixture was used for power in manufacturing BOS components, except for the aluminium products, for which a more site-specific grid mixture was used. Conventional diesel fuel was used for rail and truck transport. The PV plant utilization/administrative functions are a pick-up truck for plant maintenance, and energy to heat, cool and power office facilities for the staff. The energy sources were petrol for vehicles, and electricity and natural gas for office facilities. The disposal of field PV plant components is based on carrying them away a distance of 160 km by heavy truck. The distribution and disposal of concrete is based on transporting it 50 km. The energy to shred PV plant components is 0,4 MJ/kg.[26] The energy source for dismantling, transporting and shredding the PV components is assumed to be conventional diesel fuel.

Sensitivity analyses were also conducted with variations in the secondary aluminium content of the BOS components (100 % and 0 %), and lifetimes of inverters (10 years and 15 years without replacing parts). In addition, to verify our findings, we employed GREET,[27] a model and database widely used for LCA in the automobile cycle, in conjunction with data on the production of materials from Weiss et al[28] and from Environdec's environmental product declarations.[29]

All energy values are reported in terms of primary energy in units of GJ/kW of rated peak DC electricity output, or MJ/m^2 of installed PV modules. Primary energy is the total fuel-cycle energy and accounts for that expended to extract, refine and deliver fuels. Energy values are reported at their higher heating value. The electricity estimates are based on a US average fuel mix and power-plant efficiency, corresponding to a conversion efficiency of 33 %.[23] The GHG emissions are carbon dioxide, nitrous oxide, methane, sulphur hexafluoride, perfluorocarbons (PFCs) and chlorofluorocarbons (CFCs), which are reported in kg of CO_2 equivalencies per kW of installed PV.

Table A.53 Energy use and greenhouse gas (GHG) emissions for BOS production for reference case (33 % secondary aluminium and a 30-year lifetime of inverters, with 10 % part replacement every ten years)

Balance of system (BOS)	Mass (kg/MW)	Percentage of total	Energy (GJ/MW)	Percentage of total	GHG emissions (t CO_2 eq./MW)	Percentage of total
BOS PV support structure	16 821	10,3	699	18,7	47	23,2
PV module interconnections	453	0,3	53	1,4	2	1,2
Junction boxes	1 385	0,8	51	1,4	4	1,9
Conduits and fittings	6 561	4,0	328	8,8	20	9,8
Wire and grounding devices	5 648	3,4	769	20,6	35	17,2
Inverters and transformers	28 320	17,3	1321	35,3	55	27,1
Grid connections	1 726	1,1	127	3,4	5	2,6
Office facilities	20 697	12,6	90	2,4	8	4,1
Concrete	76 417	46,6	66	1,8	10	5,1
Miscellaneous	5 806	3,5	236	6,3	16	7,9
Total	163 834	100,0	3 740	100,0	204	100,0
Frame	18 141		2 650		184	

Life-cycle energy and GHG emissions payback times

The concept of payback time is used to evaluate the time it takes to recover the life-cycle energy and GHG emissions embodied in PV installations. Payback time is based on the assumption that the AC electricity produced by PV plants displaces an equal quantity of electricity generated by the current US energy mixture. The energy payback time (EPT) is calculated by dividing the life-cycle energy requirements of the BOS components (converted to kWh of equivalent electricity) with the actual annual electricity output of the system, after all losses, at the grid connection (that is, 1 730 kWh AC-el per kW). The actual performance and the most likely values of input parameters comprise our 'reference case'.

We also present, as the 'US average case', the expected performance under average US insolation conditions.

A.6.3.3 Results

Reference case

Table A.53 and Figures A.115 and A.116 show the results of calculating BOS life-cycle energy and GHG emissions by categories of BOS components. The total primary energy in the BOS life cycle is 543 MJ/m^2 of installed PV modules. This finding contrasts sharply with the previous central PV plant BOS estimate (that is, 1 850 MJ/m^2) that was based on the Serre plant, and reveals the energy savings from eliminating concrete foundations and roads. PV support structures account for only 12 % of the total BOS energy, while that of inverters and transformers combined accounts for 31 %. This is followed by electrical connections and PV module frames at 20 % and 19 % of total BOS energy, respectively.

Using the average US energy conversion efficiency of 33 %,[23] this gives an electricity equivalent of 49 kWh/m^2 that, after annualizing the administrative and disposal consumptions, results in an EPT of 0,21 years.

The Springerville site combines high insolation (for example, ~2 100 kWh/m^2/year) with relatively low ambient temperatures, which increases the system's efficiency; the measured system efficiency of the mc-Si PV modules at Springerville is 83,5 %. The assessments of installations under US average conditions are based on 1 800 kWh/m^2/yr insolation, a rated module efficiency of 12,2 % and a system efficiency of 80 %. This corresponds to annual electricity production of 1 420 kWh/kW installed PV. The corresponding EPT of the BOS for an average US installation is 0,37 years.

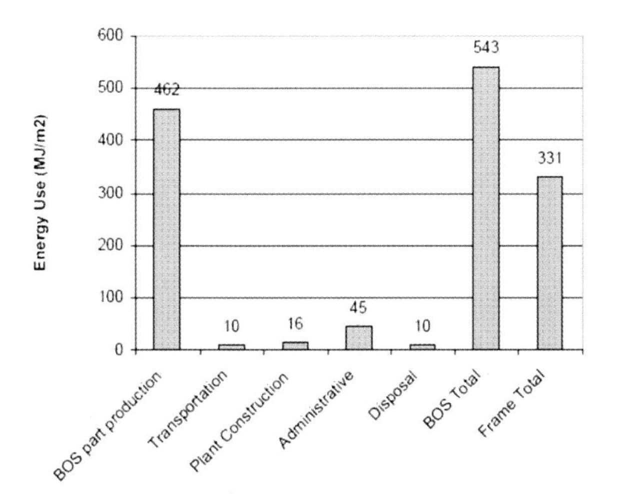

Figure A.115 Life-cycle energy consumption of BOS: Reference case

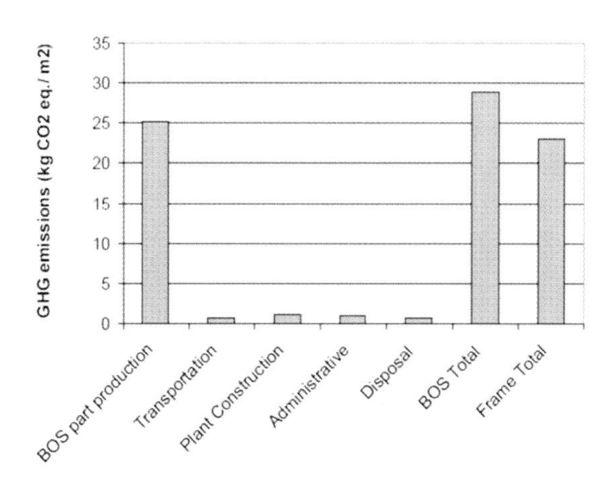

Figure A.116 Life-cycle GHG emissions of BOS: Reference case

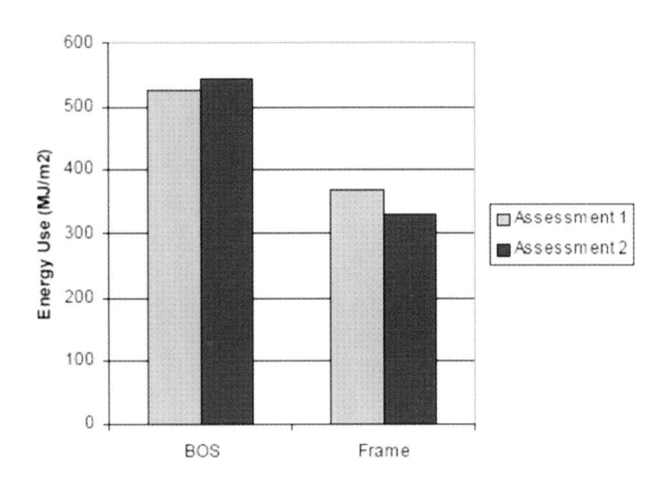

Figure A.117 Comparison of life-cycle energy use between assessments 1 and 2

Source: Assessment 1 is based on data from GREET 1.6; Weiss et al; EPD SEMC.

Assessment 2 is based on data from Franklin; NREL US LCI; Aluminium Association LCI; Ecoinvent; Annual Energy Review, EIA; DOE LCI of Biodiesel; ETH-ESU.

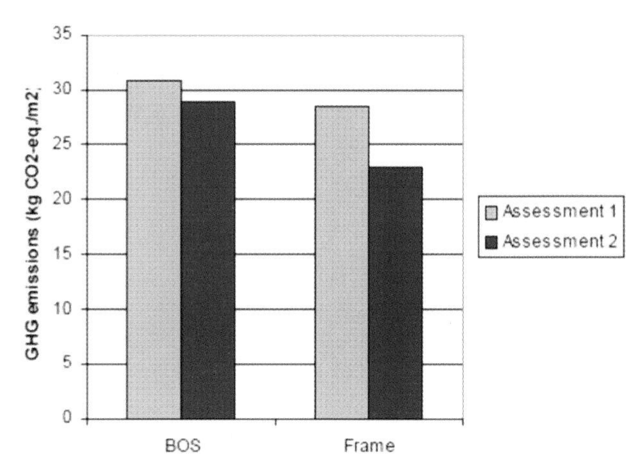

Figure A.118 Comparison of GHG emissions between assessments 1 and 2

Source: Assessment 1 is based on data from GREET 1.6; Weiss et al; EPD SEMC.

Assessment 2 is based on data from Franklin; NREL US LCI; Aluminium Association LCI; Ecoinvent; Annual Energy Review, EIA; DOE LCI of Biodiesel; ETH-ESU.

Comparisons of Figures A.115 and A.116 also show the energy consumption and corresponding GHG emissions of the aluminium frames, part of the mc-Si PV modules used at Springerville. The calculated energy consumption for the aluminium frames under reference conditions (33 % recycled Al) was 331 MJ/m^2; producing them consumes 99 % of this energy. As discussed earlier, the energy consumption if all of the aluminium is from primary sources is 457 MJ/m^2. These estimates agree with previous publications attributing 400 MJ/m^2 to Al frames, which did not, however, cite the fraction, if any, of recycled aluminium.[6] This comparison shows that the environmental impact from the life cycle of the frames is of the same magnitude as that from the total BOS components and their installation. This highlights the need for frameless PV modules to reduce the environmental impacts of the whole (modules + BOS) PV plant.

Verification analysis

A common exercise to verify the results from a multifaceted assessment, like this one, is for different analysts to assess the same system with different tools. An earlier assessment was based on GREET,[27] along with data on materials production from Weiss et al[28] and from Environdec's environmental product declarations.[29] This is labelled assessment 1 and the previously discussed reference case is labelled assessment 2.

A comparison of the results from the two assessments is shown in Figures A.117 and A.118. According to assessment 1, the BOS life-cycle energy is 526 MJ/m^2, which is 3 % lower than the reference case, whereas the corresponding GHG emissions are 31 kWh CO_2-eq./m^2, which is 7 % higher than the reference case. As shown in these figures, the differences between the two assessments are greater for the frame than for the BOS. These differences are mainly due to the different energy intensity and emission factors adopted in each assessment for aluminium part production. Energy intensity data of primary aluminium production from Weiss et al (220 MJ/kg)[28] used in assessment 1 is 5 to 10 % higher than data from other studies, including the current assessment 2. The GHG emission factors in the earlier assessment were based on the average US grid mix, while assessment 2 uses emission factors of the electricity grid mixture specifically employed by the US aluminium industry. Thus, the GHG emissions of assessment 2 reflect the actual energy consumption of the aluminium industry, which primarily relies on hydroelectric power as the main energy source.[17] Further analysis shows that differences in other assumptions, including emissions factors of transportation and part fabrication, had a negligible impact.

Sensitivity analysis

This assessment is based on actual field performance data and accurate records for the BOS components and their installation. However, some of our assumptions carry uncertainty that needs to be quantified. It pertains to the life expectancy of the inverters and the fraction of recycled aluminium used in the BOS components. The results presented below are based on the assessment 2 databases (that is, Franklin; NREL US LCI; Aluminium Association LCI; Ecoinvent; Annual Energy Review, EIA; DOE LCI of Biodiesel, and ETH-ESU[30]).

As discussed earlier, the open-frame utility-grade inverters used in Springville are expected to have the same life as utility transformers; industry data on the latter show a 30-year useful lifetime. Electronic compo-

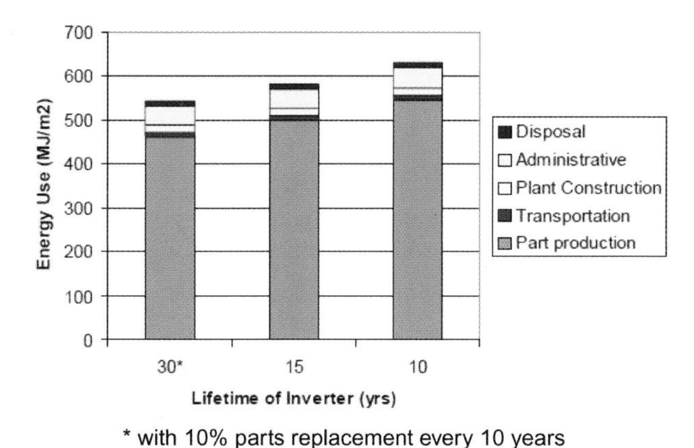

Figure A.119 Life-cycle energy consumption of BOS: Impact of inverters' life expectancy

Note: * with 10 % parts' replacement every ten years.

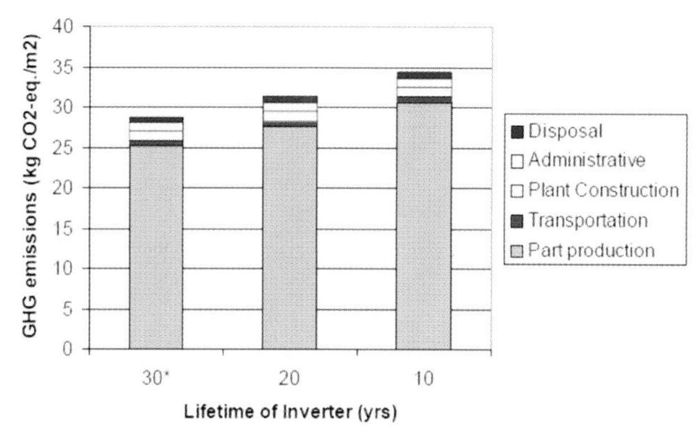

Figure A.120 Life-cycle GHG emissions of BOS: Impact of inverters' life expectancy

Note: * with 10 % parts' replacement every ten years.

nents may need to be replaced earlier; hence, we adopted a 10 % replacement of parts every ten years in our reference case. However, the integrated inverters commonly used in small installations typically are assumed to last 10 to 15 years. In our sensitivity analysis, we explored the impact on our estimates of such shorter lives. Figures A.119 and A.120 give the resulting life-cycle energy and GHG factors. For the plant at Springerville, the EPTs of the BOS increase to 0,23 and 0,25 years, corresponding to 15-year and 10-year inverter lives. For US average conditions, the EPTs increase to 0,40 and 0,43 years correspondingly.

Both global and US production of aluminium includes approximately one third from secondary (recycled) metals; this mixture was used in our reference case. Aluminium from ore (primary source) uses 10 to 20 times more energy than that from recycled metal. In the following, we also examined the impact of using 100 % primary Al (0 % recycled) and 100 % recycled Al (see Figures A.121 and A.122). The impact on the BOS components is very small since only small amounts of Al are embedded in transformers, inverters and supports. However, there is a significant impact of using either totally primary or totally secondary aluminium for producing PV module frames.

Data uncertainty analysis

As shown by the comparison of assessments 1 and 2, a choice of data sources can produce slightly different results. An interesting example of the impact of different values in a component's life-cycle inventory is the case with the transformer oil. The transformer oil used in the Springerville PV power plant consists of mostly soybean oil (> 98,5 %).[31] The life-cycle GHG emissions from the PV BOS using Ecoinvent LCI data for soybean oil is 6 % higher than the reference case based on the US Department of Energy's LCI study of soybean oil.[24] These two LCI studies adopt different assumptions on the methods of soy agriculture and in the allocations rule of energy and emissions between the soybean oil and soy meal, a co-product. However, the difference in the life-cycle energy consumption for the two cases is negligible (see Figure A.123).

A.6.4 Conclusions

The Environmental Portfolio Standard programme has proven to be a significant stimulus to increasing the installed capacity of PV systems in Arizona. The funds provided through the programme have allowed TEP to install over 5 MW of PV systems since 2000, while the kWh AC criterion has focused TEP's efforts on innova-

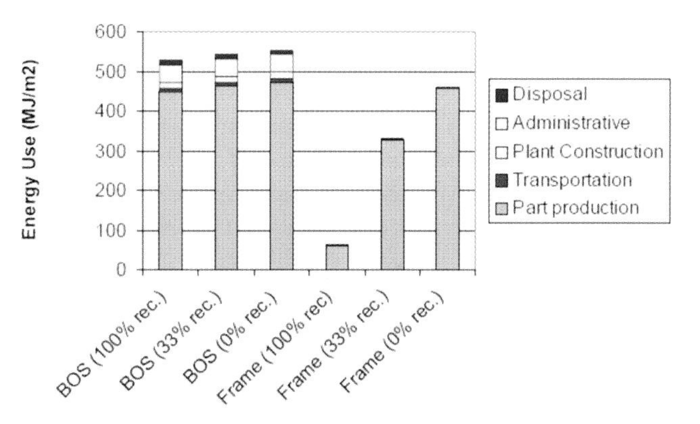

Figure A.121 Life-cycle energy consumption of BOS: Impact of recycled aluminium

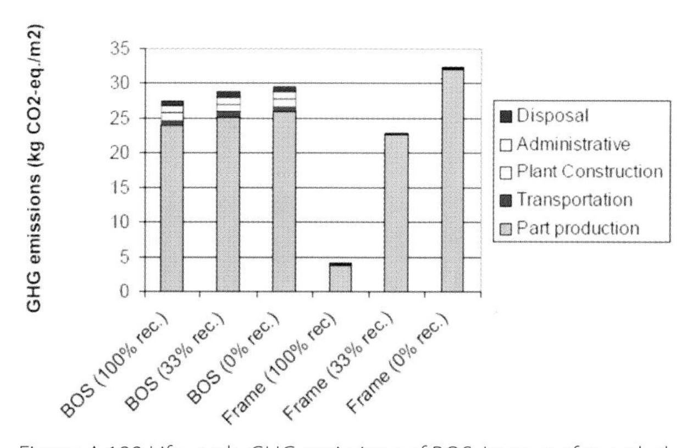

Figure A.122 Life-cycle GHG emissions of BOS: Impact of recycled aluminium

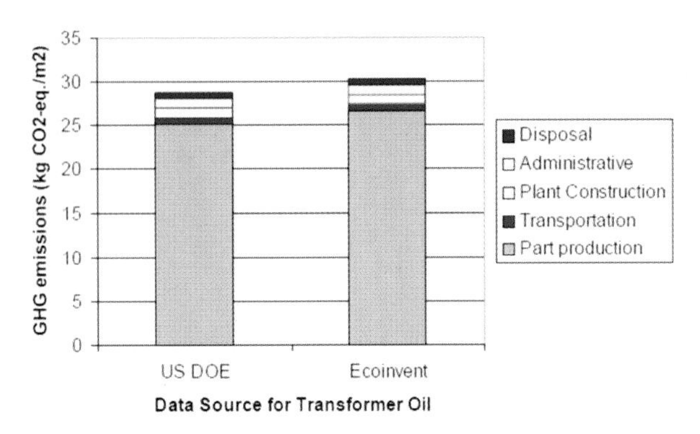

Figure A.123 Life-cycle GHG emissions of BOS: Impact of Life Cycle Inventory (LCI) data

tive low-cost PV energy generation. The energy data, system cost and maintenance experience with the Springerville crystalline silicon systems provide a treasury of information that establishes a benchmark for current utility-scale fixed flat-plate PV systems. This section has identified a number of findings, including:

- Average annual AC system energy output in 2004 is 1 720 kWh AC per kW DC of array.
- Average annual AC system power in 2004 is 0,79 of the array DC nameplate rating.
- Innovative approaches, including standardized array designs, low-cost array field BOS and bulk hardware purchases, have resulted in an installed system cost of 5,40 USD/W DC.
- Average annual O&M cost is 0,16 % of initial system installed capital cost, not including rebuild/replacement cost of the inverter.
- The mean time between unscheduled maintenance service per system is 6,9 months of operation.
- The average annual capacity factor for all systems in 2004 was 19,2 %.
- The LEC cost calculated by TEP (no financing costs) is 0,096 USD/kWh AC, which meets the 2015 SEIA baseline goal for PV generation.

The Springerville mc-Si field PV plant achieves two important advances in field PV plants: a reduction in BOS life-cycle energy and GHG emissions and a decrease in the cost of the installation.

The total primary energy for the BOS life cycle was estimated, by using different databases and analyses, to be only 526–543 MJ/m^2, which is 71 % lower than the previously published estimates based on a the Serre plant. The main difference is due to design optimization that eliminated reinforced cement foundations and decreased the need for expensive metal supports. For the Springerville site, the actual energy payback is 0,21 years; for US average conditions, the estimated EPT is 0,37 years. The GHG emissions during the life cycle of the BOS are 29 to 31 kg CO_2eq./m^2. This study indicates that PV plants potentially may approach near zero GHG emission values with the development of advanced PV technologies and installations. The total installed cost of the BOS components is 940 USD/kWp, representing a 45 % reduction from the previously reported lowest estimate for field PV plants, and brings the cost of PV electricity a step closer to being a cost-competitive source of distributed electricity generation. Undoubtedly, more decreases in the life-cycle energy, GHG emissions and costs of field PV plants will occur with advances in PV manufacturing technologies, the large-scale manufacture of standardized BOS components and utility-scale inverters, and the development of more effective installation techniques.

REFERENCES

1. Arizona Corporation Commission (2001) *Environmental Portfolio Standard R14-2-1618*, Phoenix, Arizona, www.cc.state.az.us/utility/electric/R14-2-1618.
2. Moore, L., Post, H., Hayden, H., Canada, S. and Narang, D. (2005) 'Photovoltaic power plant experience at Arizona Public Service: A 5-year assessment,' *Progress in Photovoltaics: Research and Applications*, vol 13, pp353–363
3. Alsema, E. A. (2000) 'Energy pay-back time and CO_2 emissions of PV systems', *Progress in Photovoltaics: Research and Applications*, vol 8, no 1, pp17–25
4. Frankl, P., Masini, A., Gamberale, M. and Toccaceli, D. (1998) 'Simplified life-cycle analysis of PV systems in buildings: Present situation and future trends', *Progress in Photovoltaics: Research and Applications*, vol 6, no 2, pp137–146
5. Alsema, E. A., Frankl, P. and Kato, K. (1998) 'Energy pay-back time of photovoltaic energy systems: Present status and prospects', Presented at the Second World Conference on Solar Energy Conversion, Vienna, 6–10 July 1998
6. Alsema, E. A. (2003) 'Energy pay-back and CO_2 emissions of PV systems', in T. Markvart and L. Castaner (eds) *Practical Handbook of Photovoltaics Fundamentals and Applications*, Elsevier, Oxford, UK, Chapter V-2, pp869–886
7. See www.tucsonelectric.com
8. Tucson Electric Power Company (2004) *Demand-Side Management and Renewables Data for Mid-Year 2004*, Semi-annual report to the Arizona Corporation Commission, www.greenwatts.com/Docs/ACCMidYear04.pdf
9. Hansen, T. (2003) 'The systems driven approach to solar energy: A real world experience', *Proceedings of Solar Energy Systems Symposium*, Albuquerque, New Mexico, 15–17 October, www.sandia.gov/pv
10. IEC (1998) *Photovoltaic System Performance Monitoring – Guidelines for Measurement, Data Exchange and Analysis, IEC Standard 61724*, IEC, Geneva, Switzerland

11. Marion, B., Adelstein, J., Boyle, K., Hayden, H., Hammond, B., Fletcher, T., Canada, B., Narang, D., Kimber, A., Mitchell, L., Rich, G. and Townsend, T. (2005) 'Performance parameters for grid-connected PV systems', *Proceedings of 31st IEEE Photovoltaic Specialists Conference*, Lake Buena Vista, Florida, 3–7 January 2005
12. Thomas, M., Post, H. and DeBlasio, R. (1999) 'Photovoltaic systems: An end-of-millennium review', *Progress in Photovoltaics: Research and Applications*, vol 7, pp1–19
13. Moore, L. (2001) 'Sandia's PV reliability database: Helping business do business', *Quarterly Highlights of Sandia's Solar Programs*, vol 1, www.sandia.gov/pv
14. SEIA (2005) *Our Solar Power Future: The US Photovoltaics Industry Roadmap through 2030 and Beyond*, January 2005, www.seia.org
15. Little, R. and Nowlan, M. (1997) 'Crystalline silicon photovoltaics: The hurdle for thin films,' *Progress in Photovoltaics: Research and Applications*, vol 5, pp309–315
16. Mitchell, R., Witt, C., King, R. and Ruby, D. (2002) 'PVMaT advances in the photovoltaic industry and the focus of future photovoltaic manufacturing R&D', *Proceedings of 29th IEEE Photovoltaic Specialists Conference*, New Orleans, LA, pp1444–1447
17. Hansen, T. N. (2003) 'The promise of utility scale solar photovoltaic (PV) distributed generation', Paper presented at POWER-GEN International, Las Vegas, Nevada, 10 December 2003
18. Zweibel, K. (2003) ' PV as a major source of global electricity', Paper presented at the University of Toledo, 24 February 2004, www.nrel.gov/ncpv/thin_film/docs/zweibel_t_club_feb_04_pv_global_energy_needs.ppt
19. Franklin Associates (1998) *USA LCI Database*, Franklin Associates, Prairie Village, Kansas, US
20. Althaus H.-J., Blaser S., Classen M. and Jungbluth N. (2003) *Life Cycle Inventories of Metals*, Final report of Ecoinvent 2000, vol 10, Swiss Centre for LCI, EMPADU, Dübendorf
21. National Renewable Energy Laboratory (NREL) (2004) *US Life-Cycle Inventory Database Project Data*, NREL, www.nrel.gov/lci
22. Aluminium Association (1998) *Life Cycle Inventory Report for the North American Aluminium Industry*, Aluminium Association, Arlington, VA
23. Energy Information Administration (EIA) (2003) *Annual Energy Review 2003*, DOE/EIA-0384, EIA, www.eia.doe.gov/aer
24. US DOE (1998) *Life Cycle Inventory of Biodiesel and Petroleum Diesel for Use in an Urban Bus*, NREL/SR-580-24089, NREL, Boulder, CO
25. Frischknecht (1996) *Öko-inventare von Energiesystemen, ETH-ESU*, 3rd edition
26. Automotive Engineering (1997) *Progress in Recycling Specific Materials from Automobiles*, Automotive Engineering, Warrendale, PA, August, pp51–53
27. Wang, M. (2001) *GREET Version 1,6*, Center for Transportation Research, Argonne National Laboratory, University of Chicago, Chicago, Illinois
28. Weiss, M. A., Heywood, J. B., Drake, E. M., Schafer, A. and AuYeung, F. F. (2000) *On the Road in 2020: A Life Cycle Analysis of New Automobile Technologies*, Energy Laboratory Report no MIT EL 00-003, Energy Laboratory, Massachusetts Institute of Technology, Cambridge, Massachusetts, www.fee.mit.edu/publications/PDF/el00-003.pdf
29. Environdec (2004) *Environmental Product Declarations*, Swedish Environmental Management Council, Stockholm, Sweden, www.environdec.com
30. Mason, J. E., Fthenakis, V. M., Hansen, T. and Kim, H. C. (2006) 'Energy pay-back and life cycle CO_2 emissions of the BOS in an optimized 3,5 MW PV installation', *Progress in Photovoltaics: Research and Applications*, vol 14, pp179–190
31. EPA (2002) *Joint Verification Statement: The Environmental Technology Verification Program*, Department of Toxic Substance Control, EPA, US, www.epa.gov/etv/library.htm

A.7 PHOTOVOLTAIC AND SOLAR THERMAL SYSTEMS: SIMILARITIES AND DIFFERENCES

A.7.1 Introduction

A photovoltaic (PV) system converts light directly into electricity. By contrast, a solar thermal (ST) system produces heat as an intermediate step before using that to produce electricity. The different physical processes that underlie these two approaches to electricity production lead to differences in energy threshold and response time for the two kinds of systems. For example, a PV panel responds 'instantly' (for all practical purposes) to incident photons and, in many cases, even the lowest of light intensities will suffice to produce some current. On the other hand, the heat transfer fluid in an ST collector will take a finite amount of time until its net intake of solar energy minus its heat losses to its surroundings is sufficient to enable it to produce electricity. If the heat transfer fluid does not reach some minimum temperature threshold, then the solar energy that it will have absorbed will not be convertible to electricity.

One practical consequence of this difference is that ST systems usually require a fossil-fuel back-up system to enable them to start up within a reasonable amount of time in the morning and to shut down safely at night time. Some kind of short-term storage capacity may also be incorporated within the system design in order to obviate the effect of passing clouds, or to shift noontime production peaks closer to late afternoon demand peaks. Of course, PV systems may also incorporate storage for such purposes; but an essential difference is that, unlike ST systems, PV can deliver all the energy that they generate without the need for a back-up system.

There are three reasons why the existence of ST systems should be taken into consideration in any study of the potential of VLS-PV systems. First, ST systems already exist that are an order of magnitude larger than the largest PV systems that have been built.[1] We refer, particularly, to the two 80 MW parabolic trough systems that were constructed by the Luz Corp at Harper Lake, California, during the 1980s. This means that ST more credibly lends itself to a 'VLS-ST' size upgrade than does PV at the present state of the latter's art.

Second, existing ST systems, particularly the five 30 MW Luz plants at Kramer Junction, California (which have probably been more carefully maintained than other ST systems), have proved to be very reliable in their long-term operation, having already clocked up nearly 20 years of operation time.

Third, the annual average efficiency of these parabolic trough plants is typically around 10% – equal to or higher than the efficiency of PV plants of comparable vintage.

For these reasons we need to discuss carefully the advantages and disadvantages of the two kinds of technology in order to decide whether ST or PV should be the basis for very large solar power systems.

A.7.2 Solar thermal (ST) systems

From a conceptual viewpoint, ST systems come in two varieties. One variety consists of 'relatively small' individual units, in which a single solar collector provides heat energy for its own custom-designed power generator. Examples of this type would include a variety of small parabolic dish systems, incorporating Stirling-cycle generators of typically 10 kW electrical power, that were tested in various parts of the world,[1] and up through the 0,25 km^2 solar pond that was constructed by Ormat Corp at the Dead Sea in Israel. The latter provided thermal energy for a specially designed 5 MW low-temperature organic-fluid Rankine-cycle generator.[2]

The second variety of ST systems employs a field of solar collectors in order to provide heat input for a 'relatively large' central power unit of conventional design. Examples here range in size from the fields of heliostats, each with its associated central receiver tower, that were constructed during the 1980s with power outputs of several MW, and up through the previously mentioned parabolic trough systems that were constructed in California to provide heat for steam turbines of up to 80 MW capacity each.

In spite of the large variety of ST technologies that have been tested, so far only one has succeeded in penetrating the commercial power production market: the parabolic trough type. It is therefore this type that we will consider in greater detail, not only because of its technological success, but also because of its extremely efficient use of land area. It is accordingly this type of ST plant that must be considered as the only serious rival to the VLS-PV concept.

A.7.3 Parabolic trough ST systems

The Luz parabolic trough power plants consist of a collector field that provides steam for a conventional turbo-generator. There is also a gas-fired back-up boiler to provide energy during cloudy or rainy days. The collectors actually heat oil rather than water (in order to eliminate the problem of having to support high pressures over several km of pipeline). A heat exchanger then transfers the heat from the oil to water in order to generate steam.

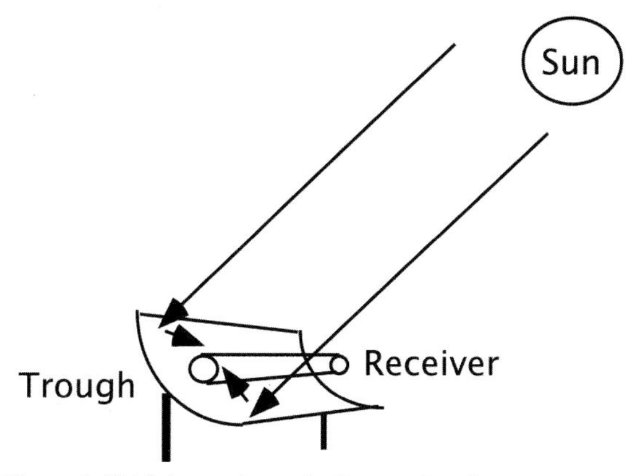

Figure A.124 Schematic parabolic trough collector

Figure A.124 shows a schematic diagram of a collector trough. Each collector unit consists of a linear mirrored trough with a parabolic cross-section. Along the focal line is a receiver tube. In the large Luz power plants, the mirrors rotate from east to west about a horizontal axis that has a north–south orientation. The receiver tube consists of a steel inner tube that is coated with a spectrally selective absorber in order to minimize radiative heat losses. This tube is surrounded by an evacuated (borosilicate) glass tube in order to minimize conductive and convective heat losses.

Long lines of these collectors ensure that, for most of its passage, the heat transfer fluid remains within the collector tubes rather than in external lengths of piping that would be more prone to heat losses. In the case of the 30 MW power plants at Kramer Junction, California, each so-called 'LS-2' collector trough is 47,1 m in length and about 5,0 m in width at its aperture. The 80 MW plants at Harper Lake, California, employed larger 'LS-3' collectors with dimensions of 95,2 m × 5,8 m. The 30 MW 'SEGS5' plant at Kramer Junction, California, has 992 collector troughs of type LS-2, with a total aperture area of 233 000 m^2. It occupies approximately 0,78 km^2 of land.

Detailed computer simulations[3] indicate that ST plants of the parabolic trough variety should have annual average efficiencies of 12 to 13 %. However, owing to a variety of maintenance problems in this first generation of commercial ST plants (for example, broken vacuum tubes and soiled mirrors), actual observed annual average efficiencies were closer to 10 %.

Table A.54 Potential electrical yield of types of VLS-PV technology in the Negev

VLS-PV yield	System type			
	Static, 30° tilt	One-axis tracking	Two-axis tracking	CPV
GWh y^{-1} GW^{-1}	1 644	2 071	2 279	1 754
kWh y^{-1} m^{-2} aperture	189	238	262	385
GWh y^{-1} km^{-2} land area	65,4	69,6	33,2	48,7

A.7.4 Comparison of potential large scale ST and PV plants

In our previous study,[4] we compared the land requirements for various kinds of VLS-PV technologies: specifically, static; one-axis tracking; two-axis tracking; and concentrator photovoltaic (CPV) systems. The results obtained are reproduced in Table A.54.

For this discussion, we draw attention to the bottom row of Table A.54 – namely, annual energy output per km^2 of land usage. For the static case, we calculated a value of 65,4 GWh per year. This assumed a land–aperture requirement of 2,89:1, as appropriate for a 30° tilted system in the Negev.[5] For the one-axis tracking situation, we calculated an annual land productivity value of 69,6 GWh/km^2, based on the land–aperture ratio of 5:3 employed by the 1 MW SMUD plant at Rancho Seco, California. For the two-axis situation, we calculated an annual land productivity value of only 33,2 GWh/km^2, based on the 8:1 land–aperture ratio employed by the 1 MW SCE 'Lugo' plant at Hesperia, California.

When the Lugo plant was designed, it was considered important that the individual trackers would almost never shade one another. However, as indicated in our 'top-down' study, this land–aperture ratio could, in principal, be reduced to 3:1 for an annual shading energy loss of only 10 % *provided that each PV cell has a bypass diode*. Hence, the 33,2 GWh/km^2 value hitherto employed for two-axis tracking could be revised upward to 79,7 GWh/km^2 for specially manufactured PV modules. Finally, the 48,7 GWh/km^2 figure for the annual land productivity of a CPV system should, for the same reason, also be revised upward to 116,9 GWh/km^2. Here, the bypass diode provision is also important; but it has already been incorporated in various early experiments with dense arrays of CPV cells, and would certainly constitute a requirement for future mass-produced CPV modules.

We may now ask the question: how does a parabolic trough ST system compare with PV? Taking an annual average system efficiency of 12,5 % for ST, and the same solar irradiance database[6] that was employed for the PV calculations in Table A.54, the expected annual average output from the ST system would be 262,4 kWh/m^2 aperture. If we now fold in a land–aperture ratio of 10:3, as employed for the Kramer Junction 30 MW plants, we arrive at an annual land productivity value of 78,7 GWh/km^2 for ST systems.

A.7.5 Conclusions

In terms of land productivity, we find that a very large ST power plant of the parabolic trough variety could be expected to be:

- more productive than a static or one-axis VLS-PV system by 20 % and 13 %, respectively;
- comparable in productivity to that of a two-axis tracking VLS-PV plant provided that all cells of the latter are specially manufactured with bypass diodes;
- less productive than a VLS-PV CPV system by 33 %.

These conclusions are based on purely energetic considerations for perfectly working systems. However, it is also important to take two other factors into consideration for a fuller appreciation of their relative advantages and disadvantages.

First, ST systems are relatively easy to match to existing fossil-fuelled power plants since they merely add a field of solar collectors to supplement the plant's heat generation requirements. The actual power-producing components already exist, of course, so no special training is required for the personnel who operate them. This may be counted as an advantage of ST compared to PV.

Second, at the present state-of-the-art for thermal storage, *ST plants require a fossil-fuelled back-up system*. This means that, pending the development of efficient thermal storage, ST is not a technology for introduction into desert areas that do not already have fossil-fuelled plants. This is a clear advantage for PV because, for obvious environmental reasons, it is important not to increase the current global use of fossil fuel.

Thus, in the short term, ST technology may be a method for enabling the existing power plants in sun-rich industrialized states, such as California, to reduce their dependence upon fossil fuel. But, in the longer term, ST will not be able to compete with VLS-PV technology, particularly of the CPV variety.

REFERENCES

1. Winter, C.-J., Sizmann, R. L. and Vant-Hull, L. L. (1991) (eds) *Solar Power Plants: Fundamentals, Technology, Systems, Economics*, Springer-Verlag, Berlin
2. Tabor, H. Z. and Doron, B. (1990) 'The Beith Ha'Arava 5 MW(e) solar pond power plant (SPPP) – Progress report', *Solar Energy*, vol 45, p247
3. Price, H. (2003) *A Parabolic Trough Solar Power Plant Simulation Model*, NREL/CP-550-33209, January
4. Kurokawa, K. (ed) (2003) *Energy from the Desert: Feasibility of Very Large Scale Photovoltaic Power Generation (VLS-PV) Systems*, James and James, London, p149
5. Kurokawa, K. (ed) (2003) *Energy from the Desert: Feasibility of Very Large Scale Photovoltaic Power Generation (VLS-PV) Systems*, James and James, London, p91
6. Faiman, D., Feuermann, D., Ibbetson, P., Zemel, A., Ianetz, A., Israeli, A., Liubansky, V. and Seter, I. (1999) *Data Processing for the Negev Radiation Survey: Fifth Year: Part 3 – Typical Meteorological Year v 2.1*, Israel Ministry of Energy and Infrastructure Publication RD-15-98, Jerusalem, March

A.8 IMPACT ASSESSMENT OF VLS-PV ON THE GLOBAL CLIMATE

A.8.1 Introduction

Our understanding of natural physical processes and their impact has considerably lagged behind the pace of development. However, recent advancements in the field of environmental science have made it possible to assess, mitigate and project the environmental impacts of man-made changes with a fair degree of accuracy. Today, it is well known that the capacity of the Earth and its ecosystems to mitigate the effects of man-made changes has been declining ever since the onset of civilization. Land surface properties, too, have changed markedly during the past several centuries, affecting local and global climate decisively. The genesis of climatological changes in temperate latitudes, for example, is traced to the large scale conversion of forests and grassland into highly productive cropland and pasture. This has influenced, beyond doubt, the interaction of our energy and water balance with the Earth's terrestrial biosphere and atmosphere. An increase in the number and intensity of thunderstorms over the years has been regarded as a direct consequence of an increase in thermal energy derived from sensible heating at the Earth's surface and condensation/freezing of water vapour.

Weather forecasters use a variety of parameters, derived from the vertical profile of thermodynamic variables (local temperature, wind circulation, available potential energy, evaporation, transpiration, etc.), to assess the potential for rainfall. Several studies were undertaken to precisely establish various correlations between different physical processes in the environment, particularly the interaction between vegetation cover and climate. These, by and large, indicate that initial changes in landscape had a cooling effect owing to a gradual increase in albedo (vegetation morphology). However, a turning point in the world's climate has been observed and further land cover changes are likely to produce a warming effect, governed by physiological mechanisms rather than by vegetation morphology, especially in the tropics and sub-tropics. Therefore, the environmental impact of large scale land-use change must be ascertained before embarking upon technologically driven mega projects. Nevertheless, a proper mix of strategies has the potential to mitigate the adverse impact on the environment and to reverse negative consequences.

A.8.2 Early investigations on climatological changes by land surface property

Early investigations in this field by Charney[1] and Charney et al[2] revealed evidence of strong linkages between land surface albedo and atmospheric circulations. Another land surface property – surface roughness – was also found by Deardorff[3] and Sud et al[4] to play an important role in energy and water fluxes to the atmosphere. These developments opened up new approaches and opportunities into understanding land–atmosphere interactions. A plethora of scientific activities followed. These created immense interest in allied investigations, such as linking leaf area index to climate change. Chase et al[5] even used the leaf area index to show that human disturbance of land cover can change atmospheric circulation and affect climate in areas distant from the actual disturbance. Newly developed techniques have been further assimilated into boundary layer modelling. These developments have provided deep insight by Stull,[6] Kain et al,[7] Huang et al[8] and Arya[9] into the linkages between the landscape and deep cumulus convection. Initial forays by Mintz[10] were primarily on land-use change-induced global climate changes. These were followed by investigations on regional or meso-scale changes. Researchers such as Bonan et al,[11] Copeland et al,[12] Stohlgren et al,[13] Liang et al,[14] and Lyons[15] invariably examined the effects of extra-tropical land cover on temperature and hydrology and of land cover change on climate.

However, most recent efforts have undoubtedly concentrated more on the regional effects of tropical deforestation and regional desertification because of the vulnerability of these ecosystems and their importance to human populations (for example, Xue et al,[16] Xiu et al[17] and Berg et al[18]). Gradually, it is being accepted that anthropogenic land cover changes of the type already observed can significantly influence global climate. Chase et al[5] made concerted efforts at simulating and analysing the effects of realistic global green leaf area changes to explain winter climate anomalies. They correlated the phenomenon with tropical deforestation affecting low-latitude convection and global-scale circulations. Zheng and Eltahir[19] further explored and simulated mid-latitude tele-connections between complete tropical deforestation and a linear wave model. They concluded that appropriate conditions existed for the propagation of tropical waves into the extra-tropics. Other significant developments by Sud et al[20] and others included simulation of isolated extra-tropical effects due to physical or physiological changes in tropical vegetation in order to understand biosphere–precipitation relationships. The efforts of Narisma and Pitman[21] to simulate the impact of 200 years of land cover change on the Australian near-surface climate are a significant step forward.

Changes in land surface albedo can result from land-use alterations and thus be tied to an anthropogenic cause. Hansen and Nazarenko[22] estimated that a forcing of –0,4 Wm^{-2} has resulted so far, approximately half of which occurred during the industrial era only. The largest effect is estimated to be at the high latitudes, where snow-covered forests that had a lower albedo have been replaced by snow-covered deforested areas.

Some researchers have pointed out that the albedo of a cultivated field is affected more by a given snowfall than the albedo of an evergreen forest. These simulations were based upon pre-industrial vegetation replaced by current land-use patterns. It was also found that global mean forcing could be –0,21 Wm^{-2}, with the largest contributions coming from deforested areas in Eurasia and North America. Betts et al[23,24] estimated an instantaneous radiative forcing of –0,20 W/m^2 by surface albedo change, replacing natural vegetation by present-day land use.

Kalnay and Cai[25] reported an evaluation of a second influence of land use (an influence other than urbanization) on climate – namely, the role of agriculture and irrigation. They used the reanalysed upper-air data to examine trends over the last 50 years, a method independent of surface-based measurements, to separate urbanization from non-urbanization effects. They concluded, in their follow-up report of corrected measurement, that the trend in daily mean temperature due to land-use changes is an increase of 0,35°C per century.[26]

Thus, landscape pattern and its average condition exert a major control on weather and climate according to Pielke[27,28] at the meso- as well as global scale. The direct surface heating anomalies and subsequent changes in circulation associated with land-use change are permanent. A tele-connection pattern is observed since land-use changes are found to be a major factor, capable of governing the temperature and precipitation distributions worldwide.

Therefore, mega-technological projects, such as the very large scale photovoltaic power generation (VLS-PV) installation in deserts proposed by Kurokawa,[29] are capable of changing entire landscapes. An initial investigation using green area modelling and meso- as well as global-scale modelling indicates that such projects are likely to increase convective rainfall under certain conditions. At this point, the land cover change is manifested as a change in surface greenness or a change in vegetation pattern, not actual VLS-PV installation due to time constraints and data availability. The analysis, basic assumptions and projections are dealt with in the following sections.

A.8.3 Preliminary modelling of VLS-PV installation

Changes in land cover influence climate and, more specifically, the hydrological cycle through the partitioning of incoming solar radiation into turbulent sensible and latent heat fluxes, as shown in Figure A.125. The energy and moisture budgets at the surface can be written as Equations A.20 and A.21 according to Pielke[27] as follows:

$$R_N = Q_G + H + L\,(E + T) \qquad [A.20]$$

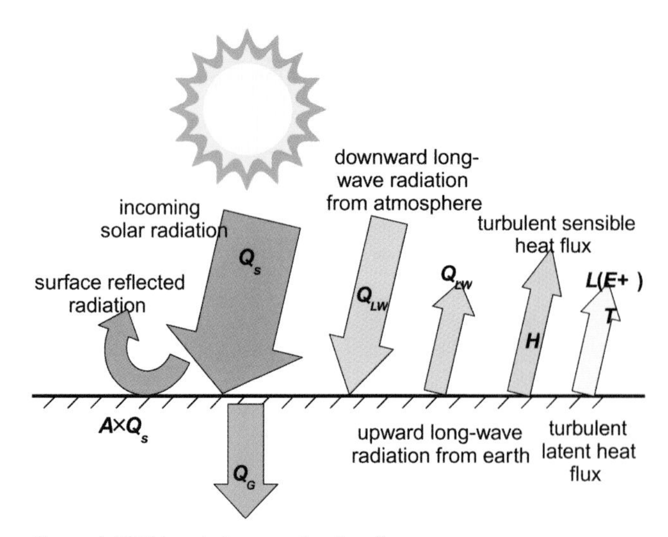

Figure A.125 Heat balance at land surface

$$R_N = Q_S\,(1 - A) + Q_{LW}\,(\downarrow) - Q_{LW}\,(\uparrow) \qquad [A.21]$$

where R_N represents the net radiative fluxes; Q_s is insolation; A is albedo; Q_G is the soil heat flux; H is the turbulent sensible heat flux; $L(E + T)$ is the turbulent latent heat flux; L is the latent heat of vaporisation; E is evaporation (conversion of liquid water from the soil surface into water vapour); T is transpiration (phase conversion to water vapour through stoma on plants); $Q_{LW}\,(\downarrow)$ is downwelling long-wave radiation; $Q_{LW}\,(\uparrow)$ is upwelling long-wave radiation.

The water balance at the land surface, as illustrated in Figure A.126, is:

$$P = E + T + RO + I \qquad [A.22]$$

where P is the precipitation; RO is runoff; and I is infiltration.

$Q_{LW}\,(\uparrow)$ and $Q_{LW}\,(\downarrow)$ in Equation A.21 are specified as follows:

$$Q_{LW}\,(\uparrow) = (1 - \varepsilon)\,Q_{LW}\,(\downarrow) + \varepsilon\,\sigma\,T_s^4 \qquad [A.23]$$

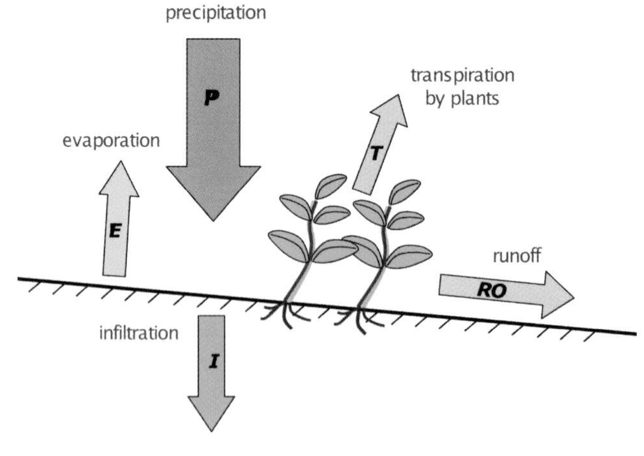

Figure A.126 Water balance at land surface

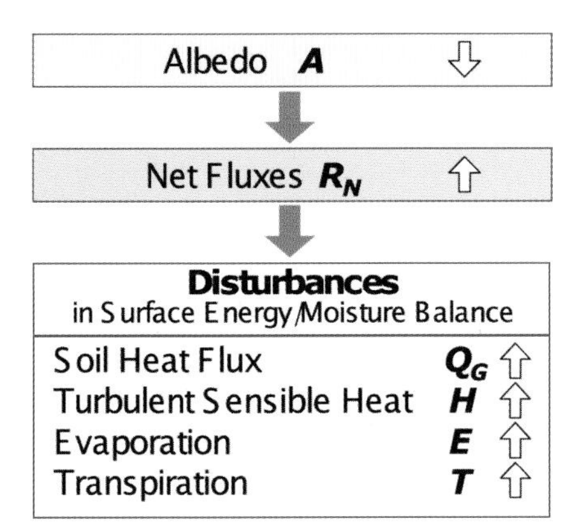

Figure A.127 An example of land surface disturbances induced by a decrease in albedo

where ε is the surface emissivity; σ is Stefan's constant; and T_s is the surface temperature.

Any land-use change that alters one or more of the variables in Equations A.20, A.21 or A.22 will directly affect the potential for rainfall. For example, as illustrated in Figure A.127, a decrease in albedo (A) would increase R_N, thus making more heat energy available for Q_G, H, E and T. If the surface is dry and bare, all of the heat energy would go into Q_G and H. The heat that goes into H increases the temperature potential. Thus, changes in the Earth's surface can induce significant disturbances in the surface energy and moisture budgets. These would influence the heat and moisture fluxes within the planetary boundary layer, convective available potential energy (CAPE) and other fundamentals of deep cumulus cloud activity.

A.8.4 Choice of model

The standard version of the National Centre for Atmospheric Research (NCAR) CCM3, coupled with the Land Surface Model (LSM) by Bonan[30] is used for simulation in the current investigation. The LSM is a general circulation model (GCM)-scale parameterization of atmospheric–land surface exchanges, and accounts for vegetation properties as functions of 1 of 24 basic vegetation types. The LSM includes a single-level canopy (the lake model) and calculates averaged surface fluxes due to sub grid-scale vegetation types and hydrology. Two phonological properties, leaf area index (LAI) and stem area index (SAI), are interpolated between prescribed monthly values. Albedos are calculated as a function of LAI and SAI among other factors. All other vegetation properties are considered to be seasonally constant.

The Community Land Model (CLM) was developed as a collaborative venture between various land modelling groups and combines the best features of three well-documented and modular land models, the LSM of Bonan,[31] the Biosphere–Atmosphere Transfer Scheme (BATS) of Dickinson et al[32] and the 1994 version of the Chinese Academy of Sciences, Institute of Atmospheric Physics LSM (IAP94) by Dai and Zeng[33] and Zeng et al.[34] CLM is preferred over other land modelling schemes such as RAMS by Cotton et al,[35] VIC and NOAH because CLM makes better use of a two-leaf canopy model by Dai et al.[36] CLM has also better soil depth parameterization (soil temperature). It simulates using a heat diffusion equation in ten soil layers, which is important as feedback and memory of the water table, because 'deep' ground processes are important to the surface water and energy budgets. It has a thinner and more realistic top soil layer (1,75 cm), which gives more accurate results for the estimation of surface soil water fluxes and the diurnal cycle of surface soil temperature. An improved annual cycle of runoff water and, consequently, evapotranspiration (because of improved soil depth parameterizations) is a significant improvement over other models.

A.8.5 Methodology

It is assumed that the PV system covers the land surface horizontally and not inclined at any angle for the entire area of the Sahara, Gobi and Thar deserts. The net energy flux at the ground surface is the sum of beam and diffuse solar radiation and long wavelength radiation from the Earth and its atmosphere, specified by Equation A.21. The sensible heat flux and latent heat flux were calculated by the following equations:

$$H = \rho_{atm} C_p (\theta_{atm} - T_g)/r_{ah} \quad [A.24]$$

where ρ_{atm} is air density; C_p is the specific heat of the air; θ_{atm} is potential temperature; T_g is ground temperature; and r_{ah} is aerodynamic resistance to sensible heat flux and water vapour transfer between atmosphere at height z_1. The potential temperature is again a function of ambient temperature, surface pressure and atmospheric pressure. The latent heat flux can be obtained from the following equation:

$$\lambda E = \rho_{atm} C_p (e_{atm} - e^*(T_g)/ \gamma (r_{aw} - r_{srf}) \quad [A.25]$$

where r_{aw}, r_{srf} are aerodynamic resistances to sensible heat flux and water vapour transfer between the atmosphere at height z_2 and z_3; e_{atm} is vapour pressure; $e^*(T_g)$ is saturation vapour pressure at ground surface; and γ is psychrometric constant.

A CLM was run for one year in two cases of the assumed albedo for the whole area of the Gobi Desert ($162,8 \times 10^4$ km^2), the Sahara Desert (743×10^4 km^2) and the Thar Desert ($44,3 \times 10^4$ km^2). In accordance with the ordinary ranges of ground albedo, as shown in Table A.55, assumed values were specified for this preliminary study (see Table A.56).

Table A.55 Ordinary ranges of ground albedo

Type	Albedo
Soil	0,05 ~ 0,4
Desert	0,2 ~ 0,45
New snow	0,9
Forest	0,1 ~ 0,2

Table A.56 Assumed ground albedo for case studies

	Visible radiation	Infrared radiation
Case A	0,20	0,20
Case B	0,10	0,20

In case B, albedo in the visible radiation range is assumed to be affected by the full installation of a large number of VLS-PV systems over the Gobi, Sahara and Thar deserts as an extreme case. The input global data was obtained from the CCM3 website (www.cgd.ucar.edu/cms/ccm3/) to run the model in off-line mode for one-year projections.

A.8.6 Model results and discussion

The surface albedo is of utmost importance in determining the absorption of solar energy, although large variations are possible due to the type and nature of the vegetation. Albedo also generally decreases as the surface wetness increases. The range of albedo for snow-free conditions is from about 0,1 for tropical forest to about 0,4 for some dry sandy surfaces. With incident mean daily solar fluxes typical of the tropics (300 to 400 Wm^{-2} in cloudless conditions), a variation of about 100 Wm^{-2} is possible. However, this would be an upper limit to spatial variations and is only conceivable at one place with an extreme climatic change or extensive human intervention, such as PV installation, deforestation or irrigation on highly reflective soil.

VLS-PV reduces albedo considerably compared to dry sandy desert soils. Therefore, it is invariably supposed to increase sensible heat flux (H) and surface heat flux (Q_G) substantially. However, complex interactions between various parameters seem to undermine this hypothesis according to the results obtained from a preliminary study of VLS-PV installation in the Thar, Gobi and Sahara deserts. This is very positive indication for VLS-PV installation. Initial results for single-year simulation of CLM indicate a very small increase in the global sensible heat flux (H) and surface heat flux (Q_G). This implies that installation of VLS-PV in deserts would have a positive radiative force due to surface albedo change replacing normal desert ecosystems.

The projected increase in daily global mean temperature due to this land-use change in the Sahara, Gobi and Thar deserts is approximately 0,001°C in one year. Considering that the Gobi Desert can supply enough world energy for the world's needs and the Sahara Desert can supply ten times as much as the world needs, this case study can be considered sufficient in demonstrating the capacity of VLS-PV to fulfil global requirements.

This rise of 0,001°C in one year can be translated into a global mean temperature rise of 0,1°C in one century if this trend is considered to be not in thermal equilibrium for at least one century. As reviewed previously, Kalnay and Cai concluded that the trend in daily mean temperature due to land-use changes, including urbanization and non-urbanization, over the last 50 years is an increase of 0,35°C per century.[26] The results in this preliminary study are similar.

The most significant finding of this study is that a proper green area modelling, coupled with VLS-PV, would have a positive impact on convective rainfall, rather than a negative impact due to a rise in sensible and surface heat flux. These green areas would further decrease dust storms, reduce CO_2 emissions and mitigate several ill effects of fossil-fuel energy use. They would attract monsoons due to evapo-transpiration characteristics and would regulate temperature. Although this study is based on land cover change, manifested as a change in surface greenness or a change in vegetation patterns, more realistic assumptions and meso-scale modelling with longer periods of simulation are needed to reveal the actual impact of VLS-PV on the global climate.

REFERENCES

1. Charney, J. G. (1975) 'Dynamics of deserts and drought in the Sahara', *Quarterly Journal of the Royal Meteorological Society*, vol 101, pp193–202
2. Charney, J. G., Quirk, W. J., Chow, S. H. and Kornfield, J. (1977) 'A comparative study of the effects of albedo change on drought in semi arid regions', *Journal of Atmospheric Science*, vol 34, pp1366–1385
3. Deardorff, J. W. (1972) 'Parameterization of the planetary boundary layer for use in general circulation models', *Monthly Weather Review*, vol 100, pp83–106
4. Sud, J. C. and Smith, W. E. (1984) 'Ensemble formulation of surface fluxes and improvement in evapotranspiration and cloud parameterization in a GCM', *Boundary Layer Meteorology*, vol 29, pp185–210
5. Chase, T. N., Pielke, R. A., Kittel, T. G. F., Nemani, R. and Running, S. W. (1996) 'The sensitivity of a general circulation model to large-scale vegetation changes', *Journal of Geophysical Research*, vol 101, pp7393–7408
6. Stull, R. B. (1988) *An Introduction to Boundary Layer Meteorology*, Kluwer Academic Publishers, Dordrecht
7. Kain, J. S. and Fritsch, J. M. (1993) 'Convective parameterization for mesoscale models: Kain–Fritsch scheme', in A. Emanuel and D. J. Raymond (eds) *The Representation of Cumulus Convection in Numerical Models*, American Meteorological Society, Boston, MA, pp165–170
8. Huang, X., Lyons, T. J. and Smith, R. C. G. (1995) 'Meteorological impact of replacing native perennial

vegetation with annual agricultural species', in J. D. Kalma and M. Sivapalan (eds) *Scale Issues in Hydrological Modelling*, Advanstar Communications, Chichester, UK, pp401–410
9. Arya, S. P. (2001) *Introduction to Meteorology*, Academic Press, San Diego, California
10. Mintz, Y. (1984) 'The sensitivity of numerically simulated climates to land–surface boundary conditions', in J. Houghton (ed) *The Global Climate*, Cambridge University Press, Cambridge, UK, pp79–105
11. Bonan, G. B., Pollard, D. and Thompson, S. L. (1992) 'Effects of boreal forest vegetation on global climate', *Nature*, vol 359, p716
12. Copeland, J. H., Pielke, R. A. and Kittel, T. G. F. (1996) 'Potential climatic impacts of vegetation change: A regional modeling study', *Journal of Geophysical Research*, vol 101, pp7409–7418
13. Stohlgren, T. J., Chase, T. N., Pielke, R. A., Kittel, T. G. F. and Baron, J. (1998) 'Evidence that local land use practices influence regional climate and vegetation patterns in adjacent natural areas', *Global Change Biology*, vol 4, pp495–504
14. Liang, X. and Xie, Z. (2001) 'A new surface runoff parameterization with sub grid-scale soil heterogeneity for land surface models', *Advances in Water Resources*, vol 24, pp1173–1193
15. Lyons, T. J. (2002) 'Clouds prefer native vegetation', *Meteorology and Atmospheric Physics*, vol 80, pp131–140
16. Xue, Y. and Shukla, J. (1993) 'The influence of land surface properties on Sahel climate. Part I: Desertification', *Journal of Climate*, vol 6, pp2232–2245
17. Xiu, A. and Pleim, J. E. (2001) 'Development of a land surface model, Part I: Application in a mesoscale meteorological model', *Journal of Applied Meteorology*, vol 40, pp192– 209
18. Berg, A. A., Famiglietti, J. S., Walker, J. P. and Houser, P. R. (2003) 'Impact of bias correction to reanalysis products on simulations of North American soil moisture and hydrological fluxes', *Journal of Geophysical Research*, vol 108 (D16), p4490
19. Zheng, A. and Eltahir, E. A. B. (1997) 'The response to deforestation and desertification in a model of West African monsoons', *Geophysical Research Letters*, vol 24, pp155–158
20. Sud, Y. C., Lau, K. M., Walker, G. K. and Kim, J. H. (1995) 'Understanding biosphere precipitation relationships: Theory, model simulations and logical inferences', *Mausam*, vol 46, pp1–14
21. Narisma, G. T. and Pitman, A. J. (2003) 'The impact of 200 years' land cover change on the Australian near-surface climate', *Journal of Hydrometeorology*, vol 4, pp424–436
22. Hansen, J. and Nazarenko, L. (2003) 'Soot climate forcing via snow and ice albedos', *Proceedings of the National Academy of Science*, vol 101, pp423–428
23. Betts, R. A., Cox, P. M., Lee, S. E. and Woodward, F. I. (1997) 'Contrasting physiological and structural vegetation feedbacks in climate change simulations', *Nature*, vol 387, pp796–799
24. Betts, A. K., Viterbo, P., Beljaars, A., Pan, H.-L., Hong, S.-Y., Goulden, M. and Wofsy, S. (1998) 'Evaluation of land–surface interaction in ECMWF and NCEP/NCAR reanalysis models over grassland (FIFE) and boreal forest (BOREAS)', *Journal of Geophysical Research*, vol 103 (D18), pp23079–23085
25. Kalnay, E. and Cai, M. (2003) 'Impact of urbanization and land-use change on climate', *Nature*, vol 423, pp528–531
26. Kalnay, E. and Cai, M. (2003) 'Impact of urbanization and land-use change on climate – Corrigenda', *Nature*, vol 425, p102
27. Pielke, R. A., Sr. (2001) 'Influence of the spatial distribution of vegetation and soils on the prediction of cumulus convective rainfall', *Reviews of Geophysics*, vol 39, pp151–177
28. Pielke, R. A., Sr. (2002) *Mesoscale Meteorological Modeling*, 2nd edition, Academic Press, San Diego, California
29. Kurokawa, K. (ed) (2003) *Energy from the Desert: Feasibility of Very Large Scale Photovoltaic Power Generation (VLS-PV) Systems*, James and James, London
30. Bonan, G. B., Oleson, K. W., Vertenstein, M., Levis, S., Zeng, X., Dai, Y., Dickinson, R. E. and Yang, Z. L. (2002) 'The land surface climatology of the Community Land Model coupled to the NCAR Community Climate Model', *Journal of Climate*, vol 15, pp3123–3149
31. Bonan, G. B. (1996) *The NCAR Land Surface Model (LSM version 1.0) Coupled to the NCAR Community Climate Model*, Technical Report NCAR/TN-429, NCAR, Boulder, CO
32. Dickinson, R. E., Henderson-Sellers, A. and Kennedy, P. J. (1993) *Biosphere Atmosphere Transfer Scheme (BATS) Version 1e as Coupled to the NCAR Community Climate Model, NCAR Technical Note*, NCAR, Boulder, CO
33. Dai, Y. and Zeng, Q. (1997) 'A land surface model (IAP94) for climate studies, Part I: Formulation and validation in off-line experiments', *Advances in Atmospheric Science*, vol 14, pp443–460
34. Zeng, X. B., Dickinson, R. E., Barlage, M., Dai, Y., Wang, G. and Oleson, K. (2005) 'Treatment of under-canopy turbulence in land models', *Journal of Climatology*, vol 18, no 23, pp5086–5094
35. Cotton, W. R., Pielke, R. A., Sr., Walko, R. L., Liston, G. E., Tremback, C. J., Jiang, H., McAnelly, R. L., Harrington, J. Y., Nicholls, M. E., Carrió, G. G. and McFadden, J. P. (2003) 'RAMS 2001: Current status and future directions', *Meteorology and Atmospheric Physics*, vol 82, pp5–29
36. Dai, Y., Dickinson, R. E. and Wang, Y. P. (2004) 'A two-big-leaf model for canopy temperature, photosynthesis, and stomatal conductance', *Journal of Climate*, vol 17, no 12, pp2281–2299

APPENDIX B

Future possibilities towards world major energy

B.1 LONG-TERM PERSPECTIVE OF ENERGY AND RENEWABLES

B.1.1 Introduction

Since the Industrial Revolution in the 18th century, world energy demand has been increasing in proportion to economic and population growth. The current worldwide primary energy supply is almost three times as much as that in the 1960s. Figure B.1 shows the trends in primary energy supply by fuels.[1] Fossil fuels, such as coal, oil and natural gas, have been the main energy resources so far.

Fossil fuels are exhaustible, however, and cannot be used forever. Table B.1 shows the latest statistics on proven fossil fuel and uranium reserves.[1,2] The concerns for fossil fuel depletion were heatedly discussed after the two oil crises during the 1970s. Renewable energy is expected to be the one possible and reliable option to overcome these problems. Renewable energy also has an advantage in terms of global warming issues. However, the current share of renewable energy in the primary energy supply is still negligible except for hydropower, although renewable energy technologies have been growing drastically.

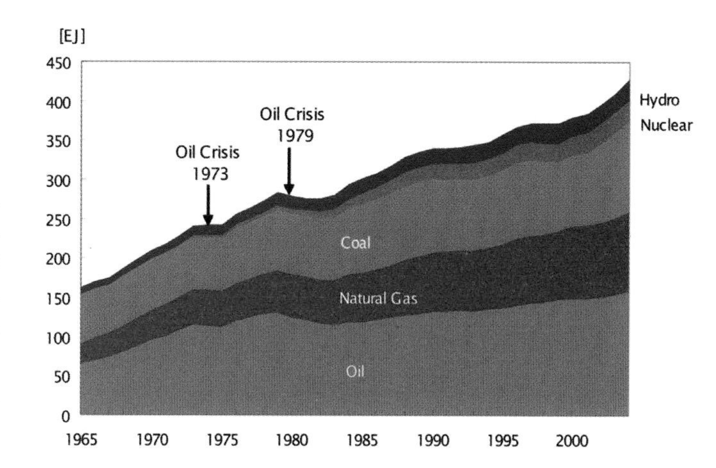

Figure B.1 Trends in total energy consumption by fuel

B.1.2 Long-term future energy perspective

With regard to the world's future energy resources, many reports have been published using different assumptions and models.

B.1.2.1 *World Energy Outlook, 2004*

In the International Energy Agency (IEA) report[3] *World Energy Outlook, 2004*, the world primary energy

Table B.1 Proved reserves of fossil fuels and uranium

	Oil[1] (billion bbl)	Natural gas[1] (10^3 billion m^3)	Coal[1] (billion tonnes)	Uranium[2] (million tonnes)
Proved reserves	1 188,6	179,5	909,1	3,95
North America	5,1 %	4,1 %	28,0 %	17,8 %
Central and South America	8,5 %	4,0 %	2,2 %	6,3 %
Europe and Eurasia	11,7 %	35,7 %	31,6 %	2,8 %
Middle East	61,7 %	40,6 %	5,6 %	23,0 %
Africa	9,4 %	7,8 %	0,0 %	18,7 %
Asia-Pacific	3,5 %	7,9 %	32,7 %	31,4 %
Annual production	29,3	2,7	5,5	0,035
R/P (years)	40,5	66,7	164	64,2*

Note: * Reserves/production ratio (R/P) of uranium is obtained by dividing its proven reserve by annual demand (0,062 million tonnes) because annual production is less than annual demand.

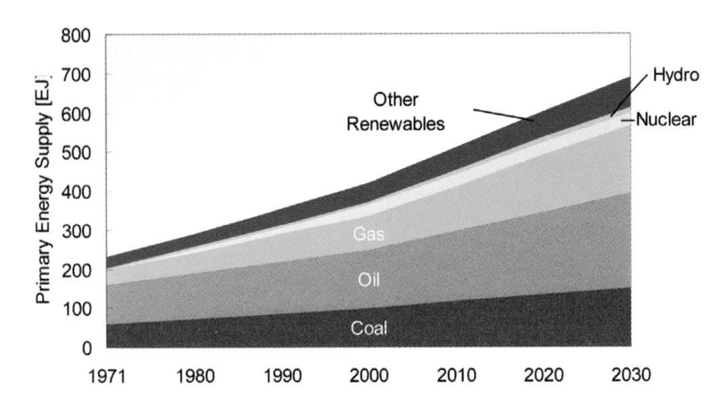

Figure B.2 World primary energy supply projection by fuel (*World Energy Outlook*), 1971–2030

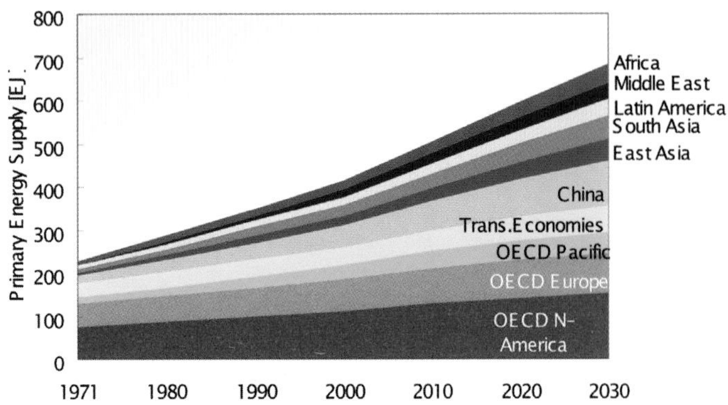

Figure B.3 World primary energy supply projection by region (*World Energy Outlook*), 1971–2030

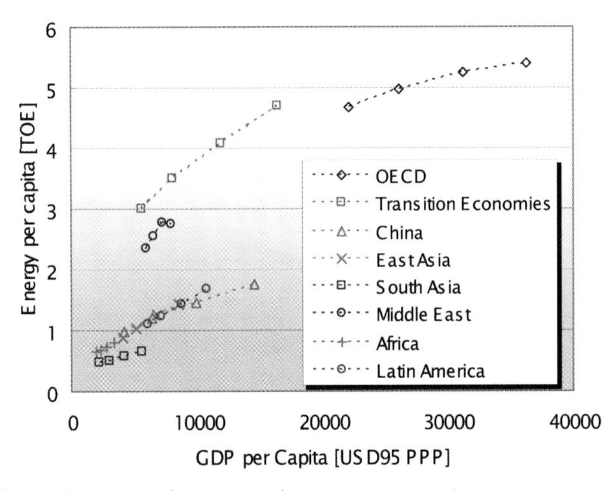

Figure B.4 Regional energy and economy per capita

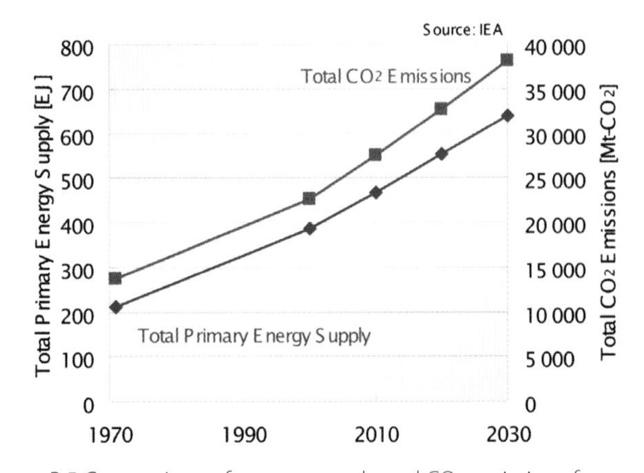

Figure B.5 Comparison of energy supply and CO_2 emissions from 1970–2030

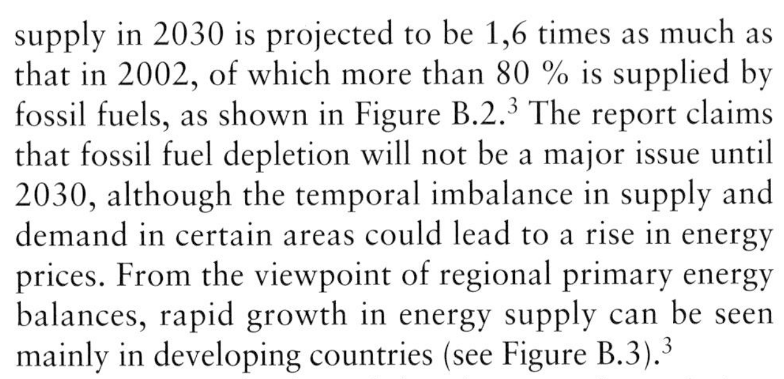

supply in 2030 is projected to be 1,6 times as much as that in 2002, of which more than 80 % is supplied by fossil fuels, as shown in Figure B.2.[3] The report claims that fossil fuel depletion will not be a major issue until 2030, although the temporal imbalance in supply and demand in certain areas could lead to a rise in energy prices. From the viewpoint of regional primary energy balances, rapid growth in energy supply can be seen mainly in developing countries (see Figure B.3).[3]

It is commonly claimed that there is a close relationship between energy consumption and economic growth. Figure B.4[3,4] illustrates relationships between energy demand per capita and gross domestic produce (GDP) by region (power purchasing party, or PPP). As shown in Figure B.4, Organisation for Economic Co-operation and Development (OECD) countries, mainly consisting of developed countries, have been consuming more energy per capita than any other regions. These countries are expected to achieve economic growth with relatively small increases in energy consumption in the future. Energy consumption of developing countries will continue to increase with their economic growth even beyond 2030.

In addition to fossil fuel depletion, global warming is another urgent issue related to energy that needs to be addressed. It is generally recognized that global warming is mainly caused by CO_2 emissions from fossil fuel consumption, as shown in Figure B.5.[3]

B.1.2.2 *International Energy Outlook (*Energy Information Administration)

The Energy Information Administration (EIA) is an official US organization that specializes in energy-related statistics. The EIA publishes forecast-type future energy scenarios every year[5] with several intervention scenarios. According to the latest scenario from the organization, world energy consumption is projected to increase by 57 % from 2002 to 2025, reaching 679 EJ in the reference case (see Figure B.6).[5]

In terms of fossil fuels, total energy consumption of coal, oil and natural gas in the projected period is expected to increase from 318 EJ in 2002 to 592 EJ in 2025. Specifically, the projection for oil consumption is slightly lower than that of the previous year,[4] mainly because of the unexpected rise in oil prices these days (see Figure B.7).[5]

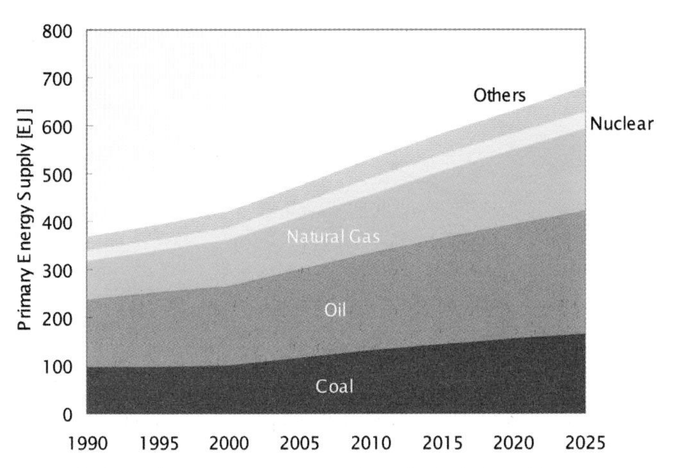

Figure B.6 World primary energy supply by fuels (International Energy Outlook reference case), 1990–2025

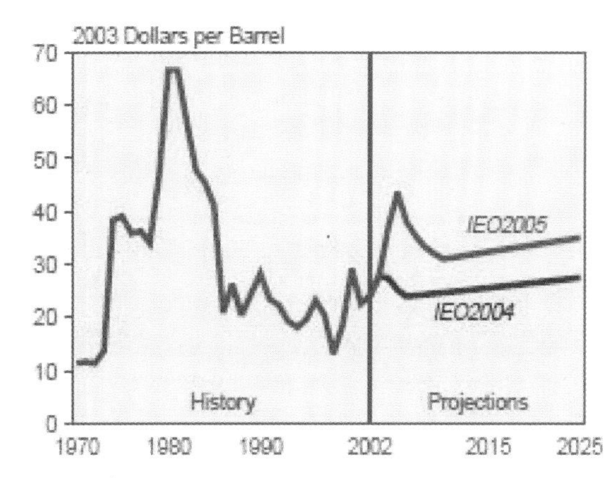

Figure B.7 Oil price projections, 1970–2025

B.1.2.3 *World Energy, Technology and Climate Policy Outlook* (European Commission)

The energy scenario from the European Commission (EC) is another future scenario with great authority.[6] According to the *World Energy, Technology and Climate Policy Outlook*, the total primary energy supply in 2030 will be approximately 721 EJ in the reference case (see Figure B.8).[7]

In this scenario, fossil fuels still dominate in the future. It is obvious that some actions are needed to avoid serious impacts related to fossil fuel depletion and global warming. As an intervention scenario, a carbon abatement case (CA case) is also developed in the study. The impact of a regional carbon tax or carbon trading scheme on total energy consumption is analysed. The results from the case study show that an 11 % reduction in total energy consumption is possible with assumed carbon prices.

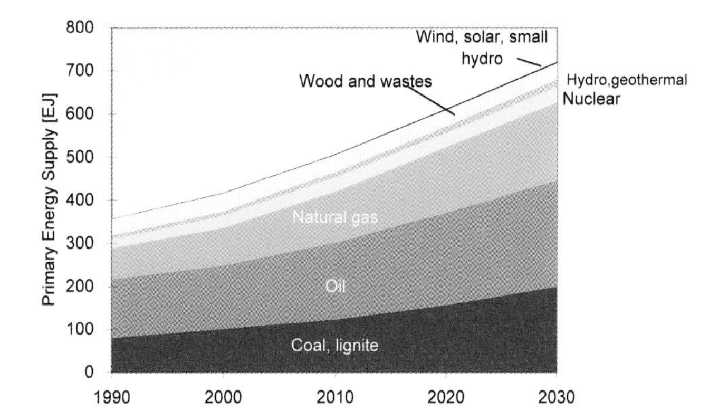

Figure B.8 World primary energy supply by fuel (*World Energy, Technology and Climate Policy Outlook* reference scenario), 1990–2030

B.1.2.4 *Energy to 2050: Scenarios for a Sustainable Future* (IEA)

The *World Energy Outlook* scenario presented earlier is a forecast scenario, which is often called a business-as-usual scenario. In this type of scenario, future energy demand and supply are calculated based on assumptions with high probability. Since the results from this report show that the future energy conditions are not desirable, another study[5] was carried out and an alternative scenario (the sustainable development scenario) was developed. The aim of the scenario is to discuss policy actions and strategies needed to be taken for a better future. Figure B.9[5] illustrates the primary energy supply of the sustainable development scenarios presented in the report.

In *Energy to 2050: Scenarios for a Sustainable Future*, the contribution of other renewables is 18,9 % (34,6 % with biomass) in 2050, while the share of coal, oil and natural gas decreases (9,8 %, 17,9 % and 26,4 %, respectively). Although the primary energy demand

Table B.2 Comparison of results between reference and carbon abatement (CA) cases

	Unit	1990	Reference case 2030	CA case 2030	Percentage change
CO_2 emissions	($GtCO_2$)	20,8	44,5	35,3	–21
Total consumption	(Gtoe)	8,7	17,1	15,2	–11
Coal, lignite	(Gtoe)	2,2	4,7	2,7	–42
Oil	(Gtoe)	3,1	5,9	5,4	–8
Natural gas	(Gtoe)	1,7	4,3	4,3	0
Nuclear	(Gtoe)	0,5	0,9	1,2	36
Renewable	(Gtoe)	1,1	1,4	1,8	35

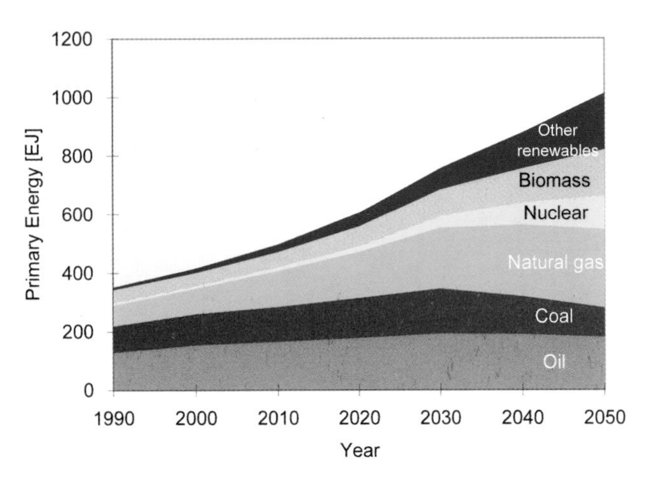

Figure B.9 World primary energy supply by fuel (sustainable development scenario), 1990–2050

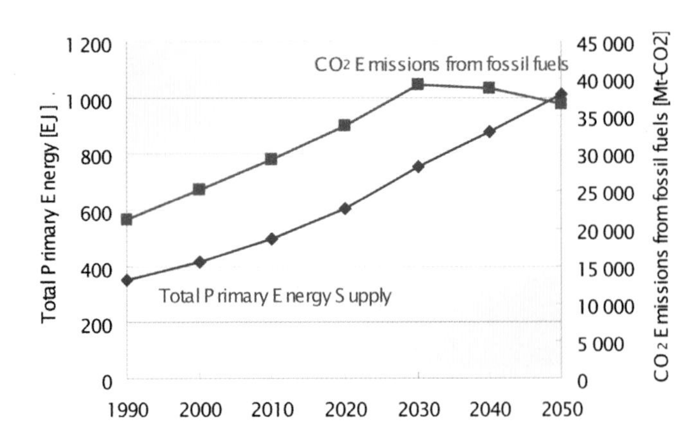

Figure B.10 World primary energy supply and CO_2 emissions (sustainable development scenario), 1990–2025

increases continuously, world CO_2 emissions decrease after 2030 as the share of alternative energy increases (see Figure B.10).[5]

B.1.2.5 *Energy Needs, Choices and Possibilities* (Shell)

In Shell's report, two different scenarios are developed and paths to sustainable energy systems are explored from a different approach.[8]

The first scenario is called *dynamic as usual* and the characteristics of the society are expressed as intense competition, conflicting interests, a wide rage of maturing and emerging technologies, etc. It is assumed that oil becomes scarce in 2040 and the share of renewable energy reaches one third of world energy in 2050. Regarding solar energy, solar photovoltaic (PV) would attract consumers with rapid cost reductions after the commissioning of a 200 MW solar PV plant in 2007.

The second scenario is called *spirit of the coming age*. The scenario assumes the development of fuel cells from an early stage and a decrease in the need for oil for transportation. Many large scale renewable plants are built in order to supply hydrogen for fuel cell vehicles.

B.1.2.6 *World in Transition towards Sustainable Energy Systems* (German Advisory Council on Global Change)

The scenario presented from the German Advisory Council on Global Change (WBGU) gives a clue to turning the energy system towards sustainable systems.[9] The report concludes that transforming the energy system over the next 100 years is feasible in technological and economical terms if appropriate actions are taken immediately. The four key components presented in the report in order to achieve sustainability are as follows:

1 major reduction in the use of fossil energy sources;
2 phase-out of the use of nuclear energy;
3 substantial development and expansion of new renewable energy sources, notably solar; and
4 improvement of energy productivity far beyond historical rates.

In the scenario, the share of renewable energy reaches approximately 83,6 % in 2100. It is particularly worth noting that the contribution of solar electricity becomes more than 64 % and the energy supply from solar energy is much more than the total energy supply in

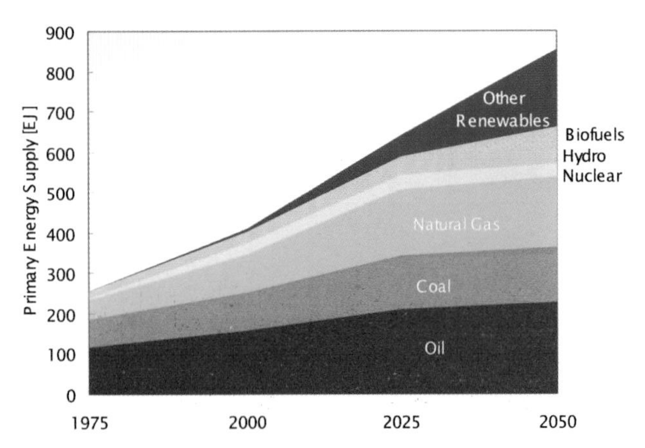

Figure B.11 World energy supply by fuel (Shell, dynamic as usual), 1975–2050

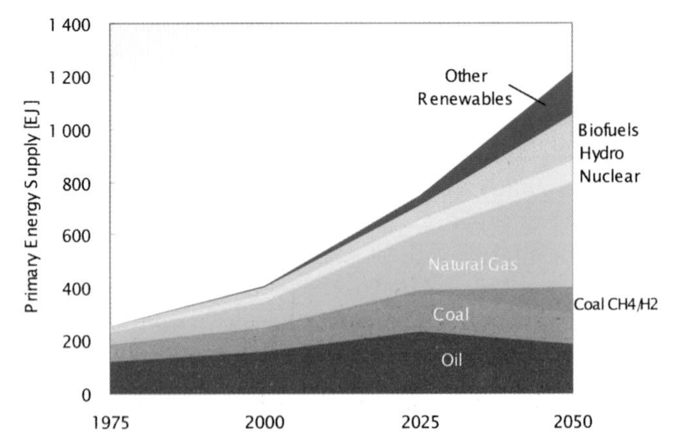

Figure B.12 World energy supply by fuel (Shell, spirit of the coming age), 1975–2050

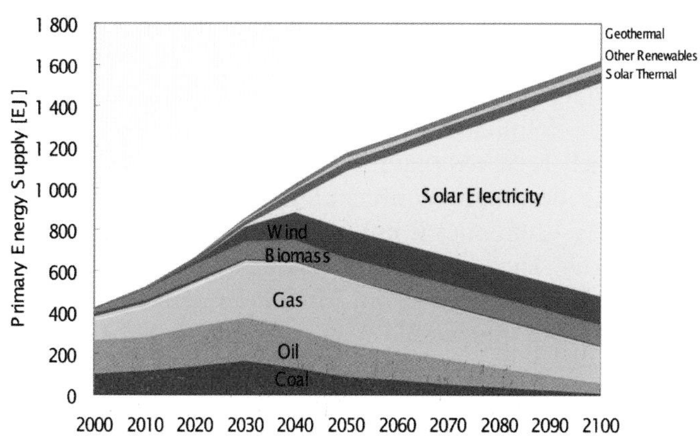

Figure B.13 World primary energy supply by fuel (German Advisory Council on Global Change), 2000–2100

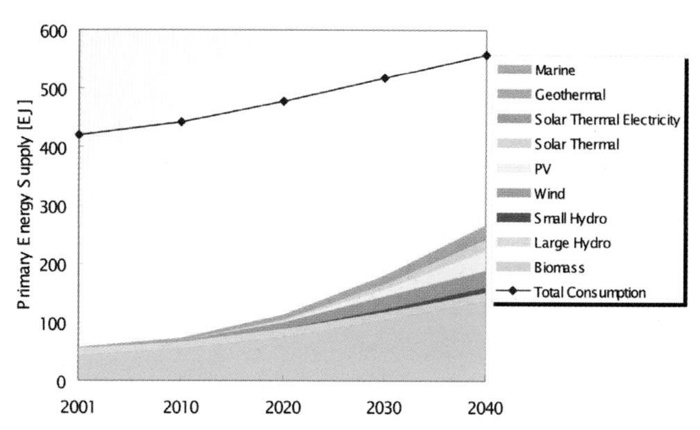

Figure B.15 World renewable energy supply by fuel (European Renewable Energy Council), 2001–2040

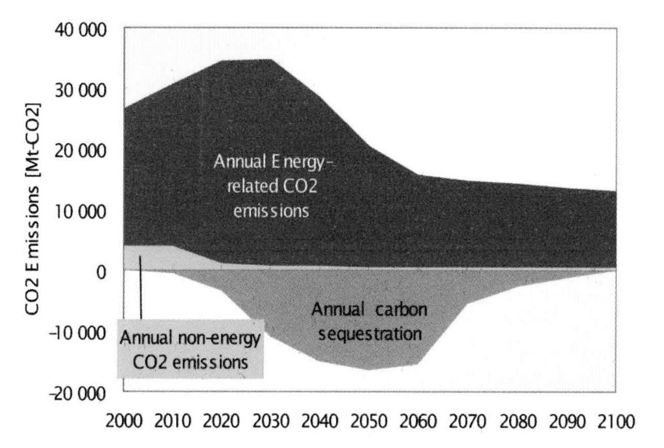

Figure B.14 Annual CO_2 emissions and carbon sequestration (German Advisory Council on Global Change), 2000–2100

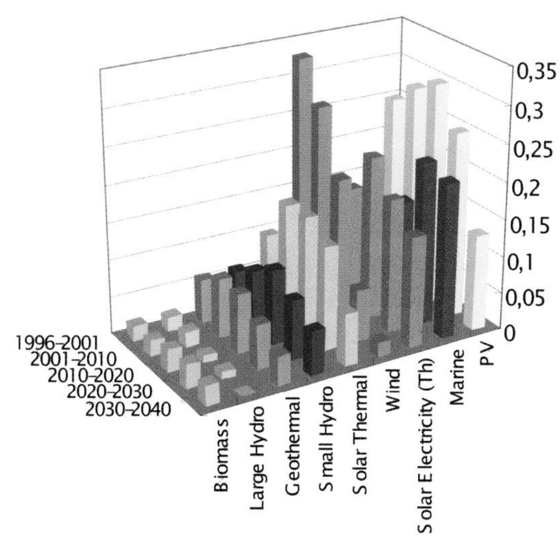

Figure B.16 Growth in renewable energy technologies in each period

2000. With this great contribution of renewable energy, energy-related CO_2 emissions decrease drastically (see Figure B.14).[9]

B.1.2.7 *Renewable Energy Scenario to 2040* (European Renewable Energy Council)

The European Renewable Energy Council (EREC) also advocates the need for renewable energy and claim that 50 % of energy could be supplied by renewables. The report also insists that appropriate policy measures are needed in order to achieve this goal. The contribution of each renewable source is shown in Figure B.15 and the growth in renewable energy technologies in each period is given in Figure B.16.

As often discussed among researchers, the intermittency of some renewable energy technologies, such as solar PV and wind power, could be major constraints when the shares of these technologies are significant. However, *Renewable Energy Scenario to 2040* maintains that other renewable energy sources with stable outputs, such as small hydro or marine technologies, will complement the system by that time.

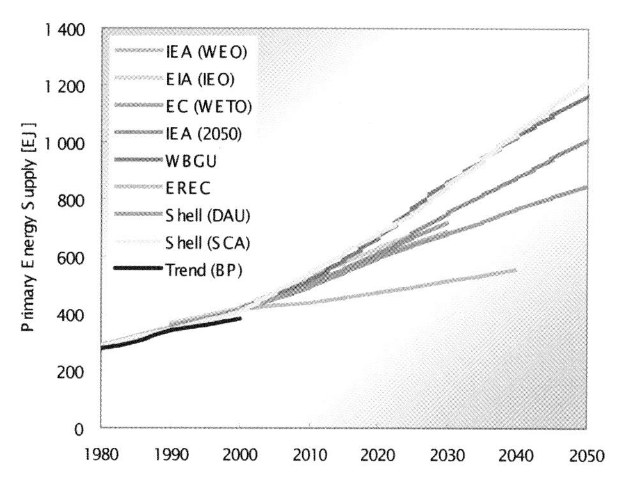

Figure B.17 Comparison of world energy scenarios, 1980–2050

B.1.3 Conclusions

Figure B.17 compares the future worldwide primary energy supply of each scenario introduced in this section. As shown, future energy demand will vary depending upon social and economic conditions. However, it is obvious that world energy demand in the next 50 years will continue to increase, and solar PV technology could contribute to energy issues as a clean and substantial energy source.

REFERENCES

1. BP (2005) *BP Statistical Review of World Energy*, BP, www.bp.com
2. OECD/NEA (1999) *IAEA Uranium 1999*, OECD, Paris, France
3. IEA (2004) *World Energy Outlook, 2004*, IEA, Paris, France
4. IEA (2004) *Key World Energy Statistics from the IEA*, IEA, Paris, France
5. EIA (2005) *International Energy Outlook, 2005*, EIA, www.eia.doe.gov/oiaf/ieo/index.html
6. EC (2003) *World Energy, Technology and Climate Policy Outlook*, EC, Brussels
7. IEA (2003) *Energy to 2050: Scenarios for a Sustainable Future*, IEA, Paris, France
8. Shell (2001) *Energy Needs, Choices and Possibilities*, Shell, www.cleanenergystates.org/CaseStudies/Shell_2050.pdf
9. WBGU (2003) *World in Transition towards Sustainable Energy Systems*, WBGU, Berlin, Germany
10. EREC (2004) *Renewable Energy Scenario to 2040*, EREC, Brussels

B.2 INTEGRATING VLS-PV SYSTEMS WITHIN AN ELECTRICAL GRID

B.2.1 Introduction

The concept of very large scale photovoltaic power generation (VLS-PV) systems is but one facet of utility-scale renewable energy generation systems. Because VLS-PV systems are of a size that can influence the performance and stability of the grid to which they are connected, it is important to consider such systems in the wider context of an electrical grid that is influenced or even dominated by renewable generation sources. It is widely accepted that the current reliance on fossil fuel as the dominant source of electrical generation in modern economies is not a sustainable approach in the long run. The only point of contention among experts is when the transition from fossil-dominated generation systems to renewable energy-dominated generation systems is likely to begin. A subject that is rarely discussed is what a renewable energy-dominated system will look like and how it will operate. In this section we discuss, from a broad perspective, some of the issues that arise when one considers an electrical grid where renewable sources constitute significant or even dominant sources of generation. The fact is that renewable energy-dominated grid systems will be much different and more complex than the current systems. The principal reason for this is that renewable energy sources are either intermittent or strongly location dependent, or both. Table B.3 illustrates these characteristics for the most commonly discussed renewable sources.

It is useful to divide renewable energy sources into two categories: intermittent sources and non-intermittent sources. As Table B.3 illustrates, those sources that are not intermittent tend to be strongly location dependent. In addition, the two most ubiquitous sources, solar and wind, are both intermittent. In a situation where intermittent renewable generation sources dominate the supply to an electrical grid, their intermittent nature leads to a very different grid management paradigm than is employed in current electrical grids, where it is assumed that centralized energy sources can respond to electrical demand within a short time and dispatchable power can be realized simply through careful anticipation of the load over relatively short time periods. For cases where intermittent sources, such as solar or wind, are a significant fraction of available generation, ensuring grid stability and the availability of dispatchable power is more complex than for our current systems. One cannot control or, in most cases, even predict with any accuracy the availability of energy from intermittent

Table B.3 Characteristics of renewable energy sources

Renewable energy source	Intermittent?	Location dependent?
Solar	Yes	No (but economic viability varies with location)
Wind	Yes	Yes (requires certain ranges of wind velocity)
Big hydro (large reservoir)	No (capacity limited in some regions)	Yes (requires suitable topography and river flow)
Small hydro (run-of-river, no reservoir)	Yes (precipitation dependent)	Yes (requires suitable topography and river flow)
Geothermal	No	Yes
Biomass	No (landfill gas can be highly variable)	Yes (requires ready availability of sources of biomass)
Tidal	Yes (but predictable)	Yes (requires suitable tidal current environment)
Wave	Yes	Yes (requires a sea-coast location with frequent wave activity)

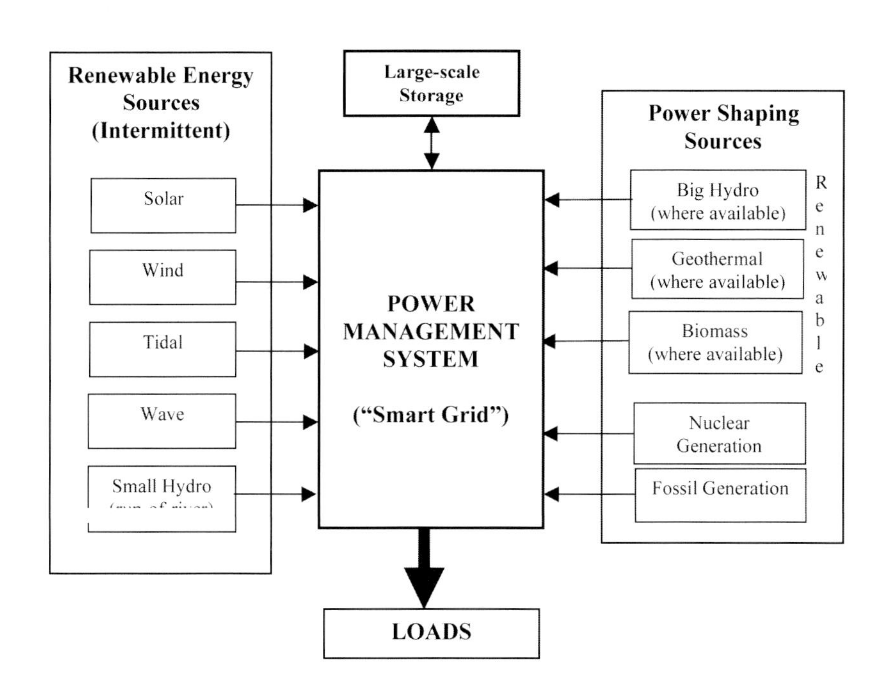

Figure B.18 Generation sources in a renewable energy-dominated grid

sources. In such cases a 'use it while you have it' approach coupled with large scale energy storage and/or non-intermittent sources for power shaping provides a viable solution.

B.2.2 Renewable energy-dominated grids

In the renewable energy-dominated electrical grids of the future, generation sources will be a mixture of a wider variety of technologies than is the case today. One can divide these sources into two categories: intermittent sources that are employed when available and non-intermittent sources (where available), as well as large scale storage, used for power-shaping to ensure that power can be delivered to the loads when required. This is illustrated in Figure B.18.

B.2.3 Blended generation

Because the renewable sources originate in nature, the mix of available energy sources is regionally dependent. Therefore, each region will have its own mix of sources and its unique way of managing the power grid. Intermittent sources, while they cannot be easily controlled, in many cases do complement one another. Figure B.19 shows an example of complementary sources for the Midwest US where the blending of wind and solar generation gives a better overall annual performance than either of these two sources taken alone. This is due to the fact that, in this region, wind is generally strong in the winter and solar is generally strong in the summer. A blend of the two sources therefore delivers a more constant supply of energy year round than would be possible from each of these sources by itself.

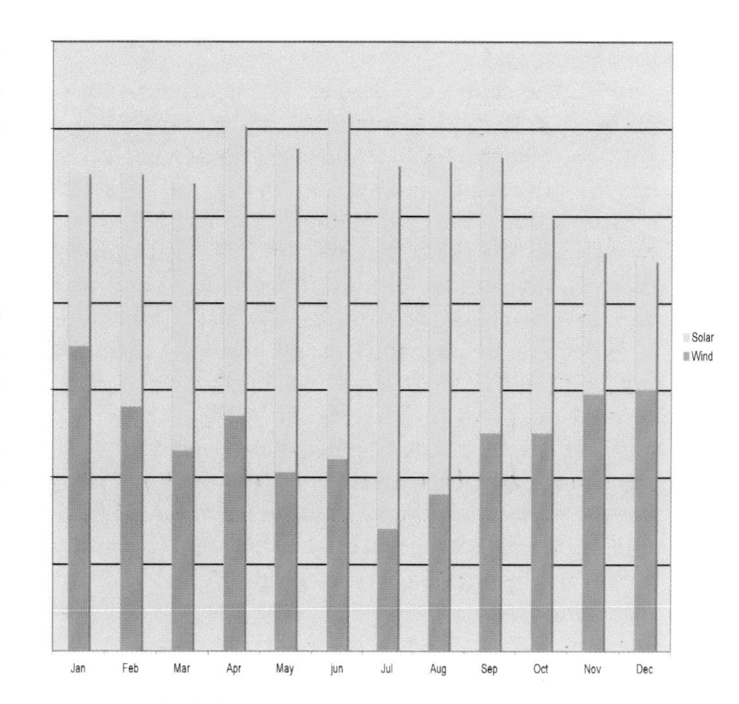

Figure B.19 Blended generation: Wind and solar

B.2.4 Shaping the power in renewable energy-dominated grids

While blended generation improves the performance of a renewable energy-dominated grid structure, the intermittent nature of these energy sources demands the availability of a suitable energy storage facility and/or non-intermittent energy sources, such as big hydro, geothermal, biomass or non-renewable sources (for example, fossil fuel or nuclear sources), to supply power when it is not available from the intermittent sources. In instances where large hydro-generation facilities are available, switching the paradigm from hydro as a main source of base power to a storage and power-shaping role is relatively straightforward. In some places available hydro may not be enough to service the full load; but it may be sufficient to serve in a power-shaping role, along with an assortment of other renewable sources. Using increasingly expensive non-renewable fossil and nuclear sources in a power-shaping role instead of primary generation would tend to minimize the impact of increasing costs of these sources as we move into the future.

B.2.5 Large scale energy storage

The capability to store and recover electrical energy is an important component of a renewable energy-dominated grid. At this point in time there is no universally satisfactory technology for accomplishing this function. Table B.4 lists some parameters of the storage technologies that are more commonly considered, in addition to the use of large hydroelectric facilities in 'storage mode'. None of these are optimal for all purposes.

Pumped hydro (pumping water uphill during periods of surplus generation) and compressed-air storage (compressing air during surplus generation) have high capacity and are among the most practical and attractive means of large-scale storage; but in both cases (as in the use of large hydro in 'storage mode') they are dependent upon local geological conditions and are therefore not universally useful methods. Hydrogen is also a potential means of storing energy for longer periods. The energy can be recovered either through fuel cells or by combustion. The capacity of fuel cell-based storage systems can be quite long depending upon the system's hydrogen storage capacity; but the need to replace fuel cells more frequently than other potential storage technologies leads to considerably higher maintenance costs. Production of hydrogen from hydrocarbon sources, the usual method of hydrogen production today, negates the reasons for implementing renewable energy systems in the first place, so future hydrogen production for large scale storage is likely to be through the electrolysis of water, employing energy from renewable sources. The electrolysis process is about 65 % efficient, so hydrogen as an energy storage mechanism will always be considerably more expensive than the energy used to produce it. In addition, the current state of technology for storing hydrogen through compression or binding to solid metals is not particularly economically efficient today. In spite of these drawbacks, hydrogen storage is potentially one of the more promising technologies for large scale universally applicable energy storage.

The remaining methods shown in Table B.4 tend to be more expensive and have limited capacity, but are useful for short-term load balancing and shaping applications in renewable energy-dominated systems. This is particularly true for flow batteries, such as the vanadium redox battery due to its ability to handle short- to medium-term fluctuations caused by such things as clouds drifting across the sun or variations in wind power. The vanadium redox battery is also suitable for extending the 'peak shaving' function of solar energy systems into the evening hours when air-conditioning loads can still be high.

B.2.6 'Smart' grids

The concept of a so-called 'smart grid' is central to the management of renewable energy-dominated electrical systems. Today's electrical grid is primarily composed of central-generating stations and electromechanical power delivery systems operated from control centres. In contrast to this, the grids of the future will integrate a multitude of distributed renewable energy sources, and use solid state electronics and modern communication technologies to manage and deliver quality power, employing computer-based automated control systems. Smart grids can support the widespread integrated use of distributed generation and go so far as to better control 'smart' appliances in homes and businesses, thereby exercising some control over load. Standardized power and communications interfaces will allow interconnection of components, including generation sources, on a 'plug-and-play' basis.

Table B.4 Electrical energy storage technologies

Technology	Capital costs (USD/kW) Low	Capital costs (USD/kW) High	Discharge period	Efficiency %	Scale (MW) Current	Scale (MW) Planned
Capacitors – high power	200[1]	600[1]	10 seconds[5]	95[1,5]	0.1–0.45[5,6]	2[7]
Capacitors – long duration	100[1]	500[1]	<1 hour[1]	95[1,5]		
Flow batteries	650[1]	2500[1]	>10 hours[1]	75–85[1]	12[2,3]	20–100[1]
Compressed air	425[5]	1000[1]	Days[5]	80–85[1,9]	50–300[9]	2700[3]
Pumped hydro	600[1,5]	1400[1]	>12 hours[5]	75–87[1,5]	2100[3]	2100[3]
Superconducting magnetic ES		300[5]	5 hours[5]	97–98[5]	30[3]	100–1000[4]
Flywheels – high power	230[1]	5501	3 – 120 seconds[5]	90[5]	1,75	201
Flywheels – long duration	3300[1]	10000[1]	~ 1 hour[5]	93[5]		
NaS batteries	259[1]	2500[1]	~1 hour[5]	85[5]	0.3–66,10	20–1001
Hydrogen fuel cell	500[5]		As needed	59[5]	0,2[5]	500[8]
Hydrogen	1700[11]		As needed	45–80[9]	0,2[12]	350[11]

This type of automated control and distribution structure is especially important in enabling renewable energy-dominated grids to manage power flow from a wide variety of intermittent and non-intermittent generation sources, together with storage centres to balance power availability and demand on a second-by-second basis. Indeed, introduction of significant quantities of VLS-PV generation into a grid structure, whether that generation is distributed or centralized, is not feasible unless the grid management structure is able to respond in an automated fashion to intermittent and/or highly variable generation sources. Some of the important aspects of such a grid management structure are:

- *Real-time monitoring*: the use of sensors embedded throughout the grid structure from generators and power lines through to sub-stations and feeders. The information flow from these sensors enables functions such as rapid diagnosis and response to anomalies in the grid, monitoring transmission lines to determine if they are reaching capacity, thus enabling more effective use of these lines, and response to fluctuations in renewable generation sources.
- *Automation of transmission and distribution*: this allows faster adjustments to changing conditions, an essential ingredient for a grid with significant generation from solar and wind sources. Automated networks would provide capacities not available in traditional grid structures, such as channelling electricity to essential services (for example, hospitals) during periods of outage or stress on the grid. Automated grid management systems also have the advantage of being able to respond rapidly to unexpected conditions, reallocate resources and soften the effect of these conditions (thus preventing blackout situations that would otherwise occur) and enabling faster recovery.
- *Communications networks*: the infrastructure of smart grids includes fibre-optics networks, cables, power-line carriers and wireless communication systems, as well as distributed computing systems that manage information flow among smart devices, allowing them to respond to grid conditions and to interact in a two-way manner with the grid management system. These networks would be protected by information security, privacy and authentication software in order to protect the integrity of the grid and its participants. Indeed, the communication infrastructure of smart grids is just as vital a part of future electrical supply systems as the wires that carry the power.
- *Demand response*: two-way communications between providers and customers opens the way to flexibly adjust power demand as well as supply to match grid conditions. This 'demand response' contrasts with the traditional utility system in which supply adjusts to meet all demands. Under this regime customers receive financial incentives to allow the grid management system to automatically cut their demand (in defined ways that they can override) at times when the grid is under pressure. This functionality is enabled by the use of such devices as 'smart appliances' that are interfaced to the grid's communications infrastructure.

These properties, infrastructure and functionality are essential to the reliable operation of an electrical grid structure where intermittent renewable sources are a significant fraction of available generation.

B.2.7 Conclusions

This section has considered, in a general way, the major issues confronting electrical power grids where VLS-PV and other major renewable energy sources are introduced. As time goes on, grids will evolve towards the type of structure discussed here. Things will not change suddenly. Rather, as substantial new generation technologies are introduced into the grid, traditional fossil and nuclear sources will gradually convert from primary generation to power shaping and back-up roles and will gradually be decommissioned as their economic viability decreases and environmental pressures increase. Depending upon the rates of change in the economics of fossil generation and environmental issues, as well as the economic feasibility of renewable sources, it is possible that nuclear generation will enjoy resurgence in the near term. In the long term, however, our generation sources will be dominated by renewable energy technologies. As these changes take place, the grid itself and its management systems will also evolve towards a 'smart grid' infrastructure. The rates of change are uncertain; but there is little doubt about the eventual general structure of the system. VLS-PV will play a pivotal role in this process.

REFERENCES

1. Electricity Storage Association (no date) www.electricitystorage.org/tech/technologies_technologies.htm
2. 19th World Energy Congress (2004) 'Vanadium redox battery and fuel cell for large-scale energy storage', 19th World Energy Congress, Sydney, Australia, 5–9 September
3. Energy Pulse (no date) www.energypulse.net
4. US DOE (no date) www.eere.energy.gov/EE/power_energy_storage.html
5. Sandia National Labs (2001) *Report 2001-0785: A Study by the DOE Energy Storage Systems Program*, March
6. DOE Energy Storage Program News Release (2004) *Joint California Energy Commission Supercapacitor/Wind Project*, January

7. University of California (no date) www.physorg.com/news3056.html
8. *Fuel Cell Magazine* (no date) *Market Reports*, www.fuelcell-magazine.com/f-bkstr.htm
9. Energy Scotland (no date) www.esru.strath.ac.uk/EandE/Web_sites/03-04/marine/tech_storage.htm
10. Tokyo Electric Power Company, Japan (no date) www.electricitystorage.org/pubs/2001/IEEE_PES_Winter2001/wm01nas.pdf
11. ConocoPhillips, Shell and Scottish and Southern Energy (no date) 'Large scale H_2 from natural gas in UK for 350MW electric generation project'
12. US DOE (no date) 'Small scale H_2 production', www.eere.energy.gov/hydrogenandfuelcells/pdfs/annual04/iia3_aaron.pdf

B.3 SOLAR HYDROGEN SCENARIO

B.3.1 Introduction

In recent years, the use of hydrogen has been positively and widely proposed and developed in conjunction with fuel cell (FC) technology, such as automobile applications, natural gas co-generation in residential and industrial uses and small mobile power supplies. There are many big projects among industrialized countries and motor companies aiming at the commercialization of fuel cells as a clean energy supply.

The usage of hydrogen itself is purely a CO_2 emission-free system since hydrogen is used as a fuel and results only in H_2O without CO_2, unlike in fossil fuel engines. However, since hydrogen itself is virtually non-existent naturally and is a secondary energy carrier, it has to be produced from other primary energy resources. Today, it is usually made from hydrocarbon fuel, such as natural gas, petroleum or coal. This is the case in most existing FC projects. This means that CO_2 is emitted in the process of hydrogen production:

$$2C_mH_n + 4mH_2O \rightarrow (n+4m)H_2 + 2mCO_2 \qquad [B.1]$$

This difference is also illustrated in Figure B.20. In order to realize clean and sustainable energy systems by strict definition, renewable energy resources should be used to produce hydrogen. Solar and wind energy are two prime examples. Coupling either of these two

Table B.5 A chronology of the main hydrogen and fuel cell events since 1766

Year	Event
1766	Discovery of hydrogen by Henry Cavendish
1783	First hydrogen balloon flight by Jacques Alexander Cesar Charles
1794	First hydrogen generator built
1800	Electrolysis discovered by the English scientists William Nicholson and Sir Anthony Carlisle
1839	Discovery of the fuel cell effect by Christian Friedrich Schoenbein
1845	Invention of the 'gas battery' (later known as the fuel cell) by William Grove
1874	Jules Verne's 'Mysterious Island' and its implications for a future hydrogen energy system
1920s and 1930s	Interest in hydrogen as a fuel in Germany, England and Canada
1920s and 1930s	Development of hydrogen dirigibles
1920	First electrolyser shipped by Canada's Electrolyser Corporation
1930s and 1940s	Rudolfo Erren – German engineer who built trucks, buses, submarines and ICEs running on H_2
1937	Hindenburg accident
1950s	Francis T. Bacon developed the first practical hydrogen-air fuel cell
1960s	The National Aeronautics and Space Administration (NASA) uses fuel cells in space programme
1970	The term 'hydrogen economy' was coined at General Motors by John Bockris
1973	Interest in fuel cell development sparked by Organization of Petroleum Exporting Countries (OPEC) oil embargo
1974	First major International Hydrogen Conference (THEME)
1974	Creation of the International Association of Hydrogen Energy (IAHE)
1988	Soviet Union converted one of its Tupolev TU-154 commercial jets to run partially on liquid hydrogen (LH_2)
1988	William Conrad – first person to fly an airplane running exclusively on LH_2
1988	German trials of fuel cell (FC)-powered submarine (Siemens)
1989	Creation of National Hydrogen Association (NHA)
1989	Creation of International Organization for Standardization (ISO) Committee for Technical Standards of Hydrogen Energy
1990	World's first solar-powered hydrogen production plant in southern Germany
1993	World's first proton exchange membrane (PEM) fuel cell bus built by Ballard Power Systems in Vancouver
1993	Japan Launched its World Energy Network (WE-NET) project
1993	Daimler-Benz and Ballard Power cooperate to build FCs for cars and buses
1997	NASA engineer Addison Bain challenged the belief that hydrogen was the cause of the Hindenburg accident
1997	Daimler-Benz and Ballard Power announced 300 MUSD spending on research on FC for transportation
1999	Royal Dutch/Shell set up a hydrogen division
1999	Europe's first hydrogen gas stations were opened in Hamburg and Munich
1999	Iceland announced its plan for the world's first hydrogen economy
2000	Ballard unveiled the world's first production-ready PEMFC for automotive use at the Detroit Auto Show

resources with hydrogen generation is a meaningful approach to achieve sustainability. In addition, hydrogen storage can act as the buffer between the availability of the resource and utilization.

Hydrogen systems have been proposed in the past, especially after the first oil crisis, as one candidate of oil-alternative energy. The main hydrogen and fuel cell events since 1766 are summarized in Table B.5.[1] A society mainly driven by hydrogen systems was named the 'hydrogen economy' by John Bockris in 1970. It should also be noted that the Apollo Space Programme achieved enormous progress in hydrogen fuel-cell technology a decade before the oil crisis.

The key driver for hydrogen technology developments during the 1970s was energy prices. In spite of a number of efforts in many countries, the fossil fuel sector was still stronger than ever and there was no clear sign for a reversal of this trend. However, over the last decade, the driver for seeking a cleaner and sustainable environment from a longer-term view has become global climate change. The first trial of solar hydrogen production appeared in 1990 in Germany, as shown in Table B.5.

Table B.6 Various theoretical means of hydrogen production

Principle	Fossil driven	Renewable driven
Hydrolysis*	Oil refining process Reformed from natural gas Coal gasification	Biomass gasification Thermochemistry by solar Thermal collector
Thermochemical process	Medium temperature process by nuclear** (high temperature gas) reactor	
Thermolysis	Thermolysis***	Thermolysis by solar furnace
Electrolysis, including high temperature water vapour electrolysis	Oil-burning power Natural gas power Nuclear power	Solar thermal or PV; wind power; hydropower; geothermal; or tidal
Thermodynamic electrolysis	Oil-burning power Natural gas power Nuclear power	Solar PV assisted by thermal collector
Photochemistry	–	Solar photochemistry
Photocatalysis	–	Oxide semiconductor Catalyst Dye catalyst
Photosynthesis	–	Chlorophyll, algae, enzyme, artificial photosynthesis
Photolysis	–	Solar direct photolysis
Micro-organism	–	Biomass fermentation

Notes: * 200°C to 600°C; ** 600°C to 1 000°C ; *** 2 500°C to 4 000°C

B.3.2 Theoretical possibilities of hydrogen production

Although there appear to be various means of producing hydrogen, current major technology is a by-product of the oil refining process. Table B.6 roughly summarizes theoretically possible means for producing hydrogen. They may be categorized into four major paths through the principles of:

1 thermal reactions;
2 electrochemical reactions;
3 photochemical reactions; and
4 micro-organic reactions.

Table B.6 also classifies two major paths through driving energy:

1 fossil fuel-driven reactions; and
2 renewable energy-driven reactions.

In this appendix, solar hydrogen production by renewable energy is of special interest. Table B.7 provides examples in three categories:

1 electrolysis by solar PV electricity;
2 thermal reactions; and
3 photon-induced reactions.

At this point in this Task 8 study, the principle aim is not to survey and discuss all of the possibilities of hydrogen production. Instead, items for further work are proposed in which longer-term technical directions are highlighted for the purpose of VLS-PV deployment as a major world energy supply.

An example of a new approach to hydrogen production (a proposal for a high temperature solid oxide electrolyser combined with high concentration PV) is described in section B.3.4. This method can also utilize thermal energy from the infrared region effectively.

First, however, the IEA's hydrogen programme is given a brief overview.

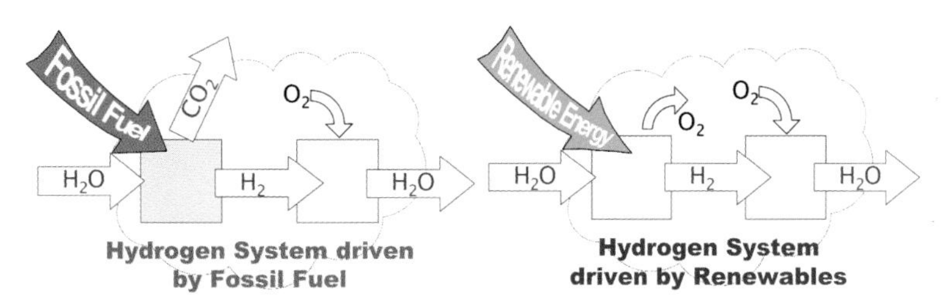

Figure B.20 Comparison of different hydrogen systems driven by fossil fuel and renewable energy, respectively

Table B.7 Possible means of solar hydrogen production

Means	Remarks	References
Solar PV power and electrolyser	Proton exchange membrane (PEM) electrolyte	2, 3
Concentrator PV and high temperature electrolyser	High temperature solid oxide electrolysis cell (SOEC)	4
High temperature electrolysis	SOEC	5, 6
Solar thermal power generation and electrolyser		
Solar thermal power generation and high temperature electrolyser		
Thermochemistry by solar thermal collector		7
Solar direct thermal decomposition of water		
Solar photochemistry		
Photocatalysis by nano-semiconductor		8
Dye catalyst		
Photosynthesis by micro-organism	Chlorophyll, algae, enzyme, artificial photosynthesis	

B.3.3 International Energy Agency (IEA) hydrogen programme[9]

B.3.3.1 Objectives

For more than 20 years, the IEA has supported collaborative activities focused on the advancement of hydrogen technology. In order for hydrogen to become a competitive energy carrier, experience and operating data need to be collected and disseminated through demonstration projects. A growing number of integrated hydrogen energy demonstrations are being carried out around the world. The IEA hydrogen programme has undertaken the task of working with demonstration project leaders to develop comprehensive reports on various demonstrations and to make this information available to the hydrogen community. Each hydrogen-based energy system was critically evaluated and compared, with system performance measurement as the central focus. The reports include project goals, a description of the main components, a representative set of experimental results, technical challenges, permitting issues, safety and public interactions.

IEA hydrogen programme members recognize that a long-term research and development (R&D) effort is required to realize the significant technological potential of hydrogen energy. This effort can help to create competitive hydrogen energy production and end-use technologies, and supports the development of the infrastructure required for its use. Attention should be given to the entire system. In particular, all of the key elements should be covered, either with new research or with common knowledge.

If the technological potential of hydrogen is realized, it will contribute to the sustainable growth of the world economy by facilitating a stable supply of energy and by helping to reduce future emissions of carbon dioxide. Cooperative efforts among nations can help to speed effective progress towards these goals. In addition, because hydrogen is in a pre-commercial phase, it is particularly suited to collaboration since there are fewer proprietary issues than may normally surface in many energy technologies.

Today, hydrogen is primarily used as a chemical feedstock in the petrochemical, food, electronics and metallurgical processing industries, but is rapidly emerging as a major component of clean sustainable energy systems. It is relevant to all energy sectors – transportation, buildings, utilities and industry. Hydrogen can provide storage options for intermittent renewable technologies, such as solar and wind, and, when combined with emerging de-carbonization technologies, can reduce the climate impacts of continued fossil fuel utilization. Hydrogen is truly the flexible energy carrier for our sustainable energy future.

As we enter the new millennium, concerns about global climate change and energy security create the forum for mainstream market penetration of hydrogen. Ultimately, hydrogen and electricity, our two major energy carriers, will come from sustainable energy sources, although fossil fuel will likely remain a significant and transitional resource for many decades.

B.3.3.2 Current research tasks in the IEA hydrogen programme

Hydrogen production technologies

Biological organisms can produce hydrogen directly from sunlight and water. In addition, semiconductor-based systems similar to photovoltaics (PV) can be used for hydrogen production. Hydrogen can also be produced indirectly via thermal processing of biomass or fossil fuels. Global environmental concerns are leading to the development of advanced processes to integrate sequestration with known reforming, gasification and partial oxidation technologies for carbonaceous fuels. These production technologies have the potential to essentially produce unlimited quantities of hydrogen in a sustainable manner:

- Task 14: photo-electrolytic production of hydrogen;
- Task 15: photo-biological production of hydrogen;
- Task 16: hydrogen from carbon-containing materials;
- Task 18: integrates systems evaluation;
- Task 19: safety;
- Task 20: hydrogen from water-photolysis.

Hydrogen storage technologies

Storage of hydrogen is an important area for cooperative research and development, particularly when considering transportation as a major user and taking into account the need for efficient energy storage for intermittent renewable power systems. Although compressed gas and liquid hydrogen storage systems have been used in vehicle demonstrations worldwide, issues of safety, capacity and energy consumption have resulted in a broadening of the storage possibilities to include metal hydrides and carbon nano-structures. Stationary storage systems that are highly efficient and that have quick response times will be important for incorporating large amounts of intermittent PV and wind into the grid as base-load power:

- Task 17: solid and liquid state hydrogen storage materials.

Hydrogen utilization technologies

Achieving the vast potential benefits of a hydrogen system requires careful integration of production, storage and end-use components with minimized cost and maximized efficiency, and a strong understanding of environmental impacts and opportunities. System models combined with detailed life-cycle assessments provide the platform for standardized comparisons of energy systems for specific applications. Individual component models form the framework by which these system designs can be formulated and evaluated:

- Task 13: design and optimization of integrated systems (completed).

Tasks in development

- Hydrogen production from low temperature processes.
- Hydrogen production from high temperature processes.

B.3.4 A new approach: Concentrator photovoltaics (CPV) combined with high temperature electrolyser

This section describes an example of the future possibility of solar hydrogen options.[4] The generation of electrolytic hydrogen with solar energy may be critically important to the world's long-term energy needs for several reasons. Feedstock (water) and solar energy are both carbon free, having no adverse impacts on global warming. The solar resource is extensive, with the potential to generate hydrogen near markets, thus minimizing transportation costs. In the past, the principal criticisms for considering PV for generating hydrogen were the high cost of PV electricity and the low efficiency of the PV system. Recently, however, concentrator PV systems have demonstrated the potential for generating lower cost electricity, primarily due to the development of solar cells approaching 40 % efficiency and reliable concentrators that can be used with them. Another critical factor favouring concentrator PV systems is the generation of by-product heat, normally dissipated to the environment, which can augment the electrolysis of water by using a high temperature solid oxide electrolysis cell.

This heat boost – 40 % measured by one of the authors at temperatures above 1 100°C – in hydrogen production leads to potential solar-to-hydrogen conversion efficiencies of 40% in the near term (the next few years), whereas efficiencies of 50 % and higher are realistic targets within five to ten years. These efficiencies exceed those of any other methods previously considered for producing electrolytic hydrogen from solar electricity. This section describes the experiments, presented in two early patents of this approach, that first demonstrated a 40 % boost in hydrogen production above that associated with the electrical output alone from solar cells. We also provide efficiency and cost analyses for generating hydrogen using today's high-efficiency solar cells in the hybrid solar concentrator. These results, based on the long-term potential for concentrator PV (CPV) systems to be mass produced at costs of less than 1 USD/W, lead to hydrogen production costs comparable with the energy costs of gasoline – recognizing that 1 kg of hydrogen has the energy equivalent of 1 US gallon of gasoline. Further development and demonstration will be needed to realize the potential of this innovative solar concentrator for generating hydrogen.

B.3.4.1 System description

Dish concentrator PV system

Solar Systems Pty Ltd in Australia has developed a dish CPV system over the past 15 years. Figure B.21 shows several of its dish concentrators operating on aboriginal lands near Alice Springs, Australia. The visitors at the site indicate the size of the units. Each dish produces about 20 kW using high efficiency silicon solar cells, but can also accommodate new high efficiency III-V multi-junction solar cells to achieve 30 kW.

Spectral splitter

The spectral splitter is placed near the focal point of the dish receiver and reflects infrared energy. Visible light is transmitted through to the solar cells and is converted into electricity. Figure B.22 schematically shows the transmission across the solar spectrum wavelengths.

Hybrid system

In Solar Systems Pty Ltd's patents,[10,11] reflected infrared light is gathered in a fibre-optic waveguide and transported to a high temperature solid oxide electrolysis cell. Figure B.23 depicts a system diagram of the components.

Figure B.21 Solar farm of dish concentrators on aboriginal lands near Alice Springs, Australia; each dish is nominally 20 kW (note the visitors for scale)

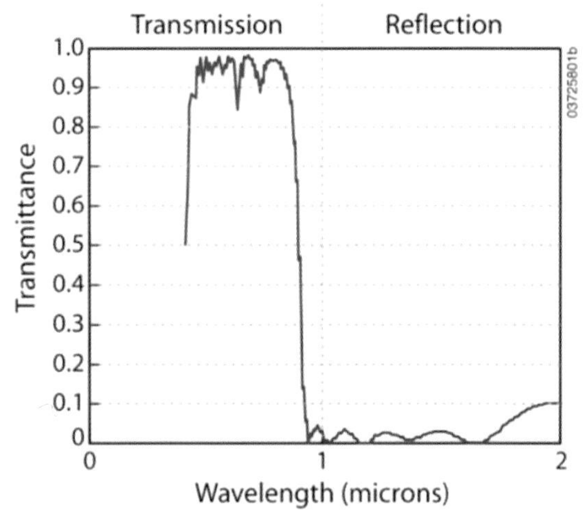

Figure B.22 Characteristics of a spectral splitter mirror

B.3.4.2 System tests

Early system tests

The testing of all components shown in Figure B.23 is described in Lasich (1997)[10] and Lasich (1999)[11]. The concentrator was a paraboloidal dish 1,5 m in diameter, arranged to track in two axes and capable of producing > 1 000 suns concentration. The full dish size was not needed and portions were appropriately shaded. The solar cell was a gallium arsenide (GaAs) PV cell with an output voltage of 1 to 1,1 V at maximum power point, with a measured efficiency of ~19 %. The voltage was an excellent match for direct connection to the electrolysis cell when operating at 1 000°C. The electrolysis cell consisted of a 5,8 cm long by 0,68 cm in diameter yttria-stabilized zirconia (YSZ) closed-end tube coated inside and out, with platinum electrodes for the tests. A metal tube surrounding the cell homogenized the solar flux over the surface of the electrolysis cell. The test rig operated above 1 000° C for more than two hours, with an excess of steam applied to the electrolysis cell. The output stream of unreacted steam and generated hydrogen was bubbled through water, and the hydrogen was collected and measured. Table B.8 shows the following results for 17 minutes of steady-state operation.

Table B.8 Electrolysis cell measurements at beginning and end of 17 minutes of system operation

Voltage (V)	Current (A)	Temperature (°C)	H_2 Production (mL)
1,03	0,67	1 020	0 (initial)
1,03	0,67	1 020	80 (17 minutes)

The ratio of the thermoneutral voltage of 1,47 V to the measured electrolysis cell voltage of 1,03 V was 1,43, corresponding to a > 40 % boost in hydrogen production due to the thermal energy input. This was also confirmed by energy balance. Combining the optical efficiencies of the concentrator dish (85 %), solar cell efficiency (with solar input of 800 W/m^2) and thermal energy boost, the total system efficiency of solar cell, electrolysis cell and optics was 22 % for conversion of solar energy to hydrogen. At the time, this efficiency was almost three times better than that recorded for a working solar plant generating hydrogen.

Expected results with multi-junction solar cells

Since the experiments for these patents were completed, solar cell efficiencies have roughly doubled from the 19 % GaAs cell efficiency used for the system tests to today's world record efficiency of 37,9 % measured at 10 suns for a GaAs-based multi-junction solar cell.[12]

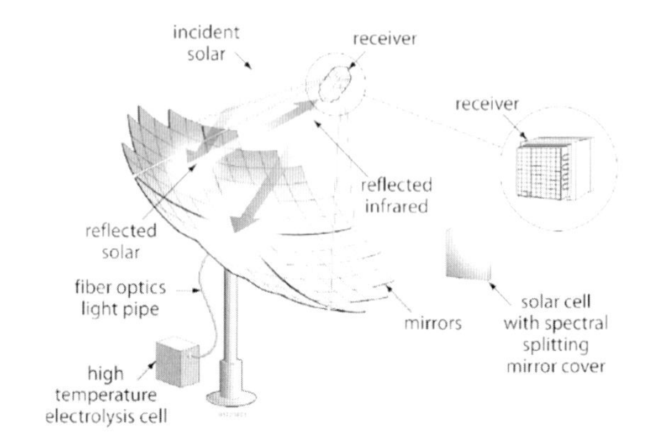

Figure B.23 Schematic of system shows sunlight reflected and focused on the receiver, with reflected infrared directed to a fibre-optics waveguide for transport to a high temperature solid oxide electrolysis cell; solar electricity is sent to the same electrolysis cell that uses both heat and electricity to split water

Similar multi-junction cells are commercially available at efficiencies above 30 %, being the state-of-the-art power source for today's space satellites. Concentrator companies around the world are working to integrate these high efficiency multi-junction cells into their system designs.[12]

Combining the observed thermal enhancement of 40 % with a multi-junction solar cell efficiency of 35 % and an optical efficiency of 85 % leads to > 40 %

conversion in the near term. A 40 % multi-junction solar cell – a result expected in the not too distant future – would yield a conversion efficiency of almost 50 %. Recent electrochemical theoretical results are consistent with these predictions based on Solar Systems Ltd Pty's early experiments.[13]

B.3.4.3 Cost analyses

Dish concentrator photovoltaic costs

The largest cost for the hybrid solar concentrator system will be for the dish concentrator and PV receiver. Algora (2004) recently completed an extensive cost analysis based on previously collected data for CPV systems.[14] Many of the costs came from installed costs for the 480 kW reflective CPV systems in Tenerife. The analysis included a wide range of parameters, including cumulative production of 10 MW for present-day systems to cumulative production of 1 000 MW for the mid-term systems where learning is incorporated. Concentrations ranged from 400 suns to 1 000 suns, with solar cell efficiencies ranging from 32 % to 40 %. Whereas module efficiencies ranged from 24,8 % to 32,2 %, the plant's AC annual efficiency ranged conservatively from 18,2 % to 23,6 %. Present-day base costs were 2,34 EUR/W (almost 3 USD/W with today's exchange rate). The lowest projected costs ranged from 0,5 to 1 EUR/W for efficiencies of 40 %, 1 000 suns concentration and cumulative production of 1 000 MW. Cost estimates for mature CPV technology are used to place this cost analysis of hydrogen generated by this hybrid solar concentrator system in a context similar to analyses completed for the electrolytic generation of hydrogen by wind systems (where cumulative production of this highly developed technology approaches 50 GW) and the conventional production of hydrogen by reforming natural gas. Cost studies for conceptual high temperature nuclear reactors (projected for mature 600 MW designs) suitable for high temperature electrolysis cells face similar problems because both the hybrid solar concentrator and high temperature nuclear reactor are in the early stages of exploration for hydrogen generation. Furthermore, high temperature solid oxide electrolysis cells will be required in large sizes (500 kW to 500 MW) for integration with nuclear reactors.[15] Units ranging up to 50 kW could be used with dish concentrators and central receivers employed for a larger scale.

Using a set of assumptions for a well-developed technology, we acquired costs in USD/kW for solid oxide electrolysis cells from a developer of solid-oxide electrolysis cells.[16] Table B.9 summarizes the cost data for a well-developed technology (1 000 MW cumulative production) for the hybrid CPV system and high temperature solid oxide electrolysis cell. Table B.10 summarizes the hydrogen production costs for a 10 MW project built with the well-developed technology. A 20 % rate of return per year was assumed and operating, storage or delivery costs were not included. Table B.11 compares these production costs with those of other hydrogen production technologies.

Table B.9 Component and system costs for 10 MW hybrid concentrator photovoltaic (CPV) project for electrolytic production of hydrogen

Component costs assuming 1000 MW technology (USD/kW)	
Concentrator PV	800
Spectral splitter	15
Fibre optics	25
Electrolysis cell	400
Total system cost	1 240

Table B.10 Hydrogen production data for mature 10 MW plant

Hydrogen cost data for mature technology	
Plant size (MW)	10
Plant cost (MUSD)	12,4
H_2 produced in one year (kg)	106
Hydrogen cost (USD/kg)	2,48

Note: The hydrogen cost of 2.48 USD/kg has considerable uncertainty (±25 %) related to technology immaturity

Table B.11 Cost comparison for the hybrid CPV production of electrolytic hydrogen

Process hydrogen production cost (USD/kg)	
Gas reformation[17]	1,15
Wind electrolysis[17]	3,10
Hybrid CPV electrolysis	2,48

Note: 1 kg of hydrogen has the energy equivalent of 1 US gallon of gasoline.

B.3.4.4 Discussion

Cost-analysis uncertainties

There are many cost analyses in the literature for hydrogen production; however, the assumptions behind them vary dramatically. The US Department of Energy (DOE) and its Hydrogen, Fuel Cells and Infrastructure Technologies Program have established a cost-analysis structure for comparing different hydrogen and fuel cell technologies within a common set of assumptions. The analysis in Table B.11 is a preliminary study needing additional work to fit within that framework. CPV systems are just beginning to enter the energy market, so cost uncertainties are significant compared with those of highly developed wind systems with a worldwide installed capacity approaching 50 GW. Nevertheless, these preliminary costs are comparable with wind electrolysis costs so that additional cost studies are warranted. Today, wind system costs are in the 800 USD/kW range – as are the estimated costs for highly developed CPV systems – whereas wind electrolysis does not have an opportunity for a heating boost in electrolysis efficiency. Assuming these cost analyses continue to be positive, we are likely to plan a larger scale electrolysis demonstration using the well-developed Solar Systems Ltd Pty's dish concentrator.

A hydrogen vision using hybrid solar concentrators

The US National Research Council and Academy of Engineering believes that one of the four most fundamental technological and economic challenges is 'to reduce sharply the costs of hydrogen production from renewable energy sources over a time frame of decades'.[18] Wind electrolysis is a strong renewable energy option, while hybrid CPV electrolysis could be another. The solar energy resource is also considered to be larger and more widely distributed than that of wind energy. And totally new system configurations may be possible with hybrid solar concentrator electrolysis. Small 50 kW systems could be part of hydrogen filling stations, reducing hydrogen distribution costs. Systems could incorporate back-up heating sources, probably natural gas in the near term, to improve the electrolysis system capacity factor.

Probably the most dramatic impact of this study has been the realization that this is a possible PV option that could provide transportation fuel on a large scale. In a scenario where hydrogen is used in fuel cell vehicles – which can have double the efficiency of standard internal combustion cars – the 'effective cost' of solar hydrogen is half (that is, 1,24 USD/kg). For customers paying 1,50 USD/litre for gasoline, this equates to 2 USD/kg, showing that the potential for a very large market clearly exists. To determine a final price of solar hydrogen to the customer, we would need to factor in the additional costs of gasoline distribution, retailing and taxes that are offset by the 'clean and renewable' value of solar hydrogen.

With the imminent market entry of CPV systems for electricity production, increasing solar cell efficiencies approaching 40 % with clearer ideas for 50 % solar cells, and the opportunity to use wasted solar heat for augmenting solar electrolysis, this is a potential 'leap frog' technology that may rapidly lower the cost of clean hydrogen.

B.3.5 Conclusions

- Hydrogen will increasingly become an energy carrier itself. It is necessary to carry out the analysis, studies, research, development and dissemination that will facilitate a significant role for hydrogen in the future.
- Hydrogen has the potential for short-, medium- and long-term applications. The steps towards realizing its potential in appropriate time frames must be understood.
- Conversion to electricity and/or hydrogen will constitute two major complementary options in the future.
- Hydrogen can assist in developing renewable and sustainable energy sources by providing an effective means of storage, distribution and conversion.
- Moreover, hydrogen can broaden the role of photovoltaics in the supply of clean fuels for transportation and heating.
- Because desert regions possess rich solar energy sources, hydrogen energy technologies offer an important potential alternative to the fossil-fuel energy supply. Hydrogen technologies can contribute to energy security, diversity and flexibility in these regions.
- As a promising example, an innovative hybrid CPV electrolysis technology was shown in this concluding section of Appendix B. Although the analysis is only preliminary, it indicates a possibility that offers a cost of hydrogen production that is lower than wind electrolysis and is in the same range as gasoline for much of the world's population.
- Further surveys during the third phase of the International Energy Agency (IEA) PVPS Programme, Task 8, on solar hydrogen options are recommended to show the future direction of massive VLS-PVs.

REFERENCES

1. Roberts, K. (2003) 'Hydrogen: Always the bridesmaid, ever the bride? EU Conference on the Hydrogen Economy: A Bridge to Sustainable Energy', Press release, Brussels, 16 June 2003
2. Aurora, P. and Duffy, J. (2005) 'Solar Hydrogen Fuel Cell System Modeling', *ISES Solar World Congress 2005*, Florida, August 2005
3. Levene, J., Mann, M., Margolis, R. and Milbrandt, A. (2005) 'Analysis of Hydrogen Production from Renewable Electricity Sources', *ISES Solar World Congress 2005*, Florida, August 2005
4. McConnell, R. D., Lasich, J. B. and Elam, C. (2005) 'A hybrid solar concentrator PV system for the electrolytic production of hydrogen', 20th EU-PVSEC Conference, Barcelona, June 2005
5. Doenitz, W., Schmidberger, R., Steinheil, E. and Streicher, R. (1978) 'Hydrogen production by high temperature electrolysis of water vapour', *Proceedings of the Second World Hydrogen Energy Conference*, Zurich, Switzerland, 21–24 August 1978, vol 1, Pergamon Press, Oxford and New York, pp403–421
6. Boehm, R., Baghzouz, Y. and Maloney, T. (2005) 'A Strategy for Renewable Hydrogen Market Penetration', *ISES Solar World Congress 2005*, Florida, August 2005
7. Sharma, R. (2005) 'Warming through Solar Thermal Power Generation', *ISES Solar World Congress 2005*, Florida, August 2005
8. Jahagirdar, A., Kadam, A. and Dhere, N. (2005) 'CIGSS Thin Films for Photoelectrochemical Water Splitting Using Multiple Bandgap Combination of Thin Film Photovoltaic Cell and Photocatalyst: Clean and Renewable Hydrogen Source', *ISES Solar World Congress 2005*, Florida, August 2005

9. IEA Hydrogen Programme (no date) www.ieahia.org/
10. Lasich, J. (1997) US Patent No 5658448, 19 August 1997
11. Lasich, J. (1999) US Patent No 5973825, 26 October 1999
12. McConnell, R. and Symko-Davies, M. (2005) *Proceedings of the 20th European Photovoltaic Solar Energy Conference*, Barcelona, Spain, 6–10 June, 2005
13. Licht, S. (2003) *Journal of Physical Chemistry B*, vol 107, pp4253-4260
14. Algora, C. (2004) Chapter 6 in A. Marti and A. Luque (eds) *Next Generation Photovoltaics*, Institute of Physics Publishing, Bristol and Philadelphia,
15. Anderson, R., Herring, S., O'Brien, J., Stoots, C., Lessing, P., Hartvigsen, J. and Elangovan, S. (2004) *Proceedings of the National Hydrogen Association Conference*, National Hydrogen Association, Washington, DC
16. J. Hartvigsen, pers comm, 2005
17. Mears, D., Mann, M., Ivy, J. and Rutkowski, M. (2004) *Proceedings of the National Hydrogen Association Conference*, National Hydrogen Association, Washington, DC
18. National Research Council and National Academy of Engineering (2004) *The Hydrogen Economy*, National Academies Press, Washington, DC